SECOND EDITION
ROTATING MACHINERY VIBRATION
From Analysis to Troubleshooting

SECOND EDITION

ROTATING MACHINERY VIBRATION

From Analysis to Troubleshooting

MAURICE L. ADAMS, JR.
Case Western Reserve University
Cleveland, Ohio

CRC Press
Taylor & Francis Group
Boca Raton   London   New York

CRC Press is an imprint of the
Taylor & Francis Group, an **informa** business

CRC Press
Taylor & Francis Group
6000 Broken Sound Parkway NW, Suite 300
Boca Raton, FL 33487-2742

© 2010 by Taylor and Francis Group, LLC
CRC Press is an imprint of Taylor & Francis Group, an Informa business

No claim to original U.S. Government works

Printed in the United States of America on acid-free paper
10 9 8 7 6 5 4 3 2 1

International Standard Book Number: 978-1-4398-0717-0 (Hardback)

This book contains information obtained from authentic and highly regarded sources. Reasonable efforts have been made to publish reliable data and information, but the author and publisher cannot assume responsibility for the validity of all materials or the consequences of their use. The authors and publishers have attempted to trace the copyright holders of all material reproduced in this publication and apologize to copyright holders if permission to publish in this form has not been obtained. If any copyright material has not been acknowledged please write and let us know so we may rectify in any future reprint.

Except as permitted under U.S. Copyright Law, no part of this book may be reprinted, reproduced, transmitted, or utilized in any form by any electronic, mechanical, or other means, now known or hereafter invented, including photocopying, microfilming, and recording, or in any information storage or retrieval system, without written permission from the publishers.

For permission to photocopy or use material electronically from this work, please access www.copyright.com (http://www.copyright.com/) or contact the Copyright Clearance Center, Inc. (CCC), 222 Rosewood Drive, Danvers, MA 01923, 978-750-8400. CCC is a not-for-profit organization that provides licenses and registration for a variety of users. For organizations that have been granted a photocopy license by the CCC, a separate system of payment has been arranged.

Trademark Notice: Product or corporate names may be trademarks or registered trademarks, and are used only for identification and explanation without intent to infringe.

Library of Congress Cataloging-in-Publication Data

Adams, Maurice L.
    Rotating machinery vibration : from analysis to troubleshooting / author, Maurice L. Adams.
        p. cm.
    "A CRC title."
    Includes bibliographical references and index.
    ISBN 978-1-4398-0717-0 (hardcover : alk. paper)
    1. Turbomachines--Vibration. 2. Rotors--Vibration. I. Title.

TJ266.A33 2010
621.406--dc22   2009041606

Visit the Taylor & Francis Web site at
http://www.taylorandfrancis.com

and the CRC Press Web site at
http://www.crcpress.com

This book is dedicated to my late parents and late brother

Maury, Libby, and George

And to my late wife

Kathy

And to my four sons

Maury, Dr. Mike, RJ, and Nate

Contents

Preface xvii
Acknowledgments xxi
Author xxv

Part I Primer on Rotor Vibration

1 Vibration Concepts and Methods 3
  1.1 One-Degree-of-Freedom Model 3
    1.1.1 Assumption of Linearity 3
    1.1.2 Unforced System 4
    1.1.3 Self-Excited Dynamic-Instability Vibrations 6
    1.1.4 Steady-State Sinusoidally Forced Systems 7
    1.1.5 Damping 8
    1.1.6 Undamped Natural Frequency: An Accurate Approximation 10
    1.1.7 1-DOF Model as an Approximation 11
  1.2 Multi-DOF Models 13
    1.2.1 Two-DOF Models 13
    1.2.2 Matrix Bandwidth and Zeros 16
    1.2.3 Standard Rotor Vibration Analyses 18
  1.3 Modes, Excitation, and Stability of Multi-DOF Models 19
    1.3.1 Modal Decomposition 19
    1.3.2 Modal Damping 24
    1.3.3 Forced Systems Decoupled in Modal Coordinates 27
    1.3.4 Harmonic Excitation of Linear Multi-DOF Models 27
    1.3.5 Dynamic Instability: The Complex Eigenvalue Problem 28
  1.4 Summary 31

2 Lateral Rotor Vibration Analysis Models 35
  2.1 Introduction 35
  2.2 Simple Linear Models 37
    2.2.1 Point–Mass 2-DOF Model 37

	2.2.2	Jeffcott Rotor Model	39
	2.2.3	Simple Nontrivial 8-DOF Model	41
		2.2.3.1 Lagrange Approach (i)	44
		2.2.3.2 Lagrange Approach (ii)	49
		2.2.3.3 Direct $F = ma$ Approach	52
2.3	Formulations for RDA Software		55
	2.3.1	Basic Rotor Finite Element	55
	2.3.2	Shaft Element Lumped Mass Matrix	57
	2.3.3	Shaft Element Distributed Mass Matrix	58
	2.3.4	Shaft Element Consistent Mass Matrix	59
	2.3.5	Shaft Element Stiffness Matrix	61
	2.3.6	Shaft Element Gyroscopic Matrix	62
	2.3.7	Addition of Nonstructural Mass and Inertia to Rotor Element	62
	2.3.8	Matrices for Complete Free–Free Rotor	63
	2.3.9	Radial-Bearing and Bearing-Support Models	64
		2.3.9.1 Bearing Coefficients Connect Rotor Directly to Ground	67
		2.3.9.2 Bearing Coefficients Connect to an Intermediate Mass	68
	2.3.10	Completed RDA Model Equations of Motion	70
2.4	Insights into Linear LRVs		70
	2.4.1	Systems with Nonsymmetric Matrices	71
	2.4.2	Explanation of Gyroscopic Effect	77
	2.4.3	Isotropic Model	79
	2.4.4	Physically Consistent Models	82
	2.4.5	Combined Radial and Misalignment Motions	82
2.5	Nonlinear Effects in Rotor Dynamical Systems		83
	2.5.1	Large Amplitude Vibration Sources that Yield Nonlinear Effects	84
	2.5.2	Journal-Bearing Nonlinearity with Large Rotor Unbalance	85
	2.5.3	Unloaded Tilting-Pad Self-Excited Vibration in Journal Bearings	94
	2.5.4	Journal-Bearing Hysteresis Loop	96
	2.5.5	Shaft-on-Bearing Impacting	97
	2.5.6	Chaos in Rotor Dynamical Systems	99
	2.5.7	Nonlinear Damping Masks Oil Whip and Steam Whirl	100
		2.5.7.1 Oil Whip Masked	100
		2.5.7.2 Steam Whirl Masked	101
	2.5.8	Nonlinear Bearing Dynamics Explains Compressor Bearing Failure	101
2.6	Summary		104

Bibliography 104
Textbooks....................................... 104
Selected Papers Concerning Rotor Dynamics Insights 105
Selected Papers on Nonlinear Rotor Dynamics 105

3 Torsional Rotor Vibration Analysis Models 111
3.1 Introduction................................. 111
3.2 Rotor-Based Spinning Reference Frames 113
3.3 Single Uncoupled Rotor 113
    3.3.1 Lumped and Distributed Mass Matrices 115
        3.3.1.1 Lumped Mass Matrix 115
        3.3.1.2 Distributed Mass Matrix 116
    3.3.2 Stiffness Matrix 117
3.4 Coupled Rotors 119
    3.4.1 Coaxial Same-Speed Coupled Rotors 120
    3.4.2 Unbranched Systems with Rigid and Flexible
        Connections 121
        3.4.2.1 Rigid Connections 122
        3.4.2.2 Flexible Connections 124
        3.4.2.3 Complete Equations of Motion 124
    3.4.3 Branched Systems with Rigid and Flexible
        Connections 126
        3.4.3.1 Rigid Connections 127
        3.4.3.2 Flexible Connections 129
        3.4.3.3 Complete Equations of Motion 129
3.5 Semidefinite Systems 130
3.6 Examples 130
    3.6.1 High-Capacity Fan for Large Altitude
        Wind Tunnel........................... 130
    3.6.2 Four-Square Gear Tester 132
    3.6.3 Large Steam Turbo-Generator Sets 134
3.7 Summary................................... 135
Bibliography 137

Part II Rotor Dynamic Analyses

4 RDA Code for Lateral Rotor Vibration Analyses 141
4.1 Introduction................................. 141
4.2 Unbalance Steady-State Response Computations 142
    4.2.1 3-Mass Rotor Model + 2 Bearings and 1 Disk 145
    4.2.2 Phase Angle Explanation and Direction
        of Rotation 149

       4.2.3 3-Mass Rotor Model + 2 Bearings/Pedestals
             and 1 Disk 152
       4.2.4 Anisotropic Model: 3-Mass Rotor + 2 Bearings/
             Pedestals and 1 Disk 155
       4.2.5 Elliptical Orbits 158
       4.2.6 Campbell Diagrams 163
  4.3 Instability Self-Excited-Vibration Threshold
       Computations 165
       4.3.1 Symmetric 3-Mass Rotor + 2 Anisotropic Bearings
             (Same) and Disk 166
       4.3.2 Symmetric 3-Mass Rotor + 2 Anisotropic Bearings
             (Different) and Disk 172
  4.4 Additional Sample Problems 173
       4.4.1 Symmetric 3-Mass Rotor + 2 Anisotropic Bearings
             and 2 Pedestals 174
       4.4.2 Nine-Stage Centrifugal Pump Model, 17-Mass
             Stations, 2 Bearings 175
             4.4.2.1 Unbalance Response 175
             4.4.2.2 Instability Threshold Speed 178
       4.4.3 Nine-Stage Centrifugal Pump Model, 5-Mass
             Stations, 2 Bearings 179
  4.5 Summary 180
  Bibliography 180

5 **Bearing and Seal Rotor Dynamics** 183
  5.1 Introduction 183
  5.2 Liquid-Lubricated Fluid-Film Journal Bearings 184
       5.2.1 Reynolds Lubrication Equation 184
             5.2.1.1 For a Single RLE Solution Point 187
       5.2.2 Journal Bearing Stiffness and Damping
             Formulations 187
             5.2.2.1 Perturbation Sizes 189
             5.2.2.2 Coordinate Transformation Properties 190
             5.2.2.3 Symmetry of Damping Array 192
       5.2.3 Tilting-Pad Journal Bearing Mechanics 192
       5.2.4 Journal Bearing Stiffness and Damping Data and
             Resources 196
             5.2.4.1 Tables of Dimensionless Stiffness and
                     Damping Coefficients 198
       5.2.5 Journal Bearing Computer Codes 199
       5.2.6 Fundamental Caveat of LRV Analyses 199
             5.2.6.1 Example 200
  5.3 Experiments to Measure Dynamic Coefficients 201

　　　　5.3.1　Mechanical Impedance Method with Harmonic
　　　　　　　Excitation 203
　　　　5.3.2　Mechanical Impedance Method with Impact
　　　　　　　Excitation 208
　　　　5.3.3　Instability Threshold-Based Approach........... 210
　　5.4　Annular Seals....................................... 212
　　　　5.4.1　Seal Dynamic Data and Resources............... 215
　　　　5.4.2　Ungrooved Annular Seals for Liquids 215
　　　　　　　5.4.2.1　Lomakin Effect 216
　　　　　　　5.4.2.2　Seal Flow Analysis Models 218
　　　　　　　5.4.2.3　Bulk Flow Model Approach............. 219
　　　　　　　5.4.2.4　Circumferential Momentum Equation 219
　　　　　　　5.4.2.5　Axial Momentum Equation 220
　　　　　　　5.4.2.6　Comparisons between Ungrooved Annular
　　　　　　　　　　　Seals and Journal Bearings.............. 222
　　　　5.4.3　Circumferentially Grooved Annular Seals
　　　　　　　for Liquids 224
　　　　5.4.4　Annular Gas Seals 225
　　　　　　　5.4.4.1　Steam Whirl Compared to Oil Whip 226
　　　　　　　5.4.4.2　Typical Configurations for Annular
　　　　　　　　　　　Gas Seals............................. 227
　　　　　　　5.4.4.3　Dealing with Seal LRV-Coefficient
　　　　　　　　　　　Uncertainties......................... 229
　　5.5　Rolling Contact Bearings 230
　　5.6　Squeeze-Film Dampers............................... 235
　　　　5.6.1　Dampers with Centering Springs 236
　　　　5.6.2　Dampers without Centering Springs 237
　　　　5.6.3　Limitations of Reynolds Equation–Based
　　　　　　　Solutions 238
　　5.7　Magnetic Bearings 239
　　　　5.7.1　Unique Operating Features of Active Magnetic
　　　　　　　Bearings.................................... 240
　　　　5.7.2　Short Comings of Magnetic Bearings............. 241
　　5.8　Compliance Surface Foil Gas Bearings 243
　　5.9　Summary... 246
　　Bibliography 246

6　Turbo-Machinery Impeller and Blade Effects............... 251
　　6.1　Centrifugal Pumps.................................. 251
　　　　6.1.1　Static Radial Hydraulic Impeller Force 251
　　　　6.1.2　Dynamic Radial Hydraulic Impeller Forces 255
　　　　　　　6.1.2.1　Unsteady Flow Dynamic Impeller Forces 255
　　　　　　　6.1.2.2　Interaction Impeller Forces 257

    6.2 Centrifugal Compressors 260
        6.2.1 Overall Stability Criteria 260
        6.2.2 Utilizing Interactive Force Modeling Similarities
              with Pumps 262
    6.3 High-Pressure Steam Turbines and Gas Turbines 263
        6.3.1 Steam Whirl 263
            6.3.1.1 Blade Tip Clearance Contribution 264
            6.3.1.2 Blade Shroud Annular Seal Contribution 265
        6.3.2 Partial Admission in Steam Turbine
              Impulse Stages 269
        6.3.3 Combustion Gas Turbines 270
    6.4 Axial Flow Compressors........................... 270
    6.5 Summary 272
    Bibliography 272

Part III Monitoring and Diagnostics

7 Rotor Vibration Measurement and Acquisition 277
    7.1 Introduction to Monitoring and Diagnostics 277
    7.2 Measured Vibration Signals and Associated Sensors 281
        7.2.1 Accelerometers 281
        7.2.2 Velocity Transducers 283
        7.2.3 Displacement Transducers 284
            7.2.3.1 Background 284
            7.2.3.2 Inductance (Eddy-Current) Noncontacting
                    Position Sensing Systems................. 285
    7.3 Vibration Data Acquisition 289
        7.3.1 Continuously Monitored Large Multibearing
              Machines 289
        7.3.2 Monitoring Several Machines at Regular Intervals ... 291
        7.3.3 Research Laboratory and Shop Test Applications.... 292
    7.4 Signal Conditioning 292
        7.4.1 Filters...................................... 293
        7.4.2 Amplitude Conventions 294
    7.5 Summary 295
    Bibliography 295

8 Vibration Severity Guidelines 297
    8.1 Introduction.................................... 297
    8.2 Casing and Bearing Cap Vibration Displacement
        Guidelines 298
    8.3 Standards, Guidelines, and Acceptance Criteria 300
    8.4 Shaft Displacement Criteria 301

Contents

    8.5 Summary 302
    Bibliography 303
    Bibliography Supplement 303

9 Signal Analysis and Identification of Vibration Causes 307
    9.1 Introduction 307
    9.2 Vibration Trending and Baselines 307
    9.3 FFT Spectrum 308
    9.4 Rotor Orbit Trajectories 310
    9.5 Bode, Polar, and Spectrum Cascade Plots 317
    9.6 Wavelet Analysis Tools 321
    9.7 Chaos Analysis Tools 325
    9.8 Symptoms and Identification of Vibration Causes 330
        9.8.1 Rotor Mass Unbalance Vibration 330
        9.8.2 Self-Excited Instability Vibrations 331
            9.8.2.1 Oil Whip 333
            9.8.2.2 Steam Whirl 333
            9.8.2.3 Instability Caused by Internal Damping
                    in the Rotor 334
            9.8.2.4 Other Instability Mechanisms 336
        9.8.3 Rotor–Stator Rub-Impacting 336
        9.8.4 Misalignment 339
        9.8.5 Resonance 340
        9.8.6 Mechanically Loose Connections 341
        9.8.7 Cracked Shafts 342
        9.8.8 Rolling-Element Bearings, Gears, and
            Vane/Blade-Passing Effects 342
    9.9 Summary 343
    Bibliography 344

Part IV Trouble-Shooting Case Studies

10 Forced Vibration and Critical Speed Case Studies 349
    10.1 Introduction 349
    10.2 HP Steam Turbine Passage through First
           Critical Speed 350
    10.3 HP–IP Turbine Second Critical Speed through
           Power Cycling 352
    10.4 Boiler Feed Pumps: Critical Speeds at Operating Speed ... 354
        10.4.1 Boiler Feed Pump Case Study 1 354
        10.4.2 Boiler Feed Pump Case Study 2 358
        10.4.3 Boiler Feed Pump Case Study 3 360
    10.5 Nuclear Feed Water Pump Cyclic Thermal Rotor Bow 361

	10.6	Power Plant Boiler Circulating Pumps	364
	10.7	Nuclear Plant Cooling Tower Circulating Pump Resonance	367
	10.8	Generator Exciter Collector Shaft Critical Speeds	367
	10.9	Summary	369
	Bibliography		370
11	**Self-Excited Rotor Vibration Case Studies**		371
	11.1	Introduction	371
	11.2	Swirl Brakes Cure Steam Whirl in a 1300 MW Unit	371
	11.3	Bearing Unloaded by Nozzle Forces Allows Steam Whirl	375
	11.4	Misalignment Causes Oil Whip/Steam Whirl "Duet"	377
	11.5	Summary	378
	Bibliography		379
12	**Additional Rotor Vibration Cases and Topics**		381
	12.1	Introduction	381
	12.2	Vertical Rotor Machines	381
	12.3	Vector Turning from Synchronously Modulated Rubs	384
	12.4	Air Preheater Drive Structural Resonances	391
	12.5	Aircraft Auxiliary Power Unit Commutator Vibration-Caused Uneven Wear	393
	12.6	Impact Tests for Vibration Problem Diagnoses	397
	12.7	Bearing Looseness Effects	398
		12.7.1 350 MW Steam Turbine Generator	398
		12.7.2 BFP 4000 hp Electric Motor	399
		12.7.3 LP Turbine Bearing Looseness on a 750 MW Steam Turbine Generator	400
	12.8	Tilting-Pad versus Fixed-Surface Journal Bearings	401
		12.8.1 A Return to the Machine of Section 11.4 of Chapter 11 Case Study	402
	12.9	Base-Motion Excitations from Earthquake and Shock	403
	12.10	Parametric Excitation: Nonaxisymmetric Shaft Stiffness	404
	12.11	Rotor Balancing	406
		12.11.1 Static Unbalance, Dynamic Unbalance, and Rigid Rotors	407
		12.11.2 Flexible Rotors	408
		12.11.3 Influence Coefficient Method	410
		12.11.4 Balancing Computer Code Examples and the Importance of Modeling	412
		12.11.5 Case Study of 430 MW Turbine Generator	418

       12.11.6 Continuous Automatic In-Service
                Rotor Balancing 419
       12.11.7 In-Service Single-Plane Balance Shot 421
   12.12 Summary 422
   Bibliography 422

Index 425

Preface

Every spinning rotor has some vibration, at least a once-per-revolution frequency component, because it is of course impossible to make any rotor perfectly mass balanced. Experience has provided guidelines for quantifying approximate comfortable safe upper limits for allowable vibration levels on virtually all types of rotating machinery. It is rarely disputed that such limits are crucial to machine durability, reliability, and life. However, the appropriate magnitude of such vibration limits for specific machinery is often disputed, with the vendor's limit usually being higher than a prudent equipment purchaser's wishes. Final payment for a new machine is occasionally put on hold, pending resolution of the machine's failure to operate below the vibration upper limits prescribed in the purchase specifications.

The mechanics of rotating machinery vibration is an interesting field, with considerable technical depth and breadth, utilizing first principles of all the mechanical engineering fundamental areas, solid mechanics, dynamics, fluid mechanics, heat transfer, and controls. Many industries rely heavily on reliable trouble-free operation of rotating machinery. These industries include power generation; petrochemical processes; manufacturing; land, sea and air transportation; heating and air conditioning; aerospace propulsion; computer disk drives; textiles; home appliances; and a wide variety of military systems. However, the level of basic understanding and competency on the subject of rotating machinery vibration varies greatly among the various affected industries. In the author's opinion, all industries reliant on rotating machinery would benefit significantly from a strengthening of their in-house competency on the subject of rotating machinery vibration. A major mission of this book is to foster an understanding of rotating machinery vibration, in both industry and academia.

Even with the best of design practices and the most effective methods of avoidance, many rotor vibration causes are so subtle and pervasive that incidents of excessive vibration in need of solutions continue to occur. Thus, a major task for the vibrations engineer is *diagnosis* and *correction*. To that end, this book is comprised of four sequential parts.

Part I: *Primer on Rotor Vibration* is a group of three chapters that develop the fundamentals of rotor vibration, starting with basic vibration concepts, followed by lateral rotor vibration and torsional rotor vibration principles and problem formulations.

Part II: *Use of Rotor Dynamic Analyses* is a group of three chapters focused on the general-purpose lateral rotor vibration PC-based code supplied with this book. This code was developed in the author's group at Case Western Reserve University and is based on the finite element approach explained in Part I. Major topics are the calculation of rotor unbalance response, the calculation of self-excited instability vibration thresholds, bearing and seal dynamic properties, and turbo-machinery impeller and blade effects on rotor vibration. In addition to their essential role in the total mission of this book, Parts I and II also provide the fundamentals of the author's graduate-level course in rotor vibration. In that context, Parts I and II provide an in-depth treatment of rotor vibration design analysis methods.

Part III: *Monitoring and Diagnostics* is a group of three chapters on measurements of rotor vibration and how to use the measurements to identify and diagnose problems in actual rotating machinery. Signal analysis methods and experience-based guidelines are provided. Approaches are given on how measurements can be used in combination with computer model analyses to optimally diagnose and alleviate rotor vibration problems.

Part IV: *Trouble-Shooting Case Studies* is a group of three chapters devoted to rotor vibration trouble-shooting case studies and topics from the author's many years of troubleshooting and problem-solving experiences. Major problem-solving cases include critical speeds and high sensitivity to rotor unbalance, self-excited rotor vibration and thresholds, unique rotor vibration characteristics of vertical machines, rub-induced high-amplitude vibration, loose parts causing excessive vibration levels, excessive support structure vibration and resonance, vibration-imposed DC motor-generator uneven commutator wear, and rotor balancing. The sizes of machines in these case studies range from a 1300 MW steam turbine generator unit to a 10 kW APU for a jet aircraft. These case studies are heavily focused on power generation equipment, including turbines, generators, exciters, large pumps for fossil-fired and nuclear PWR and BWR-powered plants, and air handling equipment. There is a common thread in all these case studies: namely, the combined use of on-site vibration measurements and signal processing along with computer-based rotor vibration model development and analyses as the optimum overall multipronged approach to maximize the probability of successfully identifying and curing problems of excessive vibration levels in rotating machinery.

The main objectives of this book are to cover all the major rotor vibration topics in a unified presentation, and to demonstrate the solving rotating machinery vibration problems. These objectives are addressed by providing depth and breadth to the governing fundamental principles plus a background in modern measurement and computational tools for rotor vibration design analyses and troubleshooting. In all engineering problem-solving endeavors, the surest way to success is to gain *physical insight* into the important phenomena involved in the problem, and that axiom is especially true in the field of rotating machinery vibration. It is the author's hope that this book will aid those seeking to gain such insight.

Maurice L. Adams, Jr.
Case Western Reserve University
Cleveland, OH

Acknowledgments

Truly qualified technologists invariably acknowledge the shoulders upon which they stand. I am unusually fortunate in having worked for several expert caliber individuals during my formative 14 years of industrial employment prior to entering academia in 1977. I first acknowledge those individuals, many of whom have unfortunately passed away over the years.

In the mid-1960s, my work in rotor dynamics began at Worthington Corporation's Advanced Products Division (APD) in Harrison, New Jersey. There I worked under two highly capable European-bred engineers, Chief Engineer Walter K. Jekat (German) and his assistant, John P. Naegeli (Swiss). John Naegeli later returned to Switzerland and eventually became the general manager of Sulzer's Turbo-Compressor Division and later the general manager of their Pump Division. My first assignment at APD was basically to be "thrown into the deep end" of a new turbo-machinery development for the U.S. Navy that even today would be considered highly challenging. That new product was comprised of a 42,000 rpm rotor with an overhung centrifugal air compressor impeller at one end and an overhung single-stage impulse steam turbine powering the rotor from the other end. The two journal bearings and the double-acting thrust bearing were all hybrid hydrodynamic–hydrostatic fluid-film bearings with water as the lubricant and running quite into the turbulence regime. Worthington subsequently sold several of these units over a period of many years. While at APD, my interest in and knowledge of centrifugal pumps grew considerably through my frequent contacts with the APD general manager, Igor Karassik, the world's most prolific writer of centrifugal pump articles, papers, and books and energetic teacher to all the then-young recent engineering graduates at APD like myself.

In 1967, having become quite seriously interested in the bearing, seal, and rotor dynamics field, I seized on an opportunity to work for an internationally recognized group at the Franklin Institute Research Laboratories (FIRL) in Philadelphia. I am eternally indebted to those individuals for the knowledge I gained from them and for their encouragement to me to pursue graduate studies part-time, which led to my engineering master's degree from a local Penn State extension near Philadelphia. The list of individuals I worked under at FIRL is almost a *who's-who list* for the field, and includes the following: Harry Rippel (fluid-film bearings), John Rumbarger (rolling-element bearings), Wilbur Shapiro (fluid-film bearings, seals, and rotor dynamics), and Elemer Makay (centrifugal pumps). I also had the privilege of working with a distinguished group of FIRL's consultants from Columbia University, specifically Professors Dudley D. Fuller, Harold G. Elrod, and Victorio "Reno" Castelli. In terms of working with the right people in one's chosen field, my 4 years at the Franklin Institute were surely the proverbial lode.

My job at Franklin Institute provided me the opportunity to publish articles and papers in the field. That bit of national recognition helped provide my next job opportunity. In 1971, I accepted a job in what was then a truly distinguished industrial research organization, the Mechanics Department at Westinghouse's Corporate R&D Center near Pittsburgh. The main attraction of this job for me was my new boss, Dr. Albert A. Raimondi, manager of the bearing mechanics section, whose famous papers on fluid-film bearings I had been using ever since my days at Worthington. An added bonus was the presence of the person holding the department manager position, A. C. "Art" Hagg, the company's then internationally recognized rotor vibration expert. My many interactions with Art Hagg were all professionally enriching. At Westinghouse, I was given the lead role on several "cutting-edge" projects, including nonlinear dynamics of flexible multibearing rotors for large steam turbines and reactor coolant pumps, bearing load determination for vertical multibearing rotors, seal development for refrigeration centrifugal compressors, and turning-gear slow-roll operation of journal bearings, developing both experiments and new computer codes for these projects. I became the junior member of an elite ad hoc trio that included Al Raimondi and D. V. "Kirk" Wright (manager of the dynamics section). Al and Kirk were the ultimate teachers and perfectionists, members of a now extinct breed of giants who unfortunately have not been replicated in today's industrial workplace environment. They encouraged and supported me in pursuing my PhD part-time, which I completed at the University of Pittsburgh in early 1977. Last, but not least, my PhD thesis advisor at Pitt, Professor Andras Szeri, taught me a deep understanding of fluid dynamics and continuum mechanics, and is also internationally recognized in the bearing mechanics field.

Upon completing my PhD in 1977, I ventured into a new employment world considerably different from any of my previous jobs, academia. Here I have found my calling. And here is where I state my biggest acknowledgment, to my students, both undergraduate and graduate, from whom I continue to learn every day as I seek to teach some of the next generation's best engineers. No wonder that even at an age closely approaching 70 years, I have no plans for retirement. It must be abundantly clear to those around me that I really enjoy mechanical engineering, since all four of my children (sons) have each independently chosen the mechanical engineering profession for themselves.

The first inspiration I had for writing a book on rotor vibration was in 1986, during the 10-week course I taught on the subject at the Swiss Federal

Acknowledgments   xxiii

Institute (ETH) in Zurich. I attribute that inspiration to the enriching intellectual atmosphere at the ETH and to my association with my host, the then ETH Professor of Turbomachinery, Dr. George Gyarmathy. I do, however, believe it is fortunate that I did not act upon that inspiration for another 10 years, because in that succeeding 10-year period I learned so much more in my field, which therefore also got into the first edition of this book. My learning in this field still continues, as the new case studies in this second edition clearly attest.

Author

Maurice L. Adams, Jr. is the founder and past president of Machinery Vibration Inc., as well as professor of mechanical and aerospace engineering at Case Western Reserve University. The author of over 100 publications and the holder of U.S. patents, he is a member of the American Society of Mechanical Engineers. Professor Adams received the BSME degree (1963) from Lehigh University, Bethlehem, Pennsylvania; the MEngSc degree (1970) from the Pennsylvania State University, University Park; and the PhD degree (1977) from the University of Pittsburgh, Pennsylvania. Prior to becoming a professor in 1977, Dr. Adams worked on rotating machinery engineering for 14 years in industry, including employment at Allis Chalmers, Worthington, Franklin Institute Research Laboratories, and Westinghouse Corporate R&D Center.

Part I

Primer on Rotor Vibration

1
Vibration Concepts and Methods

1.1 One-Degree-of-Freedom Model

The mass–spring–damper model, shown in Figure 1.1, is the *starting point* for understanding mechanical vibrations. A thorough understanding of this most elementary vibration model and its full range of vibration characteristics is absolutely essential to a comprehensive and insightful study of the rotating machinery vibration field. The fundamental physical law governing all vibration phenomena is Newton's Second Law, which in its most commonly used form says that *the sum of the forces acting upon an object is equal to its mass times its acceleration*. Both force and acceleration are vectors, so Newton's Second Law, written in its general form, yields a vector equation. For the one-degree-of-freedom (1-DOF) system, this reduces to a scalar equation, as follows:

$$F = ma \qquad (1.1)$$

where F is the sum of forces acting upon the body, m is the mass of the body, and a is the acceleration of the body.

For the system in Figure 1.1, $F = ma$ yields its differential equation of motion as follows:

$$m\ddot{x} + c\dot{x} + kx = f(t) \qquad (1.2)$$

For the system in Figure 1.1, the forces acting upon the mass include the externally applied time-dependent force, $f(t)$, plus the spring and damper motion-dependent connection forces, $-kx$ and $-c\dot{x}$. Here, the minus signs account for the spring force resisting displacement (x) in either direction from the equilibrium position and the damper force resisting velocity ($\dot{x}$) in either direction. The weight (mg) and static deflection force ($k\delta_{st}$) that the weight causes in the spring cancel each other. Equations of motion are generally written about the static equilibrium position state and then need not contain weight and weight-balancing spring deflection forces.

1.1.1 Assumption of Linearity

In the model of Equation 1.2, as in most vibration analysis models, spring and damper connection forces are assumed to be linear with (proportional

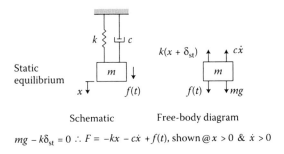

$mg - k\delta_{st} = 0 \therefore F = -kx - c\dot{x} + f(t)$, shown @ $x > 0$ & $\dot{x} > 0$

FIGURE 1.1 One-DOF linear spring–mass–damper model.

to) their respective driving parameters, that is, displacement (x) across the spring and velocity ($\dot{x}$) across the damper. These forces are therefore related to their respective driving parameters by proportionality factors, stiffness "k" for the spring and "c" for the damper. Linearity is a simplifying assumption that permeates most vibration analyses because the *equations of motion* are then made *linear*, even though real systems are never completely linear. Fortunately, the assumption of linearity leads to adequate answers in most vibration engineering analyses and simplifies considerably the tasks of making calculations and understanding what is calculated. Some specialized large-amplitude rotor vibration problems justify treating nonlinear effects, for example, large rotor unbalance such as from turbine blade loss, shock and seismic base-motion excitations, rotor rub-impact phenomena, and instability vibration limit cycles. These topics are treated in subsequent sections of this book.

1.1.2 Unforced System

The solution for the motion of the *unforced 1-DOF system* is important in its own right, but specifically important in laying the groundwork to study *self-excited instability rotor vibrations*. If the system is considered to be unforced, then $f(t) = 0$ and Equation 1.2 becomes

$$m\ddot{x} + c\dot{x} + kx = 0 \tag{1.3}$$

This is a second-order homogeneous ordinary differential equation (ODE). To solve for $x(t)$ from Equation 1.3, one needs to specify the two initial conditions, $x(0)$ and $\dot{x}(0)$. Assuming that k and c are both positive, there are three categories of solutions that can result from Equation 1.3: (i) *underdamped*, (ii) *critically damped*, and (iii) *overdamped*. These are just the traditional labels used to describe the three distinct types of roots and the corresponding three motion categories that Equation 1.3 can potentially yield when k and c are both positive. Substituting the known solution form

($Ce^{\lambda t}$) into Equation 1.3 and then canceling out the solution form yields the following quadratic equation for its roots (eigenvalues) and leads to the equation for the extracted two roots, $\lambda_{1,2}$, as follows:

$$m\lambda^2 + c\lambda + k = 0 \qquad (1.4)$$

$$\lambda_{1,2} = -\frac{c}{2m} \pm \sqrt{\left(\frac{c}{2m}\right)^2 - \left(\frac{k}{m}\right)}$$

The three categories of root types possible from Equation 1.4 are listed as follows:

Underdamped: $(c/2m)^2 \leq (k/m)$, complex conjugate roots, $\lambda_{1,2} = \alpha \pm i\omega_d$.
Critically damped: $(c/2m)^2 = (k/m)$, equal real roots, $\lambda_{1,2} = \alpha$.
Overdamped: $(c/2m)^2 \geq (k/m)$, real roots, $\lambda_{1,2} = \alpha \pm \beta$.

The well-known $x(t)$ time signals for these three solution categories are illustrated in Figure 1.2 along with the *undamped* system (i.e., $c = 0$). In most mechanical systems, the important vibration characteristics are contained in modes with the so-called *underdamped* roots, as is certainly the case for rotor dynamical systems. The general expression for the motion of the *unforced underdamped* system is commonly expressed in any one of the following four forms:

$$x(t) = Xe^{\alpha t} \begin{cases} \sin(\omega_d t + \phi_s^+) \text{ or } \sin(\omega_d t - \phi_s^-) \\ \text{OR} \\ \cos(\omega_d t + \phi_c^+) \text{ or } \cos(\omega_d t - \phi_c^-) \end{cases} \qquad (1.5)$$

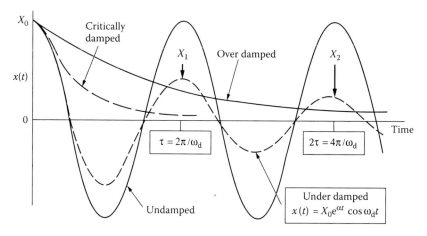

FIGURE 1.2 Motion types for the unforced 1-DOF system.

where X is the single-peak amplitude of exponential decay envelop at $t = 0$; $\omega_d = \sqrt{\omega_n^2 - \alpha^2}$, the damped natural frequency; phase angles $\phi_s^- = -\phi_s^+$, $\phi_s^+ = \phi_c^+ + 90°$, and $\phi_c^- = -\phi_c^+$ yield the same signal; $\alpha = -c/2m$, the real part of eigenvalue for underdamped system; $\omega_n = \sqrt{k/m}$, the undamped natural frequency; and $i = \sqrt{-1}$.

1.1.3 Self-Excited Dynamic-Instability Vibrations

The *unforced underdamped* system's solution, as expressed in Equation 1.5, provides a convenient way to introduce the concept of vibrations caused by *dynamic instability*. In many standard treatments of vibration theory, it is tacitly assumed that $c \geq 0$. However, the concept of *negative damping* is a convenient way to model some dynamic interactions that tap an available energy source, modulating the tapped energy to produce the so-called *self-excited vibration*.

Using the typical (shown later) multi-DOF models employed to analyze rotor-dynamical systems, design computations are performed to determine operating conditions at which self-excited vibrations are predicted. These analyses essentially are a *search* for zones of operation within which the *real part* (α) of any of the system's eigenvalues becomes positive. It is usually one of the rotor-bearing system's lower frequency corotational-orbit-direction vibration modes, at a natural frequency less than the spin speed frequency, whose eigenvalue real part becomes positive. The transient response of this mode is basically the same as would be the response of the 1-DOF system of Equation 1.3 with $c < 0$ and $c^2 < 4km$, which produces $\alpha > 0$, a positive real part for the two complex conjugate roots of Equation 1.4. As Figure 1.3 shows, this is the classic self-excited vibration case, exhibiting a vibratory motion with an exponential *growth* envelope, as opposed to the exponential *decay* envelope (for $c > 0$) shown in Figure 1.2. The widely accepted fact that safe reliable operation of rotating machinery must preclude such dynamical instabilities from zones of operation can be readily appreciated just from the graph shown in Figure 1.3.

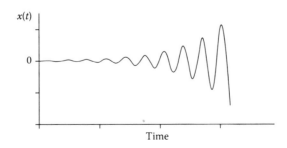

FIGURE 1.3 Initial growth of dynamical instability from an initial disturbance.

1.1.4 Steady-State Sinusoidally Forced Systems

If the system is dynamically stable ($c > 0$), that is, the natural mode is positively damped, as illustrated in Figure 1.2, then long-term vibration can persist only as the result of some long-term forcing mechanism. In rotating machinery, the one long-term forcing mechanism that is always present is the residual mass unbalance distribution in the rotor, and that can never be completely eliminated. Rotor mass unbalances are modeled by equivalent forces fixed in the rotor, in other words, a group of rotor-synchronous rotating loads each with a specified magnitude and phase angle locating it relative to a common angular reference point (*keyphaser*) fixed on the rotor. When viewed from a fixed radial direction, the projected component of such a rotating unbalance force varies sinusoidally in time at the rotor spin frequency. Without pre-empting the subsequent treatment in this book on the important topic of *rotor unbalance*, suffice it to say that there is a considerable similarity between the unbalance-driven vibration of a rotor and the steady-state response of the 1-DOF system described by Equation 1.2 with $f(t) = F_o \sin(\omega t + \theta)$. Equation 1.2 then becomes the following:

$$m\ddot{x} + c\dot{x} + kx = F_o \sin(\omega t + \theta) \tag{1.6}$$

where F_o is the force magnitude, θ is the force phase angle, and ω is the forcing frequency.

It is helpful at this point to recall the relevant terminology from the mathematics of *differential equations*, with reference to the solution for Equation 1.2. Since this is a linear differential equation, its *total solution* can be obtained by a linear superposition or adding of two component solutions: the *homogeneous* solution and the *particular* solution. For the *unforced* system, embodied in Equation 1.3, the homogeneous solution is the total solution, because $f(t) = 0$ yields a zero particular solution. For any nonzero $f(t)$, unless the initial conditions, $x(0)$ and $\dot{x}(0)$, are specifically chosen to start the system on the steady-state solution, there will be a start-up transient portion of the motion which, for stable systems, will die out as time progresses. This start-up transient is contained in the *homogeneous* solution, that is, Figure 1.2. The steady-state long-term motion is contained in the *particular* solution.

Rotating machinery designers and troubleshooters are concerned with long-term exposure vibration levels, because of material fatigue considerations, and are concerned with maximum peak vibration amplitudes passing through forced resonances within the operating zones. It is therefore only the steady-state solution, such as of Equation 1.6, that is most commonly extracted. Because this system is *linear*, only the frequency(s) in $f(t)$ will be present in the steady-state (*particular*) solution. Thus the solution of Equation 1.6 can be expressed in any of the following four steady-state solution forms, with phase angle as given for Equation 1.5 for each to represent the same signal:

$$x(t) = X \begin{Bmatrix} \sin(\omega t + \phi_s^+) \text{ or } \sin(\omega t - \phi_s^-) \\ \cos(\omega t + \phi_c^+) \text{ or } \cos(\omega t - \phi_c^-) \end{Bmatrix} \quad (1.7)$$

The steady-state *single-peak* vibration amplitude (X) and its phase angle relative to the force (let $\theta = 0$) are solvable as functions of the sinusoidally varying force magnitude (F_o) and frequency (ω), mass (m), spring stiffness (k), and damper coefficient (c) values. This can be presented in the standard normalized form shown in Figure 1.5.

1.1.5 Damping

Mechanical vibratory systems typically fall into the *underdamped* category, so each individual system mode of importance can thereby be accurately handled in the modal-coordinate space (Section 1.3 of this chapter) as the 1-DOF model illustrated in Figure 1.1. This is convenient since modern digital signal processing methods can separate out each mode's *underdamped* exponential decay signal from a total transient (e.g., impact initiated) time-base vibration test signal. Each mode's linear damping coefficient can then be determined employing the *log-decrement* method, as outlined here. Referring to Figures 1.2 and 1.4c, test data for a mode's *underdamped* exponential decay signal can be used to determine the damping

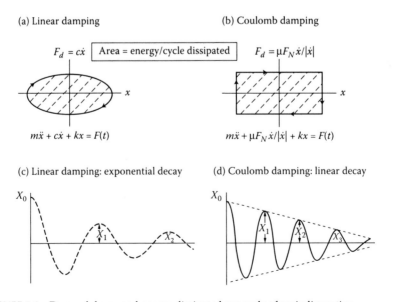

FIGURE 1.4 Damped decay and energy dissipated per cycle of periodic motion.

coefficient as follows:

$$x(n\tau) = X_0 e^{\alpha n\tau}, \quad \therefore \frac{x(n\tau)}{x(0)} = e^{\alpha n\tau}, \quad \therefore \alpha n\tau = \ln\left(\frac{X_n}{X_0}\right)$$

$$1 - \text{cycle period} \equiv \tau = \frac{2\pi}{\omega_d}$$

Recalling $\alpha = -\dfrac{c}{2m}$ yields the damping coefficient $\quad c = -\dfrac{2m}{n\tau}\ln\left(\dfrac{X_n}{X_0}\right)$

(1.8)

Vibration damping means are extremely important in nature as well as engineered devices. The standard linear model for damping is akin to a drag force proportional to velocity magnitude. But many important damping mechanisms are nonlinear, for example, Coulomb damping, internal material hysteresis damping. What is typically done to handle the modeling of nonlinear damping is to approximate it with the linear model by matching energy dissipated per cycle. This works well since modest amounts of damping have negligible effect on natural frequency. Energy/cycle dissipated by damping under single-frequency harmonic cycling is illustrated in Figure 1.4 for linear and Coulomb friction damping.

The *log-decrement* test method for determining damping was previously shown to utilize the transient decay motion of an initially displaced but unforced system. In contrast, the *half power bandwidth* test method utilizes the steady-state response to a harmonic excitation force. The steady-state linear response to a single-frequency harmonic excitation force of slowly varied frequency will correspond to a member of the family displayed in Figure 1.5a. For the single-DOF linear damped model, the following equation is applicable for low damped systems:

$$Q \equiv \frac{\text{Frequency at peak vibration amplitude}}{\omega_2 - \omega_1} = \frac{1}{2\varsigma} \quad (1.9)$$

$$Q \equiv \frac{\omega_{\text{peak}}}{\omega_2 - \omega_1} = \frac{1}{2\varsigma}$$

Referring to the underdamped plots in Figure 1.5a, $\omega_1 < \omega_{\text{peak}}$ and $\omega_2 > \omega_{\text{peak}}$ are the frequencies where a horizontal line at $0.707 \times$ *amplitude peak* intersects the particular amplitude versus the frequency plot. $\omega_{\text{peak}} \simeq \omega_n$ is the frequency at peak vibration amplitude. Q comes from the word *quality*, long used to measure the *quality* of an electrical resonance circuit. The term *high Q* is often used synonymously for *low damping*. For sources of damping other than linear, such as structural damping, the equivalent linear damping coefficient can be determined using Equation 1.9. So the

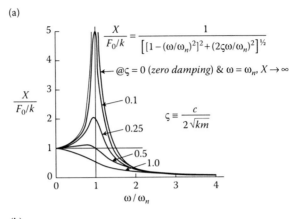

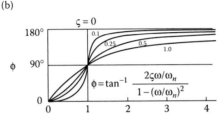

FIGURE 1.5 One-DOF steady-state response to a sinusoidal force. (a) $X/(F_0/k)$ versus ω/ω_n, (b) ϕ versus ω/ω_n, $\varsigma = 1$ at critically damped.

sharpness-of-peak of a measured steady-state plot of vibration amplitude versus frequency provides a measure of the damping present.

1.1.6 Undamped Natural Frequency: An Accurate Approximation

Because of the modest amounts of damping typical of most mechanical systems, the undamped model provides good answers for natural frequencies in most situations. Figure 1.5 shows that the natural frequency of the 1-DOF model is the frequency at which an excitation force produces maximum vibration (i.e., a *forced resonance*) and is thus important. As shown in a subsequent topic of this chapter (*Modal Decomposition*), each natural mode of an undamped multi-DOF model is exactly equivalent to an undamped 1-DOF model. Therefore, the accurate approximation now shown for the 1-DOF model is usually applicable to the important modes of multi-DOF models.

The ratio (ς) of *damping to critical damping* (frequently referenced as a percentage, e.g., $\varsigma = 0.1$ is "10% damping") is derivable as follows. Shown with Equation 1.4, the defined condition for "critically damped" is $(c/2m)^2 = (k/m)$, which yields $c = 2\sqrt{km} \equiv c_c$, the "critical damping."

Vibration Concepts and Methods

Therefore, the *damping ratio*, defined as $\varsigma \equiv c/c_c$, can be expressed as follows:

$$\varsigma \equiv \frac{c}{2\sqrt{km}} \qquad (1.10)$$

With Equations 1.4 and 1.5, the following were defined: $\omega_n = \sqrt{k/m}$ (undamped natural frequency), $\alpha = -c/2m$ (real part of eigenvalue for an underdamped system), and $\omega_d = \sqrt{\omega_n^2 - \alpha^2}$ (damped natural frequency). Using these expressions with Equation 1.10 for the damping ratio (ς) leads directly to the following formula for the damped natural frequency:

$$\omega_d = \omega_n \sqrt{1 - \varsigma^2} \qquad (1.11)$$

This well-known important formula clearly shows just how well the *undamped natural frequency* approximates the *damped natural frequency* for typical applications. For example, a generous damping estimate for most potentially resonant mechanical system modes is 10–20% of critical damping ($\varsigma = 0.1$–0.2). Substituting the values $\varsigma = 0.1$ and 0.2 into Equation 1.11 gives $\omega_d = 0.995\omega_n$ for 10% damping and $\omega_d = 0.98\omega_n$ for 20% damping, that is, 0.5% error and 2% error, respectively. For even smaller damping ratio values typical of many structures, the approximation just gets better. A fundamentally important and powerful *dichotomy*, applicable to the important modes of many mechanical and structural vibratory systems, becomes clear within the context of this accurate approximation: *A natural frequency is only slightly lowered by the damping, but the peak vibration caused by an excitation force at the natural frequency is overwhelmingly lowered by the damping.* Figure 1.5 clearly shows all this.

1.1.7 1-DOF Model as an Approximation

Equation 1.2 is an exact mathematical model for the system schematically illustrated in Figure 1.1. However, *real-world* vibratory systems do not look like this classic 1-DOF picture, but in many cases it adequately approximates them for the purposes of engineering analyses. An appreciation for this is essential for one to make *the connection* between the mathematical models and the real devices, for whose analysis the models are employed.

One of many important examples is the concentrated mass (m) supported at the free end of a uniform cantilever beam (length L, bending moment of inertia I, Young's modulus E) as shown in Figure 1.6a. If the *concentrated mass* has considerably more mass than the beam, one may reasonably assume the beam to be *massless*, at least for the purpose of analyzing vibratory motions at the system's lowest natural frequency transverse mode. One can thereby adequately approximate the *fundamental mode* by

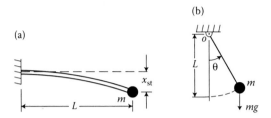

FIGURE 1.6 Two examples treated as linear 1-DOF models: (a) cantilever beam with a concentrated end mass and (b) simple pendulum.

a 1-DOF model. For *small* transverse static deflections (x_{st}) at the free end of the cantilever beam resulting from a transverse static load (F_{st}) at its free end, the equivalent spring stiffness is obtained directly from the cantilever beam's static deflection formula. This leads directly to the equivalent 1-DOF undamped system equation of motion, from which its undamped *natural frequency* (ω_n) is extracted, as follows:

$$x_{st} = \frac{F_{st}L^3}{3EI} \quad \text{(beam deflection formula)} \quad \text{and} \quad F_{st} \equiv kx_{st}$$

$$\therefore k = \frac{F_{st}}{x_{st}} = \frac{3EI}{L^3}$$

Then,

$$m\ddot{x} + \left(\frac{3EI}{L^3}\right)x = 0, \quad \therefore \omega_n = \sqrt{\frac{k}{m}} = \sqrt{\frac{3EI}{mL^3}} \quad (1.12)$$

In this example, the primary approximation is that the *beam is massless*. The secondary approximation is that the *deflections are small* enough so that simple linear beam theory provides a good approximation of beam deflection.

A second important example is illustrated in Figure 1.6b, the simple planar pendulum having a mass (m) concentrated at the free end of a rigid link of negligible mass and length (L). The appropriate form of Newton's Second Law for motion about the fixed pivot point of this model is $M = J\ddot{\theta}$, where M is the sum of moments about the pivot point "o," J (equal to mL^2 here) is the mass moment-of-inertia about the pivot point, and θ is the single motion coordinate for this 1-DOF system. The instantaneous sum of moments about the pivot point "o" consists only of that from the gravitational force mg on the concentrated mass, which is shown as follows (minus sign because M is always opposite θ):

$$M = -mgL\sin\theta, \quad \therefore mL^2\ddot{\theta} + mgL\sin\theta = 0$$

Dividing by mL^2 gives the following motion equation:

$$\ddot{\theta} + \left(\frac{g}{L}\right)\sin\theta = 0 \tag{1.13}$$

This equation of motion is obviously nonlinear. However, for small motions ($\theta \ll 1$) $\sin\theta \cong \theta$; hence it can be linearized as an approximation, as follows:

$$\ddot{\theta} + \left(\frac{g}{L}\right)\theta = 0, \quad \therefore \omega_n = \sqrt{\frac{g}{L}} \tag{1.14}$$

In this last example, the primary approximation is that the *motion is small*. The secondary approximation is that the pendulum has all its *mass concentrated* at its free end. Note that the stiffness or the restoring force effect in this model is not from a spring but from gravity. It is essential to make simplifying approximations in all vibration models, in order to have feasible engineering analyses. It is, however, also essential to understand the practical limitations of those approximations, to avoid producing analysis results that are highly inaccurate or, worse, do not even make physical sense.

1.2 Multi-DOF Models

It is conventional practice to model rotor dynamical systems with multi-DOF models, usually by utilizing standard finite element procedures. To comfortably apply and understand such models, it is helpful to first consider somewhat simpler models having more than 1-DOF.

The number of degrees of freedom (DOFs) of a dynamical system is the number of kinematically independent spatial coordinates required to uniquely and totally specify any position state the system can have. Consequently, with $F = ma$ the governing physical principle, this DOF number is also equal to the number of second-order ODEs required to mathematically characterize the system. Clearly, the 1-DOF system shown in Figure 1.1 is consistent with this general rule, that is, one spatial coordinate (x) and one ODE, Equation 1.2, to mathematically characterize the system. The 2-DOF system is the next logical step to study.

1.2.1 Two-DOF Models

As shown in the previous section, even the 1-DOF model can provide usable engineering answers when certain simplifying assumptions are justified. It is surely correct to infer that the 2-DOF model can provide usable

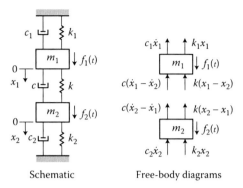

FIGURE 1.7 2-DOF model.

engineering answers over a considerably broader range of problems than the 1-DOF model. Also, first understanding the 2-DOF model is the best approach to tackling the subject of multi-DOF models. Figure 1.7 shows a common 2-DOF model. With the aid of the ever-important *free body diagrams*, application of $F = ma$ individually to each mass yields the following two equations of motion for this model:

$$m_1\ddot{x}_1 + (c + c_1)\dot{x}_1 + (k + k_1)x_1 - c\dot{x}_2 - kx_2 = f_1(t)$$
$$m_2\ddot{x}_2 + (c + c_2)\dot{x}_2 + (k + k_2)x_2 - c\dot{x}_1 - kx_1 = f_2(t) \tag{1.15}$$

With two or more DOFs, it is quite useful to write the equations of motion in matrix form, as follows for Equations 1.15:

$$\begin{bmatrix} m_1 & 0 \\ 0 & m_2 \end{bmatrix} \begin{Bmatrix} \ddot{x}_1 \\ \ddot{x}_2 \end{Bmatrix} + \begin{bmatrix} c + c_1 & -c \\ -c & c + c_2 \end{bmatrix} \begin{Bmatrix} \dot{x}_1 \\ \dot{x}_2 \end{Bmatrix}$$
$$+ \begin{bmatrix} k + k_1 & -k \\ -k & k + k_2 \end{bmatrix} \begin{Bmatrix} x_1 \\ x_2 \end{Bmatrix} = \begin{Bmatrix} f_1(t) \\ f_2(t) \end{Bmatrix} \tag{1.16}$$

For a multi-DOF system with any number of DOFs, the motion equations are typically written in the following condensed matrix notation:

$$[M]\{\ddot{x}\} + [C]\{\dot{x}\} + [K]\{x\} = \{f(t)\} \tag{1.17}$$

where [M] is the *mass matrix*, [C] is the *damping matrix*, and [K] is the *stiffness matrix*.

Note that all the three matrices in Equation 1.16 are symmetric, a property exhaustively treated in Section 2.4. Also note that Equation 1.16 is coupled through displacements and velocities, but not through accelerations. This is easily observable when the motion equations are in matrix form, as in

Vibration Concepts and Methods

Equation 1.16, noting that the mass matrix has *zeros* for the off-diagonal terms whereas the stiffness and damping matrices do not. Coupling means that these two differential equations are not independent, and thus are solvable only as a simultaneous pair. The stiffness and damping coupling of the two motion equations clearly reflects the physical model (Figure 1.7), as the two masses are connected to each other by a spring (k) and a damper (c).

Figure 1.8 shows a second 2-DOF example, the planar double-compound pendulum, which demonstrates acceleration (inertia) coupling. This example is also utilized here to introduce the well-known Lagrange equations, an alternate approach to applying $F = ma$ directly as done in all the previous examples. The Lagrange approach does not utilize the free-body diagrams that are virtually mandatory when applying $F = ma$ directly.

The Lagrange equations are derived directly from $F = ma$, and therefore embody the same physical principle. Their derivation can be found in virtually any modern second-level text on *Dynamics* or *Vibrations*, and they are expressible as follows:

$$\frac{d}{dt}\left(\frac{\partial T}{\partial \dot{q}_i}\right) - \frac{\partial T}{\partial q_i} + \frac{\partial V}{\partial q_i} = Q_i, \quad i = 1, 2, \ldots, n_{\text{DOF}} \quad (1.18)$$

The q_i's and $\dot{q}_i$'s are the *generalized coordinates* and *velocities*, respectively, T is the *kinetic energy*, V is the *potential energy*, and Q_i's are the *generalized forces*. Generalized coordinates can be either straight-line displacements (e.g., x, y, z) or angular displacements (e.g., $\theta_x, \theta_y, \theta_z$). Thus, a generalized force associated with a straight-line displacement will in fact have units of force, whereas a generalized force associated with an angular displacement will have units of moment or torque. Here, kinetic energy can be a function of both *generalized coordinates* and *velocities* whereas potential energy is a function of *generalized coordinates* only, that is, $T = T(\dot{q}_i, q_i)$ and $V = V(q_i)$. Obtaining the two equations of motion for the 2-DOF double-compound pendulum (Figure 1.8) is summarized as follows:

$$T = \tfrac{1}{2}m_1 v_1^2 + \tfrac{1}{2}m_2 v_2^2 \quad (1.19)$$

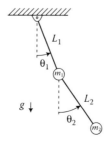

FIGURE 1.8 Planar double-compound pendulum with concentrated masses.

Here, v_1 and v_2 are the speeds of m_1 and m_2, respectively, and their squares result in the following:

$$v_1^2 = L_1^2 \dot{\theta}_1^2$$
$$v_2^2 = L_1^2 \dot{\theta}_1^2 + L_2^2 \dot{\theta}_2^2 + 2 L_1 L_2 \dot{\theta}_1 \dot{\theta}_2 (\cos \theta_1 \cos \theta_2 + \sin \theta_1 \sin \theta_2) \quad (1.20)$$
$$V = m_1 g L_1 (1 - \cos \theta_1) + m_2 g [L_1 (1 - \cos \theta_1) + L_2 (1 - \cos \theta_2)]$$

Substituting the T and V expressions into the Lagrange equations ($q_1 = \theta_1$ and $q_2 = \theta_2$) leads to the two equations of motion for the double-compound pendulum model shown in Figure 1.8. These two motion equations are nonlinear just as shown in Equation 1.13 for the simple pendulum. Therefore, they can be linearized for *small motions* ($\theta_1 \ll 1$ and $\theta_2 \ll 1$) in the same manner as Equation 1.14 was obtained from Equation 1.13, to obtain the following:

$$\begin{bmatrix} (m_1 + m_2) L_1^2 & m_2 L_1 L_2 \\ m_2 L_1 L_2 & m_2 L_2^2 \end{bmatrix} \begin{bmatrix} \ddot{\theta}_1 \\ \ddot{\theta}_2 \end{bmatrix} + \begin{bmatrix} (m_1 + m_2) g L_1 & 0 \\ 0 & m_2 g L_2 \end{bmatrix} \begin{bmatrix} \theta_1 \\ \theta_2 \end{bmatrix} = \begin{bmatrix} 0 \\ 0 \end{bmatrix}$$
$$(1.21)$$

Since Equations 1.21 are written in matrix form, it is clear from the mass matrix and the zeros in the stiffness matrix that this model has acceleration (inertia) coupling but not displacement coupling. Also, the stiffnesses or generalized restoring forces (moments) in this model are not from springs but from gravity, just like the simple pendulum model illustrated in Figure 1.6b. Damping was not included in this model. As in the previous example, the matrices in Equation 1.21 for this example are symmetric, as they must be.

1.2.2 Matrix Bandwidth and Zeros

The 4-DOF model in Figure 1.9 has a characteristic common for models of many types of vibratory structures, such as many rotor vibration models, namely, *narrow bandwidth matrices*. Specifically, this system's *mass* matrix is "diagonal" (i.e., only its *diagonal* elements are nonzero) and its *stiffness* matrix is "tri-diagonal" (i.e., only its central three *diagonals* are nonzero), as shown in Figure 1.8. Obviously, a model's matrices are essentially its equations of motion. For this model, the *diagonal* nature of the mass matrix reflects that the model has no inertia coupling, in contrast to the model in Figure 1.8. For rotors, as shown in Chapter 2, the *lumped-mass* approach gives a diagonal mass matrix, in contrast to the so-called *distributed-mass* and *consistent-mass* approaches, which are preferred over the *lumped-mass* approach since they yield better model resolution

Vibration Concepts and Methods 17

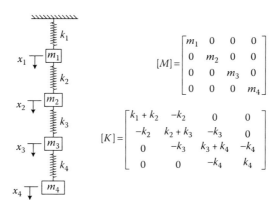

FIGURE 1.9 4-DOF lumped-mass model.

accuracy. That is, they require a smaller number of DOFs or finite elements (i.e., smaller matrices) to get the same model resolution accuracy. Both the *distributed-mass* and *consistent-mass* models yield *multi-diagonal* mass matrices. Unlike the *lumped-mass* model shown in Figure 1.9, they embody a first-order account of *inertia* coupling between adjacent masses in the model, and thus a better resolution accuracy. As in the previous examples, the matrices for the model in Figure 1.9 are symmetric, as they must be.

Rotors are essentially beams, albeit circular beams. In rotor vibration models, rotors are typically sectioned (discretized) using *circular-bar finite elements*, with the local radial and angular displacement coordinates numbered sequentially according to axial location along the rotor. *Lateral* rotor vibration models therefore will have *narrow bandwidth* motion equation matrices when the rotor model is connected directly to *ground* through each bearing's equivalent stiffness and damping elements, without intervening bearing masses and without *connections* between the bearings other than through the rotor. This is the most typical *lateral* rotor vibration model. However, when more elaborate *bearing support structure* models are employed, the total system's motion equation matrices are generally not of *narrow bandwidth*, with the resulting bandwidth depending on the coordinate numbering sequence implemented in the specific computer code. Even then, the model's matrices still contain mostly *zeros*, that is, most finite elements are connected in the model only to a limited number of their neighboring finite elements.

Similarly, *torsional* rotor vibration models have *narrow bandwidth* motion equation matrices (typically *tri-diagonal*) for single rotational drive lines. However, for two or more shafts connected, for example, by gears, the matrices will most likely not be of *narrow bandwidth*, as shown in Chapter 3, but the matrices will still contain mostly *zeros*.

The topic of "matrix bandwidth and zeros" becomes a significant computational consideration for systems having very large numbers of

DOFs. That is, with large models (large number of DOFs), one has very large matrices in which most of the elements are filled with zeros, causing computer memory and computations to be unnecessarily taxed by multiplying and storing *lots of zeros*. Special measures are typically employed in modern computational schemes to circumvent this. Fortunately, rotor vibration models do not generally have such a large number of DOFs to require such special measures, especially in light of the enormous and continuing increases in PC and Work Station memory and computational speed. *Matrix bandwidth* is simply a result of how the motion equations are sequentially ordered or numbered. So even for the simple 4-DOF model in Figure 1.9, the bandwidth could be maximized just by reordering the equations accordingly, for example, $\{x_1, x_4, x_2, x_3\}$ instead of $\{x_1, x_2, x_3, x_4\}$. General purpose *finite-element* computer codes often have user-optional algorithms for *matrix bandwidth minimization*, where the user-supplied displacement coordinates are automatically renumbered for this purpose.

1.2.3 Standard Rotor Vibration Analyses

Achieving good models for rotor vibration analyses of many single-span two-bearing rotors may require models with as many as 100 DOFs. For a multispan rotor model of a complete large steam-powered turbo-generator, models of 200 or 300 DOFs are typically deemed necessary to accurately characterize the system. Obtaining the important vibration characteristics of a machine or structure from large DOF models is not nearly as daunting as one might initially think, because of the following axiom: *Rarely is it necessary in engineering vibration analyses to solve the model's governing equations of motion in their totality.* For example, *lateral rotor vibration analyses* generally entail no more than the following three categories:

i. Natural frequencies (damped or undamped) and corresponding mode shapes.
ii. Self-excited vibration threshold speeds, frequencies, and mode shapes.
iii. Vibration over full speed range due to specified rotor mass unbalances.

None of these three categories of analyses actually entails obtaining the general solution for the model's coupled differential equations of motion. That is, the needed computational results can be extracted from the model's equations of motion without having to obtain their general solution, as later detailed.

In the next section, basic topics important to these *standard vibration analyses* are covered. Extraction of *natural frequencies* and *mode shapes* as well as *instability threshold speeds* are both embedded in the classic *Eigenvalue–Eigenvector* mathematics problem associated with linear vibratory systems. Specifically, the extraction of *natural frequencies* and corresponding *mode shapes* for multi-DOF models are explained. Standard algorithms used for these analyses are treated in later chapters. *Steady-state rotor unbalance vibration* is simply an extension of the 1-DOF Equation 1.6.

1.3 Modes, Excitation, and Stability of Multi-DOF Models

Linear vibration models are typically categorized as either *undamped* or *damped*. Although all real systems have some damping, the important vibration characteristics are usually contained in the "lightest damped" modes, well within the so-called underdamped zone (defined in Section 1.1.1 in connection with Equation 1.4). As a consequence, an *undamped* model can usually provide adequate assessment of *natural frequencies* and corresponding *mode shapes*, with simpler computations and easier-to-visualize results than with damping included. Moreover, with modern computational schemes like finite-element methods, most structures with defined geometry and material linear elastic properties can be well modeled insofar as *inertia* and *elastic* characteristics are concerned, that is, *mass* and *stiffness* matrices adequate for vibration modeling purposes are usually obtainable. Conversely, the job of characterizing a structure's damping properties can be an elusive task, relying more on experience, testing, and sometimes rough approximation. Therefore, when analyzing the influence of structural variables on *natural frequencies* and *mode shapes*, it usually makes more sense to analyze the system using the undamped model of the structure.

In stark contrast, when analyzing the steady-state *peak vibration* amplitude at a *forced resonance* or the *threshold* location of a *self-excited vibration* (dynamic instability), *damping* is an absolutely essential ingredient in the analyses. For example, as Figure 1.5 clearly shows, the nearer the forcing frequency approaches the natural or resonance frequency, the more important is the influence of damping. Obviously, what allows an excitation force to slowly pass through or "sit at" a natural frequency without potentially damaging the machine or structure is the *damping present in the system*!

1.3.1 Modal Decomposition

Each natural mode of an undamped model is exactly equivalent to an undamped 1-DOF model and is mathematically decoupled from the model's other natural modes, as observed when the motion equations are transformed into what are called the *modal coordinates*. Such a coordinate transformation is similar and mathematically equivalent to observing material *stress components* at a point in the *principal coordinate system*, wherein viewed decoupling appears, that is, all the *shear stresses disappear* and the normal stresses are the *principal stresses*. Similarly, when an undamped mutli-DOF model's equations of motion are transformed into their *modal coordinates*, both the *mass* and *stiffness matrices* become *diagonal matrices*, that is, all *zeros* except for their main diagonal elements. That is, the equations of motion become *decoupled* when they are transformed from the *physical space* into the *modal space*, as explained here.

Equations of motion for free (*unforced*) and undamped multi-DOF models can be compactly expressed in matrix form as follows, where the q_i's are the previously defined *generalized coordinates* (Section 1.2.1):

$$[M]\{\ddot{q}\} + [K]\{q\} = \{0\} \quad (1.22)$$

For a specified set of initial conditions, $\{q(0)\}$ and $\{\dot{q}(0)\}$, this set of equations is guaranteed a unique solution by virtue of applicable theorems from differential equation theory, provided both $[M]$ and $[K]$ are positive-definite matrices. Therefore, if a solution is found by any means, it must be the solution. Historically, the approach that has guided the successful solution to many problems in mechanics has been the use of good physical insight to provide the correct guess of the solution form. Such is the case for the solution to Equation 1.22.

The vibratory displacement in a multi-DOF model is a function of both time and location in the model. The correct guess here is that the complete solution can be comprised of superimposed contributory solutions, each being expressible as the product of a time function, $s(t)$, multiplied by a spatial function of the coordinates, $\{u\}$. This is the classic *Separation of Variables* method, expressed as follows:

$$q_i(t) = u_i s(t), \ i = 1, 2, \ldots, N = \text{Number of DOF} \quad (1.23)$$

Substituting Equation 1.23 into Equation 1.22 yields the following equation:

$$[M]\{u\}\ddot{s}(t) + [K]\{u\}s(t) = \{0\} \quad (1.24)$$

Each of these N equations ($i = 1, 2, \ldots, N$) can be expressed as

$$\sum_{j=1}^{N} M_{ij} u_j \ddot{s}(t) + \sum_{j=1}^{N} K_{ij} u_j s(t) = 0 \quad (1.25)$$

rearranged to have a function of time only on one side of the equation and a function of location only on the other side of the equation, as follows:

$$-\frac{\ddot{s}(t)}{s(t)} = \frac{\sum_{j=1}^{N} K_{ij} u_j}{\sum_{j=1}^{N} M_{ij} u_j} \quad (1.26)$$

Following the usual argument of the separation of variables method, for a time-only function to be equal to a location-only function they both must equal the same constant (say ω^2), positive in this case. A positive constant gives harmonic motions in time, physically consistent with having finite energy in a conservative model and contrary to the exponential solutions that a negative constant gives. The following equations are thereby obtained:

$$\ddot{s}(t) + \omega^2 s(t) = 0 \quad (1.27)$$

$$\sum_{j=1}^{N} \left(K_{ij} - \omega^2 M_{ij} \right) u_j = 0, \quad i = 1, 2, \ldots, N \quad (1.28)$$

Equation 1.27 has the same form as the equation of motion for an unforced and undamped 1-DOF model, that is, same as Equation 1.3 with $c = 0$. Therefore, the solution of Equation 1.27 can be surmised directly from Equation 1.5, as follows:

$$s(t) = X \begin{Bmatrix} \sin(\omega t + \phi_s^+) \text{ or } \sin(\omega t - \phi_s^-) \\ \cos(\omega t + \phi_c^+) \text{ or } \cos(\omega t - \phi_c^-) \end{Bmatrix} \quad (1.29)$$

Any of the four equation (Equation 1.29) forms can be used to represent the same harmonic signal. The following form is arbitrarily selected here:

$$s(t) = X \cos(\omega t - \phi) \quad (1.30)$$

Equation 1.30 indicates a harmonic motion with all the coordinates having the same frequency and the same phase angle. The information to determine the specific frequencies at which the model will satisfy such a harmonic motion is contained in Equation 1.28, which are a set of N linear homogeneous algebraic equations in the N unknowns, u_j. Determining the values of ω^2 that provide nontrivial solutions to Equation 1.28 is the classic characteristic value or eigenvalue problem. The trivial solution (all u_j's = zero) is a static equilibrium state. Equation 1.28 can be compactly shown in matrix form as follows:

$$\left[K - \omega^2 M \right] \{u\} = \{0\} \quad (1.31)$$

From linear algebra it is known that for a nontrivial solution of Equation 1.31, the determinant of equation coefficients must be equal to zero, as follows:

$$D \equiv |K - \omega^2 M| = 0 \tag{1.32}$$

Expanding D, the characteristic determinant, yields an Nth-order polynomial equation in ω^2, usually referred to as the frequency or characteristic equation, which has N roots (eigenvalues) for ω^2. These eigenvalues are real numbers because $[M]$ and $[K]$ are symmetric, and are positive because $[M]$ and $[K]$ are positive-definite matrices. Virtually any modern text devoted just to vibration theory will contain an expanded treatment on modal decomposition and rigorously develop its quite useful properties, which are summarized here.

The N roots of Equation 1.32 each provide a positive natural frequency, ω_j ($j = 1, 2, \ldots, N$), for one of the model's N natural modes. These undamped natural frequencies are typically ordered by relative magnitude, as follows:

$$\omega_1 \leq \omega_2 \leq \cdots \leq \omega_N$$

Each root of Equation 1.32 substituted into Equation 1.31 yields a solution for the corresponding eigenvector $\{u_p\}$. Since Equation 1.31 is homogeneous (all right-hand sides = zero), each $\{u_p\}$ is determined only to an arbitrary multiplier. That is, if $\{u_p\}$ is a solution with ω_p^2 then $a\{u_p\}$ is also a solution, where "a" is an arbitrary positive or negative real number. Each eigenvector (modal vector) thus contains the mode shape, or the relative magnitudes of all the physical coordinates, ($q_i, i = 1, 2, \ldots, N$), for a specific natural mode. To plot a mode shape, one usually scales the modal vector by dividing all translation displacement elements by the largest, thus maintaining their relative proportion on a 0-to-1 plot.

It can be rigorously shown that each modal vector of an N-DOF model is orthogonal (with either mass or stiffness matrix as the weighting matrix) to all the other modal vectors in an N-dimensional vector space. This is somewhat the same way the x, y, and z axes of a three-dimensional Cartesian coordinate space are orthogonal. Consequently, the total set of modal vectors forms a complete orthogonal set of N vectors in the N-dimensional vector space that contains all possible displacement states of the model. Thus, any instantaneous position state of a model can be expressed as an instantaneous *linear superposition* of the contributions from all of its natural modes. In other words, the so-called modal coordinates $\{\eta(t)\}$ contain the amount of each natural mode's contribution to the system's instantaneous position state, $\{q(t)\}$. This property of the modal vectors can be expressed by the following linear transformation, where the $N \times N$ modal matrix,

[U], is formed using each one of the $N \times 1$ modal vectors as one of its columns:

$$\{q(t)\} = [U]\{\eta(t)\} \quad (1.33)$$

Here it is convenient to scale each of the modal vectors as follows ("T" denotes *transpose*):

$$\{u_p\}^T[M]\{u_p\} = 1 \quad (1.34)$$

Then the resulting modal matrix, [U], will satisfy the following equation:

$$[U]^T[M][U] = [I] \quad (1.35)$$

Here, [I] is the identity matrix, with 1 on each main diagonal element and zeros elsewhere. Equation 1.35 is actually a linear transformation of the mass matrix into modal coordinates, with the modal vectors scaled (normalized); hence all the modal masses are equal to 1. Applying the identical transformation on the stiffness matrix also produces a diagonal matrix, with each main diagonal element equal to one of the eigenvalues ω_j^2 as follows:

$$[U]^T[K][U] = \left[\omega_{ij}^2\right] \quad (1.36)$$

Here, the array $[\omega_{ij}^2]$ is defined similar to the *kronecker delta*, as follows:

$$\omega_{ij}^2 \equiv \begin{cases} \omega_j^2, & i = j \\ 0, & i \neq j \end{cases} \quad (1.37)$$

Substituting the linear transformation of Equation 1.33 into the original equations of motion Equation 1.22 and then premultiplying the result by $[U]^T$ yield the following result:

$$[U]^T[M][U]\{\ddot{\eta}(t)\} + [U]^T[K][U]\{\eta(t)\} = 0 \quad (1.38)$$

Utilizing in Equations 1.38, 1.35, and 1.36, which express the modal vectors' orthogonality property, shows that the equations of motion are decoupled in the modal coordinate space. Accordingly, Equation 1.38 becomes

$$\{\ddot{\eta}(t)\} + \left[\omega_{ij}^2\right]\{\eta(t)\} = 0 \quad (1.39)$$

Equation 1.39 clearly shows that each natural mode is equivalent to an undamped 1-DOF model. Each natural mode's response to a set of initial conditions is therefore of the same form as for the undamped 1-DOF model, shown as follows:

$$\eta_p(t) = A_p \cos(\omega_p t - \phi_p) \tag{1.40}$$

Consequently, utilizing the linear superposition of the contributions from all the model's natural modes, the motion of a free undamped multi-DOF system is expressible as follows, where the A_p's are the single-peak amplitudes of each of the modes:

$$\{q(t)\} = \sum_{p=1}^{N} A_p \{u_p\} \cos(\omega_p t - \phi_p)$$

This can be expressed in matrix form as follows:

$$\{q(t)\} = [U]\{A_p \cos(\omega_p t - \phi_p)\} \tag{1.41}$$

1.3.2 Modal Damping

A major role of damping is to dissipate vibration energy that would otherwise lead to intolerably high vibration amplitudes at forced resonances or allow self-excited vibration phenomena to occur. As already shown for the 1-DOF model, a natural frequency is only slightly lowered by the damping, but the peak vibration caused by an excitation force at the natural frequency is overwhelmingly lowered by the damping. This clearly applies to multi-DOF models, as shown by the modal damping approach which follows.

Modeling a structure's damping properties can be an elusive task, relying more on experience, testing and occasionally rough approximation. An actual damping mechanism may be fundamentally quite nonlinear like Coulomb rubbing friction and internal material hysteresis damping. But to maintain a linear model, the dissipated vibration energy mechanism must be modeled as drag forces proportional to the velocity magnitude differences across the damper-connected elements in the model, such as shown in Figures 1.1 and 1.7. Viscous damping is a natural embodiment of the linear damping drag force model. Satisfactory models for forced-resonance vibration amplitudes and instability thresholds can be obtained when the energy-per-cycle dissipated by the actual system is commensurate with the linear damping model, even if the actual damping mechanism is nonlinear.

For a multi-DOF model, one convenient way to incorporate damping into the model is on a mode-by-mode basis. This is an optimum modeling procedure in light of modern testing and digital signal processing techniques that provide equivalent linear damping ratios $\varsigma \equiv (c/c_c)_j$ for each of the prevalent modes excited in testing. Thus, appropriate damping

Vibration Concepts and Methods

can be incorporated into the model by adding it to each relevant mode in the modal coordinate system. Accordingly, Equation 1.39 is augmented as follows:

$$\{\ddot{\eta}(t)\} + 2[\varsigma_i\omega_j]\{\dot{\eta}(t)\} + \left[\omega_{ij}^2\right]\{\eta(t)\} = 0 \qquad (1.42)$$

Here, $\varsigma_i\omega_j$ is a diagonal array defined similar to the Kronecker delta, as follows:

$$\varsigma_i\omega_j \equiv \begin{cases} \varsigma_j\omega_j, & i=j \\ 0, & i \neq j \end{cases} \qquad (1.43)$$

The often used 1-DOF version of Equation 1.42 is obtained by dividing Equation 1.3 by m and using the definition for ς given in Equation 1.10, yielding $\ddot{x} + 2\varsigma\omega_n\dot{x} + \omega_n^2 x = 0$.

Mathematically, an N-DOF model has N modes. However, the discrete model (e.g., finite-element model) should have the DOF number, N, several (like 10, "more or less") times the mode number, n, of the actual system's highest frequency mode of importance. This statement assumes usual mode numbering by ascending frequency, $\omega_1 \leq \omega_2 \leq \cdots \leq \omega_n \leq \cdots \leq \omega_N$. The underlying objective is for the discrete model to adequately characterize the actual continuous media system in the frequency range up to the maximum modal frequency of importance. *It is of fundamental modeling importance that at frequencies progressively higher the characteristics of the discrete model and those of the actual system progressively diverge.* The desired number of important modes will depend on the nature of the problem analysis. For example, to analyze forced resonances, one hopefully knows the actual maximum excitation-force frequency ω_{max}. As a rule, all modal frequencies within and somewhat above the excitation frequency range should be included even though some of these modes may be of lesser importance than others.

Consider an application in which an actual system has been tested, providing damping ratio data for the lowest frequency n modes. The first n elements ($j = 1, 2, \ldots, n < N$) of the $N \times N$ diagonal modal damping matrix will each contain its own value, $\varsigma_j\omega_j$. The modal damping matrix will otherwise consist of zeros, and thus modified from Equation 1.43 as follows:

$$\varsigma_i\omega_j = \begin{cases} \varsigma_j\omega_j, & i = j \leq n \\ 0, & i = j > n \\ 0, & i \neq j \end{cases} \qquad (1.44)$$

With each jth mode having its own decoupled equation of motion in modal coordinates, the previously stated equivalency between the damped

1-DOF model and relevant modes of a multi-DOF model is thus shown, as follows:

$$\ddot{\eta}_j + 2\varsigma_j\omega_j\dot{\eta}_j + \omega_j^2\eta_j = 0 \qquad (1.45)$$

The equations of motion in the physical coordinates are then as follows:

$$[M]\{\ddot{q}\} + [C_m]\{\dot{q}\} + [K]\{q\} = \{0\} \qquad (1.46)$$

The elements $2\varsigma_j\omega_j$ form a diagonal matrix in modal coordinates to incorporate the mode-by-mode damping model. Consequently, the transformation to physical coordinates to obtain $[C_m]$ would appear to be simply the inverse of the transformation that diagonalizes $[M]$ and $[K]$, as shown in the following equation:

$$[C_m] = \left[U^T\right]^{-1}[2\varsigma_i\omega_j][U]^{-1} \qquad (1.47)$$

However, for typical large systems the available $N \times N$ modal matrix $[U]$ often contains only n nonzero eigenvectors extracted from the finite element model of N DOFs, where $n < N$. So the available modal matrix is missing $N - n$ columns, and thus not invertible. An approach to circumvent this difficulty is to use the *Static Condensation* method to reduce the number of DOFs to the number of modes to be retained, and then solve for all the eigenvectors of the statically condensed model. The $n \times n$ $[U]$ matrix should then be invertible, allowing Equation 1.47 to be functional. Element values in the modal-based damping matrix may not benefit from the luxury of a good test on the actual system, particularly if the actual system has not yet been built. In that case, modal damping is estimated from experience and previous damping data.

Unlike $[C_m]$, an arbitrary damping matrix, $[C]$, is not diagonalized by the transformation that diagonalizes $[M]$ and $[K]$. There is an older approach called *proportional damping* that postulates a damping matrix $[C_p]$ in the physical coordinate system proportional to a linear combination of $[M]$ and $[K]$. That is, $[C_p] \equiv a[M] + b[K]$. This is done simply so that $[C_p]$ is diagonalized by the transformation into modal coordinates. Here, "a" and "b" are real numbers with appropriate dimensional units. Available prior to modern modal test methods, the proportional damping approach provides a damping model that also preserves decoupling in modal coordinates, but not as physics motivated as a modal damping model. Compared to the modal damping approach, proportional damping is not as directly related to a mode-by-mode inclusion of damping based on modal testing. Clearly, the proportional damping approach can be mathematically viewed as a pseudo modal damping, given its diagonal form within the modal coordinate space.

Vibration Concepts and Methods

1.3.3 Forced Systems Decoupled in Modal Coordinates

This important topic is shown by adding a system of external time-dependent forces to either the *modally damped* model of Equation 1.46 or the *undamped* model of Equation 1.22, $[C] = [0]$, both of which are contained within the following equation:

$$[M]\{\ddot{q}\} + [C_m]\{\dot{q}\} + [K]\{q\} = \{f(t)\} \tag{1.48}$$

Since the modal vectors span the vector space of all possible model displacement states, modal decomposition is applicable to forced systems as well. Clearly, transformation of Equation 1.48 into the modal coordinate system provides the following equivalent decoupled set of equations:

$$\{\ddot{\eta}(t)\} + 2[\varsigma_i \omega_j]\{\dot{\eta}(t)\} + \left[\omega_{ij}^2\right]\{\eta(t)\} = [U]^T\{f(t)\} \tag{1.49}$$

Here, the vector of modal forces is $\{\Phi(t)\} \equiv [U]^T \{f(t)\}$. This shows that each modal force $\Phi_i(t)$ is a linear combination of all the physical forces $f_j(t)$. And the contribution of each physical force to $\Phi_i(t)$ is in proportion to the modal matrix element U_{ji} (or U_{ij}^T), which is called the *participation factor* of the jth physical coordinate for the ith mode.

As an important example, Equation 1.49 shows that a physical harmonic force having a particular mode's natural frequency will produce its maximum resonance vibration effect if applied at the physical coordinate location having that mode's largest participation factor. Conversely, if the same harmonic force is applied in a physical coordinate with a zero-participation factor (called a "nodal point" for that mode), the force's contribution to that mode's vibration will be zero. This is particularly relevant to *rotor balancing problems*, explaining why both a rotor unbalance magnitude and its axial location are important.

1.3.4 Harmonic Excitation of Linear Multi-DOF Models

The most frequently performed type of vibration analysis is the *steady-state* response from *harmonic* excitation forces. Various single-frequency solutions at different frequencies can be superimposed to obtain a simultaneous multifrequency steady-state solution, provided the model is linear. Also, using the single-frequency case, the frequency can be varied over the desired range in a given application. Thus, the formulation and solution for the single-frequency case is the building block for vibration analyses. The generic governing equation for this case can be expressed as follows, where [C] is arbitrary and not necessarily modal:

$$[M]\{\ddot{x}\} + [C]\{\dot{x}\} + [K]\{x\} = \left\{F_j e^{i(\omega t + \theta_j)}\right\} \tag{1.50}$$

Here, x is used as the generalized coordinate symbol, and the harmonic forcing functions have individual magnitudes F_j and phase angles θ_j. Since they have a common excitation frequency ω, it is convenient to represent each harmonic excitation force as a planar vector rotating counter clockwise (ccw) at ω (rad/s) in the complex-plane exponential form. The right-hand side of Equation 1.50 represents the standard notation for this representation, $i \equiv \sqrt{-1}$. The instantaneous projection of each planar vector onto the *real axis* of the complex plane is the instantaneous physical value of the corresponding sinusoidal time-varying scalar force component.

Equation 1.50 is the multi-DOF version of the 1-DOF model representation in Equation 1.6 whose *steady-state* solution (the so-called *particular* solution) is harmonic, Equation 1.7. For the multi-DOF Equation 1.50, the steady-state solution is also harmonic, and shown as follows using the exponential complex form:

$$x_j = X_j e^{i(\omega t + \phi_j)} \tag{1.51}$$

Here, X_j is the single-peak amplitude of the jth coordinate's harmonic motion at frequency ω and phase angle ϕ_j. Substitution of this known solution form, Equation 1.51, into the equations of motion, Equation 1.50, and then dividing through by $e^{i\omega t}$, yields the following simultaneous set of complex algebraic equations:

$$[-\omega^2 M + i\omega C + K]\{X_j e^{i\phi_j}\} = \{F_j e^{i\theta_j}\} \tag{1.52}$$

In this set of equations, the known inputs are the model's M, C, and K matrices, the excitation forcing frequency ω and magnitude F_j, and the phase angle θ_j for each of the excitation forces. The outputs are the single-peak amplitude X_j and the phase angle ϕ_j for each jth physical motion coordinate of the model.

1.3.5 Dynamic Instability: The Complex Eigenvalue Problem

Consider the unforced general multi-DOF linear model, expressed as follows:

$$[M]\{\ddot{x}\} + [C]\{\dot{x}\} + [K]\{x\} = \{0\} \tag{1.53}$$

As explained in Chapter 2, the stiffness and damping coefficients for rotor vibration model elements that dynamically connect the rotor to the rest of the model (e.g., bearings, squeeze-film dampers, and seals) are often nonsymmetric arrays, especially the stiffness coefficients for journal bearings. Also, the gyroscopic effects of rotor-mounted disk-like masses add skew-symmetric element pairs to the model's [C] matrix. No symmetry

TABLE 1.1
Eigenvalue Categories and Associated Types of Unforced Motion

Eigenvalue Category	Mode Motion: $\eta(t) = Ae^{\alpha t} \cos(\omega t - \phi)$
1. $\alpha = 0 \; \omega \neq 0$	Zero damped, steady-state sinusoidal motion
2. $\alpha < 0 \; \omega \neq 0$	Underdamped, sinusoidal, exponential decay
3. $\alpha > 0 \; \omega \neq 0$	Negatively damped, sinusoidal, exponential growth
4. $\alpha = 0 \; \omega = 0$	So-called rigid-body mode, constant momentum
5. $\alpha < 0 \; \omega = 0$	Overdamped, nonoscillatory, exponential decay
6. $\alpha > 0 \; \omega = 0$	(i) Negatively damped more than "critical" amount
	(ii) Statically unstable nonoscillatory exponential growth

restrictions are made here on [K], [C], or [M]. Solutions of Equation 1.53 have the following form:

$$\{x\} = \{X\}e^{\lambda t}, \quad \text{where } \lambda = \alpha \pm i\omega \quad (1.54)$$

For rotor vibration analyses, interest is focused on machine operating zones wherein dynamic instability (self-excited vibration) is predicted to occur. In particular, the boundary location of such an operating zone is usually what is sought. With this objective in mind, it is important to first understand the relationship between the eigenvalue type of a specific mode and the mode's characteristics or motion properties. The eigenvalues for each mode can be of a variety of fundamental types, each type denoting a specific property, similar to the 1-DOF model in Section 1.1.1.

Table 1.1 provides a complete list of eigenvalue types and the corresponding mode motion properties. Referring to Figure 1.2, Section 1.1.1, the underdamped and overdamped Categories 2 and 5, respectively, in Table 1.1 are just like their 1-DOF counterparts. Thus, when Category 5 has the smallest absolute value $|\alpha|$ for which $\omega = 0$, this corresponds to the 1-DOF critically damped case. Furthermore, Category 1 is like the 1-DOF $c = 0$ undamped case. Category 3 corresponds to the 1-DOF negatively damped $c < 0$ case illustrated in Figure 1.3. Categories 4 and 6 were not explicitly discussed in Section 1.1.1, but each of these also has a 1-DOF counterpart as well. For Category 4 the 1-DOF model has $k = 0$ and $c = 0$, giving $m\ddot{x} = 0$ momentum conserved. For Category 6, recalling that $\varsigma = c/(2\sqrt{km})$ and $\omega_n^2 = k/m$ for the 1-DOF model, clearly $\varsigma < 0$ with $\varsigma^2 > 1$ corresponds to 6(i), and $k < 0$ corresponds to 6(ii).

The usual analysis application concerning self-excited rotor vibration is to predict the limits or boundaries of safe operating conditions, to predict dynamic instability thresholds. In Table 1.1, a prediction of such a threshold corresponds to finding the parameter boundary (usually rotor spin speed or machine power output) where the system transitions from Category 2 (positively damped) to Category 3 (negatively damped). Exactly on such a transition boundary, the mode in question is in Category 1 (zero damped).

Equation 1.53 is a set of N second-order ODEs. The usual approach to formulate the associated eigenvalue problem entails first transforming Equation 1.53 into an equivalent set of $2N$ first-order differential equations. To that end, the following associated vectors are defined,

$$\{y\} \equiv \{\dot{x}\}, \quad \therefore \{\dot{y}\} = \{\ddot{x}\} \quad \text{and} \quad \{z\} \equiv \begin{Bmatrix} \{y\} \\ \{x\} \end{Bmatrix}, \quad \therefore \{\dot{z}\} \equiv \begin{Bmatrix} \{\dot{y}\} \\ \{\dot{x}\} \end{Bmatrix}$$

so that Equation 1.53 is transformed into the following:

$$\begin{bmatrix} [0] & [M] \\ [M] & [C] \end{bmatrix} \{\dot{z}\} + \begin{bmatrix} [-M] & [0] \\ [0] & [K] \end{bmatrix} \{z\} = \{0\} \quad (1.55)$$

Naturally, Equation 1.53 and Equation 1.55 have solutions of the same form, Equation 1.54 as follows:

$$\{z\} = \{Z\}e^{\lambda t}, \quad \text{where } \lambda = \alpha \pm i\omega \quad (1.56)$$

A $2N \times 2N$ matrix $[A]$ is defined as

$$[A] \equiv \begin{bmatrix} [0] & [M] \\ [M] & [C] \end{bmatrix}^{-1} \begin{bmatrix} [-M] & [0] \\ [0] & [K] \end{bmatrix} \quad (1.57)$$

Compact Equation 1.55 is shown as follows:

$$\{\dot{z}\} + [A]\{z\} = \{0\} \quad (1.58)$$

Substituting Equation 1.56 into Equation 1.58 and dividing the result by $e^{\lambda t}$ yields the following *complex eigenvalue problem*:

$$[A + I\lambda]\{Z\} = \{0\} \quad (1.59)$$

Here $[I]$ is the identity matrix. For the general multi-DOF models, the eigenvalues λ_j and associated eigenvectors $\{Z\}_j$ can be mathematically complex. This is in contrast to the real eigenvalues and real eigenvectors for the undamped models treated earlier in this section. To solve both the real eigenvalue problem presented earlier and the complex eigenvalue problem covered here, modern computational methods are readily available. Application of these methods is covered in subsequent chapters as required.

1.4 Summary

This chapter is intended as a comprehensive primer on basic vibration concepts and methods. In that context, it has general application beyond this book's primary subject, rotating machinery. But the main purpose of this chapter is to provide the needed vibration fundamentals to draw upon in the remainder of the book. Therefore, throughout this chapter, frequent references are made to connect a specific vibration topic to some aspect of *Rotating Machinery Vibration*.

For anyone requiring a more detailed presentation of the material presented in this chapter, there are several excellent texts devoted entirely to *Vibration*.

PROBLEM EXERCISES

1. (a) Develop the equation of motion for the shown configuration.
   (b) Express the system's natural frequency.

$$y = \frac{Fl^3}{48EI}, \quad \text{center deflection under center load}$$

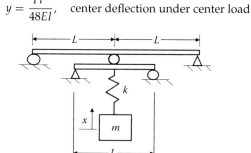

Both beams have negligible mass and the same *EI*. Only *m* has mass

2. A rigid beam of negligible mass with a concentrated mass *m* at each end is shown. Also at each end is located a spring of stiffness *k* and a dashpot of damping *c*. Assume that motions are much smaller than the configuration's dimensions. $F(t) = F_0 \sin \omega t$ develop (a) equations of motion and (b) natural frequencies (C = 0).

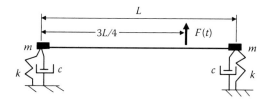

3. A two-bearing rotor operates at 1800 rpm. All parts are made of steel, comprising a solid shaft modeled as perfectly rigid, and a mid-span pump impeller modeled as a rigid disk that has the mass and transverse moment of inertia of the actual impeller. Neglect the polar moment of inertia of the shaft and the disk. The two radial bearings positioned at the ends of the rotor are modeled as being dynamically isotropic, meaning that their radial stiffness (9500 lb/in = 332,880 n/m each) is the same in any radial direction. This system has two resonance frequencies (critical speeds), one below 1800 cpm and one far enough above 1800 cpm not to matter.

Determine the minimum required isotropic radial linear damping strength C, in parallel with the stiffness at each bearing, so as to limit the radial vibration peak amplitude to within the close radial clearance (0.015 in. = 0.0381 cm) at the pump impeller leakage control rings, that is, as the rotor slowly run-up from zero to the 1800 rpm operating speed. Unbalance at the disk is 40 oz in.

Shaft length = 60 in. (152.4 cm)   Shaft outer diameter
                                          = 4 in. (10.16 cm)
Disk diameter = 17 in. (43.18 cm)   Disk width
                                          = 4 in. (10.16 cm)

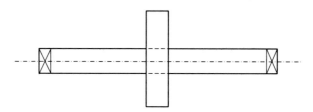

4. This problem is based on the previous Problem 3 in which the bearing damping coefficient was determined to limit the amplitude of rotor vibration at its lower unbalance excited resonance speed. Specifying an initial nonzero displacement magnitude with zero initial velocity, solve the transient (homogeneous) solution. Treat this solution as if it were an experimentally obtained time-base signal. Then use this "experimental" displacement signal with the log-decrement method to determine the damping coefficient for each bearing. Your result should return the value obtained in Problem 3 and be used here to obtain the "experimental" transient displacement decay signal.

5. The 3-DOF model, in its static equilibrium state, has three lumped masses that translate only horizontally. It has one external forcing function, two elastic beams, two linear connecting springs, and two linear connecting dampers.

   (a) Develop the equations of motion for this system.

(b) Put the equations of motion into matrix form.
Cantilever beam static end deflection under end load:
$$\delta = \frac{PL^3}{3EI}$$

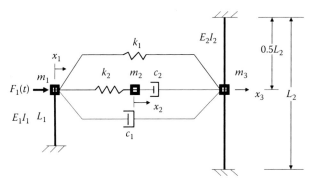

6. Shown, two mutually perpendicular elastic beams of negligible mass and a single mass m behind the simply supported beam. A pin, which is rigidly part of m, protrudes through both beams and is free to slide without friction in the two perpendicular fitted slots shown. Beams have the same EI and displacements are very small: (a) develop the equations of motion in shown coordinates (X, Y), (b) put the equations of motion into matrix form, and (c) write the equations of motion in modal coordinates and express the natural frequencies.

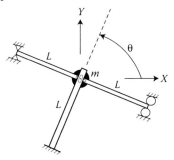

7. A uniform beam of negligible mass supports two concentrated masses. Beam EI is uniform. Develop the equations of motion for small amplitude displacements. Start by formulating the flexibility matrix, employing standard beam deflection formulas, and then invert this flexibility matrix to obtain the stiffness matrix.

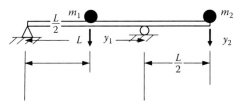

2

Lateral Rotor Vibration Analysis Models

2.1 Introduction

Lateral rotor vibration (LRV) is radial-plane orbital motion of the rotor spin axis. *Transverse rotor vibration* is used synonymously with LRV. Figure 2.1 shows the sketch of a complete steam turbine generator rotor (minus turbine blades) and a sample of its computed LRV vibration orbits, shown grossly enlarged. Actual LRV orbits are typically only a few thousandths of an inch (a few hundredths of a millimeter) across. LRV is an important design consideration in many types of rotating machinery, particularly turbo-electrical machines such as steam and combustion gas turbine generators sets, compressors, pumps, gas turbine jet engines, turbochargers, and electric motors. Thus, LRV impacts several major industries.

Usually, but not always, the potential for rotor dynamic beam-bending-type deflections significantly contributes to the LRV characteristics. The significance of LRV rotor bending increases with bearing-to-rotor stiffness ratio and with rotor spin speed. Consequently, in some rotating machines with low operating speed and/or low bearing-to-rotor stiffness ratio, the LRV is essentially of a rigid rotor vibrating in flexible bearings/supports. The opposite case (i.e., a flexible rotor in essentially rigid bearings) is also possible but rotor dynamically less desirable, because it lacks some vibratory motion at the bearings which often provide that essential ingredient, *damping*, to keep vibration amplitudes at resonance conditions within tolerable levels.

For the same reason, it is generally undesirable to have journal bearings located at *nodal points* of important potentially resonant modes, that is, the squeeze-film damping capacity of a bearing cannot dissipate vibration energy without some vibratory motion across it. Figure 2.1 is a case with significant participation of both rotor bending and relative motion at the bearings. This is the most interesting and challenging LRV category to analyze.

A rotor's flexibility and mass distributions and its bearings' flexibilities combined with its maximum spin speed essentially determine whether or not residual rotor unbalance can produce forced LRV resonance. That is, these aforementioned factors determine if the rotor–bearing system has one or more *lateral natural frequency* modes below the operating speed.

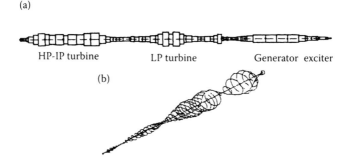

FIGURE 2.1 LRV example; vibration orbits show rotor dynamic flexibility: (a) steam turbo-generator rotor sketch (minus turbine blades) and (b) sample vibration orbits of above turbo-generator (isometric view).

If so, then the rotor must pass through the speed(s) (called "critical speeds") where the residual mass unbalances act as once-per-rev (synchronous) harmonic forces to excite the one or more natural frequencies the rotor speed traverses when accelerating to operating speed and when coasting down. Resonant mode shapes at critical speeds are also determined by the same aforementioned rotor and bearing properties. Many types of modern rotating machinery are designed to operate above one or more (sometimes several) *critical speeds*, because of demands for compact high-performance machines.

When one or more *critical speeds* are to be traversed, LRV analyses are required at the design stage of a rotating machine. These analyses generally include computations to ensure that the machine is not inadvertently designed to run continuously at or near a *critical speed*. These analyses should also include computed *unbalance rotor vibration* levels over the entire speed range, to ensure that the rotor–bearing system is adequately damped to safely pass through the critical speeds within the operating speed range. Furthermore, these analyses should include *dynamic stability* computations to ensure that there are no *self-excited vibration* modes within the combined ranges of operating speed and output of the machine. Lastly, if LRV *rotor bending* significantly contributes to the critical speeds' *mode shapes*, then the rotor must be balanced using one of the proven *flexible rotor balancing* procedures (e.g., *Influence Coefficient Method*), which are more complicated than the simpler *two-plane rigid-rotor* balancing procedure. Providing an introductory appreciation for all these is the objective of Table 2.1, which somewhat simplistically subdivides the degree of LRV complexity into three categories. The three categories in Table 2.1 could be further delineated, as made clear in subsequent chapters, with the aid of the applicable first principles covered in the remaining sections of this chapter.

Chapter 3 deals with *torsional rotor vibration* (TRV), which involves torsional twisting of the rotor. In single-rotor drive lines it is rare for significant

Lateral Rotor Vibration Analysis Models

TABLE 2.1

Three Elementary LRV Complexity Categories

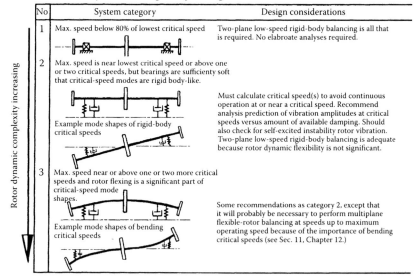

dynamic coupling to exist between *LRV* and *TRV* characteristics. Consequently, these two types of rotor vibration, while potentially coexisting to significant degrees in the same rotor, practically do not significantly interact. There are a few exceptions to this, for example, high-speed refrigerant centrifugal compressors for high-capacity refrigeration and air conditioning systems. Such compressor units are typically comprised of two parallel *rotor dynamically flexible* shafts coupled by a two-gear single-stage speed increaser. In that specific type of rotating machinery, the gear teeth forces provide a potential mechanism for coupling the LRV and TRV characteristics. Even in that exceptional application, such *lateral–torsional coupling* is generally not factored into design analyses. Near the end of Chapter 3, subsequent to the coverage of applicable first principles for both LRV and TRV, Table 3.1 is presented to show some quite interesting and important contrasts between LRV and TRV, which are not frequently articulated and thus not widely appreciated.

2.2 Simple Linear Models

2.2.1 Point–Mass 2-DOF Model

The simplest LRV model that can encompass radial-plane orbital rotor motion has 2-DOF, as shown in Figure 2.2. In this model, the rotor point

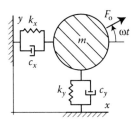

FIGURE 2.2 Simplest LRV model that can handle radial-plane orbital motion.

mass (m) is allowed to translate in a radial x–y plane. It is connected to ground through linear springs and dampers and may be excited by time-varying radial force components such as the rotating force (mass-unbalance) shown. The two equations of motion for this model with the shown rotating excitation force are easily derived from $F = ma$ to obtain the following:

$$m\ddot{x} + c_x\dot{x} + k_x x = F_o \cos \omega t$$
$$m\ddot{y} + c_y\dot{y} + k_y y = F_o \sin \omega t \tag{2.1}$$

Here, the springs and dampers can be used to include radially isotropic shaft flexibility in series with bearing parallel flexibility and damping. Note that in this model, the x-motion and y-motion are decoupled, that is, the two motion equations are decoupled. However, if the x–y axes are not chosen aligned with the springs and dampers as shown, the equations become coupled, even though the model is "physically" unchanged in an alternate x–y orientation. Therefore, the x–y *physical coordinates* shown in Figure 2.2 are also the *modal coordinates* for this model. Naturally, if a model can be configured in a set of physical coordinates that yield completely decoupled motion equations, then that set of physical coordinates are also the model's modal coordinates. For most vibration models, LRV or otherwise, this is not possible. If the conditions $k_x = k_y \equiv k$ and $c_x = c_y \equiv c$ are imposed, then the model is *isotropic*, which means the x–y coordinate system can then be rotated to a different orientation in its plane with no change to the motion equations. Therefore, such an isotropic model remains decoupled and the physical coordinates are its modal coordinates in any x–y orientation. Thus, for the isotropic model the radial stiffness, k, is the same in any radial direction, as is the radial damping, c.

Many types of radial bearings and seals have fluid dynamical features that produce significant and important LRV *cross coupling* between orthogonal radial directions. A more generalized version of the simple 2-DOF model in Figure 2.2 can incorporate such cross coupling, as shown in the following two coupled equations, written in matrix form:

$$\begin{bmatrix} m & 0 \\ 0 & m \end{bmatrix} \begin{Bmatrix} \ddot{x} \\ \ddot{y} \end{Bmatrix} + \begin{bmatrix} c_{xx} & c_{xy} \\ c_{yx} & c_{yy} \end{bmatrix} \begin{Bmatrix} \dot{x} \\ \dot{y} \end{Bmatrix} + \begin{bmatrix} k_{xx} & k_{xy} \\ k_{yx} & k_{yy} \end{bmatrix} \begin{Bmatrix} x \\ y \end{Bmatrix} = \begin{Bmatrix} F_x(t) \\ F_y(t) \end{Bmatrix} \quad (2.2)$$

As shown in considerably more detail later in this chapter and in Chapter 5, such 2×2 $[c_{ij}]$ and $[k_{ij}]$ matrices for bearings and seals are extremely important inputs for many LRV analyses, and have been the focus of extensive research to improve the accuracy for quantifying their matrix coefficients. In general, these coefficient matrices for bearings and seals cannot be simultaneously diagonalized in a single $x-y$ coordinate system, in contrast to the model shown in Figure 2.2. In fact, as explained later in this chapter, the bearing and seal stiffness coefficient matrices are often nonsymmetric and their damping coefficient matrices may also be nonsymmetric when certain fluid dynamical factors are significant (e.g., fluid inertia).

2.2.2 Jeffcott Rotor Model

A centrally mounted disk on a slender flexible uniform shaft comprises the model employed by H. H. Jeffcott [*Philosophical Magazine* 6(37), 1919] to analyze the lateral vibration of shafts in the neighborhood of the (lowest) critical speed. Figure 2.3a is a lateral planar view of this model and Figure 2.3b is its extension to include bearing flexibility. If the concentrated midspan disk mass m in these two models is treated strictly as a point mass, then both of these models fit the 2-DOF model in Figure 2.2. If bearing stiffness is included but bearing damping neglected, bearing

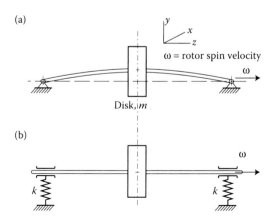

FIGURE 2.3 Jeffcott rotor model: (a) Jeffcott model and (b) modified Jeffcott model with bearing flexibility.

stiffness and half-shaft transverse bending stiffness are simply added as springs in series. To additionally include bearing damping, the bearing stiffness and damping in-parallel combination is added in series with the half-shaft transverse bending stiffness. This results in an equivalent radial stiffness and damping coefficients that are frequency dependent, but still within the 2-DOF model in Figure 2.2. However, if x and y disk angular displacements (θ_x and θ_y) are allowed, and the disk's transverse and polar moments-of-inertia (I_T and I_P) about its center are included, then the Figure 2.3 models have 4-DOFs, the generalized coordinates for the disk then being x, y, θ_x, and θ_y. If the disk is centered between two identical half shafts (same length, diameter, and material), and both the modified Jeffcott model bearings are identical with symmetric stiffness coefficients as in the model of Figure 2.2, then the 4-DOF models (undamped) are somewhat simpler than they would otherwise be. That is, there is no coupling through the [M] and [K] matrices, as shown in the following equations of motion which are then applicable:

$$\begin{bmatrix} m & 0 & 0 & 0 \\ 0 & m & 0 & 0 \\ 0 & 0 & I_T & 0 \\ 0 & 0 & 0 & I_T \end{bmatrix} \begin{Bmatrix} \ddot{x} \\ \ddot{y} \\ \ddot{\theta}_x \\ \ddot{\theta}_y \end{Bmatrix} + \begin{bmatrix} 0 & 0 & 0 & 0 \\ 0 & 0 & 0 & 0 \\ 0 & 0 & 0 & \omega I_P \\ 0 & 0 & -\omega I_P & 0 \end{bmatrix} \begin{Bmatrix} \dot{x} \\ \dot{y} \\ \dot{\theta}_x \\ \dot{\theta}_y \end{Bmatrix}$$

$$+ \begin{bmatrix} k_x & 0 & 0 & 0 \\ 0 & k_y & 0 & 0 \\ 0 & 0 & K_x & 0 \\ 0 & 0 & 0 & K_y \end{bmatrix} \begin{Bmatrix} x \\ y \\ \theta_x \\ \theta_y \end{Bmatrix} = \begin{Bmatrix} F_x(t) \\ F_y(t) \\ M_x(t) \\ M_y(t) \end{Bmatrix} \quad (2.3)$$

where I_T is the disk transverse inertia, I_P is the disk polar inertia, K_x is the x-moment stiffness, K_y is the y-moment stiffness, M_x is the x-applied moment on the disk, and M_y is the y-applied moment on the disk.

From the model represented in Equations 2.3, there still appears a [C] matrix multiplying the velocity vector, albeit 14 of the 16 elements in [C] are zero. The two nonzero elements in the [C] matrix of Equations 2.3 embody the so-called *gyroscopic effect* of the disk, which shows up as skew-symmetric components of [C]. As more fully explained later in this chapter, the gyroscopic effect is conservative (i.e., it is an inertia effect and thus dissipates no energy) even though it "resides" in the [C] matrix. Note that the *gyroscopic effect* couples the θ_x and θ_y motions.

For the simply supported Jeffcott rotor model, Figure 2.3a, the four (diagonal) nonzero stiffness matrix elements in Equations 2.3 describe the flexible shaft's radially isotropic force and moment response to the disk's four coordinates (x, y, θ_x, and θ_y). For the modified Jeffcott model, Figure 2.3b, the stiffness elements in Equations 2.3 contain the combination of the isotropic rotor flexibility in series with the bearings' symmetric

Lateral Rotor Vibration Analysis Models

flexibilities (falls into Category 3 in Table 2.1). Equations 2.3 can be applied to the model in Figure 2.3b, irrespective of whether the shaft is flexible or completely rigid. Thus, in this rigid-shaft flexible-bearing case, the stiffness elements in Equations 2.3 describe only the flexibility of the bearings (Category 2 in Table 2.1).

Slightly less simple versions of the models in Figure 2.3 arise when the disk is allowed to be located off center, or the two half shafts are not identical in every respect, since the equations of motion are then more coupled. Similarly, for a disk located outboard of the bearing span on an "overhung" extension of the shaft, the model is not as simple as that described by Equations 2.3. Early rotor vibration analysts like H. H. Jeffcott, without the yet-to-be-invented digital computer, resorted out of necessity to a variety of such models in designing the machines for the rapid electrification and industrialization during the first part of the twentieth century. Such simple models are still quite useful in honing the rotor vibration specialist's understanding and insights, and are exhaustively covered in several texts devoted to rotor vibration theory (see the Bibliography section at the end of this chapter).

Here, it is expedient to transition from the classic simple LRV models just summarized to the modern finite-element models for multibearing flexible rotors having general mass and flexibility properties. That transition step is covered in the next topic, the *Simple Nontrivial LRV Model*, an 8-DOF system whose equations of motion can be written on a single page even though it contains all the generic features of general LRV models.

2.2.3 Simple Nontrivial 8-DOF Model

Even if one understands the underlying physical principles imbedded in a computationally intensive engineering analysis computer code, it is still somewhat of a "black box" to all except the individual(s) who wrote the code. In that vein, the equations of motion for a multi-DOF system are essentially contained in the elements of the model's $[M]$, $[C]$, and $[K]$ matrices, which are "constructed and housed inside the computer" during computation. Therefore, prior to presenting the formulation and development of the Rotor Dynamic Analysis (RDA) Finite Element PC software supplied with this book, the complete equations of motion are here rigorously developed for a simple nontrivial 8-DOF LRV model using both the Lagrange and direct $F = ma$ approaches. This 8-DOF model is illustrated in Figure 2.4, and contains the following features of general purpose multi-DOF LRV models:

a. Bending of the shaft in two mutually perpendicular lateral planes.
b. Two completely general dynamically linear bearings.

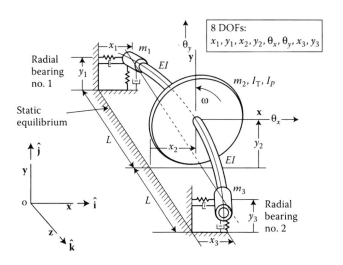

FIGURE 2.4 Simple nontrivial 8-DOF model for LRV.

c. Three concentrated masses connected by flexible shafting.
d. The central concentrated mass also has transverse and polar moments of inertia and associated angular coordinates.

In most vibration modeling, such as with finite-element formulations, the actual continuous media system is modeled by a *discrete* assemblage of $F = ma$-based ODEs. This means the governing *partial differential equation* (PDE) embodying the applicable physical principle(s) of the continuous media structure is approximated by a set of ODEs. The more *pieces* into which the structure model is subdivided, the larger the number of ODEs (equal to the number of DOFs) and the more accurately they approximate the governing PDE. The fundamental reason for doing this is because general solutions to most governing PDEs are obtainable only for the simplest of geometric shapes. The underlying objective is to model the system by a sufficient number of DOFs in order to adequately characterize the actual continuous media system in the frequency range up to ω_n, the highest natural frequency of interest for the system being analyzed. At frequencies progressively higher than ω_n, the characteristics of the discrete model and those of the actual system progressively diverge. The practical application details of these considerations are covered in Chapter 4, which is essentially a users' manual for the RDA code supplied with this book.

The system in Figure 2.4 is modeled here by three lumped masses. The two end masses (m_1 and m_3) are allowed only planar displacements in x and y, whereas the central mass (m_2) is allowed both x and y displacements plus x and y angular displacements θ_x and θ_y. With this model, the two flexible half shafts can be treated either as *massless*, or subdivided into *lumped masses* that are combined with the concentrated masses at mass stations 1, 2, and 3. The usual way of doing this is to subdivide each shaft section into two equal axial-length sections, adding the left-half mass to the left station mass and the right-half mass to the right station mass.

The equations of motion for the system in Figure 2.4 are first derived using two different variations of the Lagrange approach, followed by the direct $F = ma$ approach. The two different Lagrange derivations presented differ only as follows: (i) treating the *gyroscopic effect* as a reaction moment upon the disk using rigid-body rotational dynamics or (ii) treating the *gyroscopic effect* by including the disk's spin-velocity kinetic energy within the total system kinetic energy function, T. The second of these two Lagrange avenues is a bit more demanding to follow than the first, since it requires using the so-called *Euler angles* to define the disk's angular coordinates. For each of the approaches used here, the starting point is the rotor beam-deflection model consistent with the half shafts' bending moment boundary conditions (bending moment $\propto$ curvature) and the eight generalized coordinates employed. This deflection model is shown in Figure 2.5, where deflections are shown as greatly exaggerated.

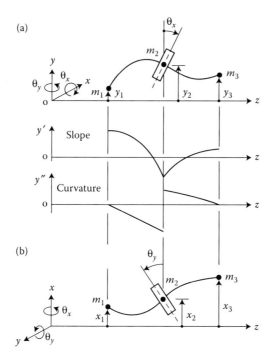

FIGURE 2.5 Rotor beam-deflection model for an 8-DOF system, with all generalized coordinates shown in their respective positive directions: (a) beam deflection, slope, and curvature in $y-z$ plane and (b) $x-z$ deflection only, but slope and curvature similar to (a).

2.2.3.1 Lagrange Approach (i)

Referring back in Section 1.2.1 to the description of the Lagrange equations, they can be expressed as follows:

$$\frac{d}{dt}\left(\frac{\partial T}{\partial \dot{q}_i}\right) - \frac{\partial T}{\partial q_i} + \frac{\partial V}{\partial q_i} = Q_i, \quad i = 1, 2, \ldots, n_{DOF} \quad (2.4)$$

where T and V are the kinetic and potential energy functions, respectively; and q_i and Q_i are the generalized coordinates and generalized forces, respectively. In this derivation, the left-hand side of Equations 2.4 is used to develop the rotor model mass and stiffness matrices. The bearings' stiffness and damping components as well as the rotor disk's gyroscopic moment are treated as generalized forces and thus brought into the equations of motion on the right-hand side of Equations 2.4.

For a beam in bending, the potential energy can be derived by integrating the strain energy over the length of the beam. Linear beam theory is used here, so the bending strain energy in two planes ($x-z$ and $y-z$) can be linearly superimposed as

$$V = \int_0^{2L} \frac{M_{xz}^2 + M_{yz}^2}{2EI} \, dz \quad (2.5)$$

where M_{xz} is the bending moment in $x-z$ plane $= EIx''$, M_{yz} is the bending moment in $y-z$ plane $= EIy''$, E is Young's modulus of the shaft material, and I is the bending area moment of inertia for the two uniform diameter half shafts.

As is evident from Figure 2.5, a linear bending curvature function satisfies the two zero-moment end-boundary conditions (at $z = 0$ and $z = 2L$) and its discontinuity at $z = L$ provides an instantaneous moment difference across the disk consistent with the disk's instantaneous dynamics. Because of the discontinuity, the integration indicated in Equation 2.5 must be performed in two pieces, $z = 0$ to L and $z = L$ to $2L$. Accordingly, each half shaft has *three generalized coordinates* (two translations and one angular displacement) to specify its deflection curve in the $x-z$ plane and likewise in the $y-z$ plane. Therefore, deflection functions with *three coefficients* and linearly varying second derivatives (i.e., curvatures) are required. Thus, a third-order polynomial can be used, but it has four coefficients; hence one term must be omitted. The second-order term is omitted because the zero-order term is needed to retain x and y rigid-body translations and the first-order term is needed to retain x and y rigid-body rotations. The following expressions follow from these requirements. First, the left half

Lateral Rotor Vibration Analysis Models

shaft is treated.

$z = 0$ to L Boundary conditions

$x = az^3 + bz + c$ $x(0) = x_1 = c$

$x' = 3az^2 + b$ $x(L) = x_2 = aL^3 + bL + x_1$

$x'' = 6az$ $x'(L) = \theta_y = 3aL^2 + b$

$\theta_x, \theta_y \ll 1$, $\therefore \tan \theta_x \cong \theta_x$ and $\tan \theta_y \cong \theta_y$

From the above simultaneous equations with boundary conditions utilized at $z = 0$ and $z = L$, the coefficient "a" is determined and results in the following expression for x–z plane curvature:

$$x'' = \frac{3}{L^3}(x_1 - x_2 + \theta_y L)z, \quad z = 0 \text{ to } L \tag{2.6}$$

Similarly, the y–z plane curvature over $z = 0$ to L is determined to be the following:

$$y'' = \frac{3}{L^3}(y_1 - y_2 - \theta_x L)z, \quad z = 0 \text{ to } L \tag{2.7}$$

For the right half shaft, the same polynomial form is used for beam deflection as for the left half shaft, except that $(2L-z)$ must be put in place of z, as follows:

$z = L$ to $2L$ Boundary conditions

$x = a(2L - z)^3 + b(2L - z) + c$ $x(2L) = x_3 = c$

$x' = -3a(2L - z)^2 - b$ $x(L) = x_2 = aL^3 + bL + x_3$

$x'' = 6a(2L - z)$ $x'(L) = \theta_y = -3aL^2 - b$

From these simultaneous equations with boundary conditions utilized (at $z = L$ and $z = 2L$), the coefficient "a" is determined and results in the following expression for x–z plane curvature:

$$x'' = \frac{3}{L^3}(x_3 - x_2 - \theta_y L)(2L - z), \quad z = L \text{ to } 2L \tag{2.8}$$

Similarly, the y–z plane curvature over $z = L$ to $2L$ is determined to be the following:

$$y'' = \frac{3}{L^3}(y_3 - y_2 + \theta_x L)(2L - z), \quad z = L \text{ to } 2L \tag{2.9}$$

The curvature expressions from Equations 2.6–2.9 are used for bending moment in the integration of strain energy expressed in Equation 2.5

(i.e., $M_{xz} = EIx''$ and $M_{yz} = EIy''$ from linear beam theory). Because of the curvature discontinuity at $z = L$, the integral for strain energy must be split into two pieces, as follows:

$$V = \frac{EI}{2}\left[\int_0^L [(x'')^2 + (y'')^2]\,dz + \int_L^{2L} [(x'')^2 + (y'')^2]\,dz\right] \quad (2.10)$$

There are obvious math steps left out at this point, in the interest of space. The obtained expression for potential energy is given as follows:

$$V = \frac{3EI}{2L^3}\Big(x_1^2 + 2x_2^2 + x_3^2 - 2x_1 x_2 - 2x_2 x_3 + 2x_1 \theta_y L - 2x_3 \theta_y L + 2\theta_y^2 L^2 \\ + y_1^2 + 2y_2^2 + y_3^2 - 2y_1 y_2 - 2y_2 y_3 - 2y_1 \theta_x L + 2y_3 \theta_x L + 2\theta_x^2 L^2\Big) \quad (2.11)$$

In this approach, the gyroscopic effect is treated as an external moment upon the disk, so expressing the kinetic energy is a relatively simple step since the disk's spin velocity is not included in T. Kinetic energies for m_1 and m_3 are just $\frac{1}{2}m_1 v_1^2$ and $\frac{1}{2}m_3 v_3^2$, respectively. For the disk (m_2), kinetic energy (T_{disk}) can be expressed as the sum of its mass center's translational kinetic energy (T_{cg}) and its rotational kinetic energy (T_{rot}) about the mass center. The kinetic energy function is thus given as follows:

$$T = \tfrac{1}{2}\Big[m_1(\dot{x}_1^2 + \dot{y}_1^2) + m_2(\dot{x}_2^2 + \dot{y}_2^2) + I_T(\dot{\theta}_x^2 + \dot{\theta}_y^2) + m_3(\dot{x}_3^2 + \dot{y}_3^2)\Big] \quad (2.12)$$

$I_T = \tfrac{1}{4}m_2 R^2$ and $I_P = \tfrac{1}{2}m_2 R^2$

The generalized forces for the bearings are perturbations from static equilibrium, and are treated as linear displacement and velocity-dependent forces, expressible for each bearing as follows:

$$f_x^{(n)} = -k_{xx}^{(n)} x - k_{xy}^{(n)} y - c_{xx}^{(n)} \dot{x} - c_{xy}^{(n)} \dot{y} \\ f_y^{(n)} = -k_{yx}^{(n)} x - k_{yy}^{(n)} y - c_{yx}^{(n)} \dot{x} - c_{yy}^{(n)} \dot{y} \quad (2.13)$$

where, n is the bearing no. = 1, 2.

Treating the gyroscopic effect in this approach simply employs the following embodiment of Newton's Second Law for rotation of a rigid body:

$$\dot{\vec{H}} = \vec{M} \quad (2.14)$$

Equation 2.14 states that the instantaneous *time-rate-of-change of the rigid body's angular momentum* ($\vec{H}$) is equal to the sum of the instantaneous moments

($\vec{M}$) upon the rigid body, both ($\vec{H}$) and ($\vec{M}$) being referenced to the same base point (the disk's center-of-gravity is used). Here, $\vec{H} = \hat{i}I_T\dot{\theta}_x + \hat{j}I_T\dot{\theta}_y + \hat{k}I_P\omega$ is the angular momentum, with the spin velocity (ω) held constant. To make the mass moment-of-inertia components time invariant, the (x, y, z) unit base vectors ($\hat{i}, \hat{j}, \hat{k}$) are defined to precess with the disk's axis of symmetry (i.e., spin axis) at an angular velocity $\vec{\Omega} = \hat{i}\dot{\theta}_x + \hat{j}\dot{\theta}_y$. Since the ($\hat{i}, \hat{j}, \hat{k}$) triad rotates at the precession velocity ($\vec{\Omega}$), the total inertial time-rate-of-change of the rigid body's angular momentum ($\vec{H}$) is expressed as follows:

$$\dot{\vec{H}} = \dot{\vec{H}}_\Omega + \vec{\Omega} \times \vec{H} \tag{2.15}$$

Using the chain rule for differentiating a product, $\dot{\vec{H}}_\Omega = \hat{i}I_T\ddot{\theta}_x + \hat{j}I_T\ddot{\theta}_y$ is the portion of $\dot{\vec{H}}$ obtained by differentiating $\dot{\theta}_x$ and $\dot{\theta}_y$, and $\vec{\Omega} \times \vec{H}$ is the portion obtained by differentiating the rotating base vectors ($\hat{i}, \hat{j}, \hat{k}$). The disk's angular motion displacements ($\theta_x, \theta_y \ll 1$) are assumed to be very small; therefore, the precessing triad ($\hat{i}, \hat{j}, \hat{k}$) has virtually the same orientation as the nonrotating x–y–z coordinate system. Thus, a vector referenced to the precessing ($\hat{i}, \hat{j}, \hat{k}$) system has virtually the same x–y–z scalar components in the nonprecessing ($\hat{i}, \hat{j}, \hat{k}$) system. Equation 2.14 then yields the following expressions for the x and y moment components that must be applied to the disk to make it undergo its x and y angular motions.

$$\begin{matrix} M_x = I_T\ddot{\theta}_x + I_P\omega\dot{\theta}_y \\ M_y = I_T\ddot{\theta}_y - I_P\omega\dot{\theta}_x \end{matrix} \quad \text{rearranged to} \quad \begin{matrix} M_x - I_P\omega\dot{\theta}_y = I_T\ddot{\theta}_x \\ M_y + I_P\omega\dot{\theta}_x = I_T\ddot{\theta}_y \end{matrix} \tag{2.16}$$

The I_T acceleration terms in Equations 2.16 are included via the Lagrange kinetic energy function (T), Equation 2.12. However, the I_P terms are not included, and these are the gyroscopic inertia components that are rearranged here to the left side of the equations, as shown, to appear as moment components (fictitious) applied to the disk. The gyroscopic moment components that are "applied" to the disk as generalized forces in Equations 2.4 are then as follows:

$$\begin{aligned} M_{\text{gyro},x} &= -I_P\omega\dot{\theta}_y \\ M_{\text{gyro},y} &= +I_P\omega\dot{\theta}_x \end{aligned} \tag{2.17}$$

Equations 2.11 and 2.12 for V and T, respectively, as well as Equations 2.13 for bearing dynamic force components upon m_1 and m_3 and Equations 2.17 for gyroscopic moment components upon the disk are all applied in Equations 2.4, the Lagrange equations. In the interest of space, the clearly indicated math steps are omitted at this point. The derived eight equations of motion for the model shown in Figure 2.4 are presented in the matrix form, as follows:

$$\begin{Bmatrix} m_1\ddot{x}_1 \\ m_1\ddot{y}_1 \\ m_2\ddot{x}_2 \\ m_2\ddot{y}_2 \\ I_T\ddot{\theta}_x \\ I_T\ddot{\theta}_y \\ m_3\ddot{x}_3 \\ m_3\ddot{y}_3 \end{Bmatrix} + \begin{bmatrix} c_{xx}^{(1)} & c_{xy}^{(1)} & 0 & 0 & 0 & 0 & 0 & 0 \\ c_{yx}^{(1)} & c_{yy}^{(1)} & 0 & 0 & 0 & 0 & 0 & 0 \\ 0 & 0 & 0 & 0 & 0 & 0 & 0 & 0 \\ 0 & 0 & 0 & 0 & 0 & 0 & 0 & 0 \\ 0 & 0 & 0 & 0 & 0 & I_P\omega & 0 & 0 \\ 0 & 0 & 0 & 0 & -I_P\omega & 0 & 0 & 0 \\ 0 & 0 & 0 & 0 & 0 & 0 & c_{xx}^{(2)} & c_{xy}^{(2)} \\ 0 & 0 & 0 & 0 & 0 & 0 & c_{yx}^{(2)} & c_{yy}^{(2)} \end{bmatrix} \begin{Bmatrix} \dot{x}_1 \\ \dot{y}_1 \\ \dot{x}_2 \\ \dot{y}_2 \\ \dot{\theta}_x \\ \dot{\theta}_y \\ \dot{x}_3 \\ \dot{y}_3 \end{Bmatrix}$$

$$+ \frac{3EI}{L^3} \begin{bmatrix} (1+\bar{k}_{xx}^{(1)}) & \bar{k}_{xy}^{(1)} & -1 & 0 & 0 & L & 0 & 0 \\ \bar{k}_{yx}^{(1)} & (1+\bar{k}_{yy}^{(1)}) & 0 & -1 & -L & 0 & 0 & 0 \\ -1 & 0 & 2 & 0 & 0 & 0 & -1 & 0 \\ 0 & -1 & 0 & 2 & 0 & 0 & 0 & -1 \\ 0 & -L & 0 & 0 & 2L^2 & 0 & 0 & L \\ L & 0 & 0 & 0 & 0 & 2L^2 & -L & 0 \\ 0 & 0 & -1 & 0 & 0 & -L & (1+\bar{k}_{xx}^{(2)}) & \bar{k}_{xy}^{(2)} \\ 0 & 0 & 0 & -1 & L & 0 & \bar{k}_{yx}^{(2)} & (1+\bar{k}_{yy}^{(2)}) \end{bmatrix}$$

$$\times \begin{Bmatrix} x_1 \\ y_1 \\ x_2 \\ y_2 \\ \theta_x \\ \theta_y \\ x_3 \\ y_3 \end{Bmatrix} = \{R\} \quad (2.18)$$

$$\bar{k}_{ij}^{(n)} \equiv \frac{L^3}{3EI} k_{ij}^{(n)}$$

$\{R\} \equiv$ vector of time-varying forces and moments applied upon the system.

In Equation 2.18 the $\{m\ddot{q}\}$ vector shown takes advantage of multiplying the diagonal mass matrix (all zeros except on the main diagonal) by the acceleration vector, thus compressing the space needed to write the full equations of motion. Properly applied, these equations of motion for the 8-DOF model are a reasonable approximation for the first and possibly the second natural frequency modes of an axially symmetric rotor on two dynamically linear bearings, especially if most of the rotor mass is located near the rotor's axial center between the bearings. It is also a worthy model on which to "benchmark" a general purpose linear LRV computer code. More importantly, this model's equations of motion layout for detailed scrutiny all the elements of the motion equation matrices, on slightly over half a page, for an 8-DOF model that has all the generic features of general multi-DOF LRV models. One can thereby gain insight into the computations that take place when a general purpose LRV code is used.

2.2.3.2 Lagrange Approach (ii)

This approach differs from the just completed previous Lagrange approach only in how the gyroscopic moment is derived; hence only that facet is shown here. Specifically, the issue is the portion of the disk's rotational kinetic energy (T_{rot}) due to its spin velocity. Using a coordinate system with its origin at the disk's mass center and its axes aligned with principal-inertia axes through the disk's mass center, the disk's kinetic energy due to rotation can be expressed as follows:

$$T_{rot} = \tfrac{1}{2}\left(I_{xx}\omega_x^2 + I_{yy}\omega_y^2 + I_{zz}\omega_z^2\right) \quad (2.19)$$

However, this expression cannot be directly used in the kinetic energy function (T) for the Lagrange equations because ω_x, ω_y, and ω_z are not the time derivatives of any three angular coordinates, respectively, that could specify the disk's angular position. The angular orientation of any rigid body can, however, be prescribed by three angles, the so-called Euler angles. Furthermore, the first time derivatives of these three angles provide angular velocity components applicable to T_{rot} for the Lagrange equations. While this approach can be applied to any rigid body, the application here is somewhat simplified because $I_{xx} = I_{yy} \equiv I_T$ and $\theta_x, \theta_y \ll 1$.

The three Euler angles are applied in a specified order that follows. $(\hat{i}, \hat{j}, \hat{k})$ is a mass-center principal-inertia triad corresponding to an $x-y-z$ principal-inertia coordinate system fixed in the disk at its center. When all the Euler angles are zero, $(\hat{i}, \hat{j}, \hat{k})$ aligns with a nonrotating triad $(\hat{I}, \hat{J}, \hat{K})$.

To "book keep" the three sequential steps of orthogonal transformation produced by the three sequential Euler angles, it is helpful to give a specific identity to the $(\hat{i}, \hat{j}, \hat{k})$ triad for each of the four orientations it occupies, from "start to finish," in undergoing the three Euler angles. These identities are given along with each Euler angle specified. It is also quite helpful at this point for the reader to isometrically sketch each of the four $x-y-z$ coordinate system angular orientations, using a common origin.

- *Initial state (all Euler angles are zero):* $(\hat{i}, \hat{j}, \hat{k})$ aligns with $(\hat{I}, \hat{J}, \hat{K})$.
- *First Euler angle:* Rotate disk θ_y about the y-axis (i.e., $\hat{i}, \hat{k}$ about $\hat{j} = \hat{J}$),

$$(\hat{i}, \hat{j}, \hat{k}) \text{ moves to } (\hat{i}', \hat{j}', \hat{k}'), \quad \text{where } \hat{j}' = \hat{j} = \hat{J}$$

- *Second Euler angle:* Rotate disk θ_x about the x-axis (i.e., $\hat{j}', \hat{k}'$ about $\hat{i}'$),

$$(\hat{i}', \hat{j}', \hat{k}') \text{ moves to } (\hat{i}'', \hat{j}'', \hat{k}''), \quad \text{where } \hat{i}'' = \hat{i}'$$

- *Third Euler angle:* Rotate the disk ϕ about the z-axis (i.e., $\hat{i}'', \hat{j}''$ about $\hat{k}''$),

$$(\hat{i}'', \hat{j}'', \hat{k}'') \text{ moves to } (\hat{i}, \hat{j}, \hat{k}), \quad \text{where } \hat{k}'' = \hat{k}$$

The following angular velocity vector for the disk is now specified in components that are legitimate for use in the Lagrange approach since each velocity component is the first time derivative of a generalized coordinate:

$$\vec{\omega}_{total} = \dot{\theta}_y \hat{j} + \dot{\theta}_x \hat{i}' + \omega \hat{k} \tag{2.20}$$

$$\omega = \dot{\phi}$$

The remaining step is to transform $\hat{J}$ and $\hat{i}'$ in Equation 2.20 into their $(\hat{i}, \hat{j}, \hat{k})$ components to obtain the disk's angular velocity components in a principal-inertia x–y–z coordinate system. This is accomplished simply by using the following associated direction-cosine orthogonal transformations:

$$\begin{Bmatrix} \hat{i}' \\ \hat{j}' \\ \hat{k}' \end{Bmatrix} = \begin{bmatrix} \cos\theta_y & 0 & -\sin\theta_y \\ 0 & 1 & 0 \\ \sin\theta_y & 0 & \cos\theta_y \end{bmatrix} \begin{Bmatrix} \hat{I} \\ \hat{J} \\ \hat{K} \end{Bmatrix}$$

$$\begin{Bmatrix} \hat{i}'' \\ \hat{j}'' \\ \hat{k}'' \end{Bmatrix} = \begin{bmatrix} 1 & 0 & 0 \\ 0 & \cos\theta_x & \sin\theta_x \\ 0 & -\sin\theta_x & \cos\theta_x \end{bmatrix} \begin{Bmatrix} \hat{i}' \\ \hat{j}' \\ \hat{k}' \end{Bmatrix} \tag{2.21}$$

$$\begin{Bmatrix} \hat{i} \\ \hat{j} \\ \hat{k} \end{Bmatrix} = \begin{bmatrix} \cos\phi & \sin\phi & 0 \\ -\sin\phi & \cos\phi & 0 \\ 0 & 0 & 1 \end{bmatrix} \begin{Bmatrix} \hat{i}'' \\ \hat{j}'' \\ \hat{k}'' \end{Bmatrix}$$

Lateral Rotor Vibration Analysis Models

Multiplying these three orthogonal matrices together according to the proper Euler angle sequence yields an equation of the following form:

$$\begin{Bmatrix} \hat{i} \\ \hat{j} \\ \hat{k} \end{Bmatrix} = [R_\phi][R_{\theta_x}][R_{\theta_y}] \begin{Bmatrix} \hat{I} \\ \hat{J} \\ \hat{K} \end{Bmatrix} \tag{2.22}$$

Equation 2.22, product of the three orthogonal transformation matrices, is also an orthogonal matrix, embodying the total orthogonal transformation from the initial state to the end state orientation following application of the three Euler angles, and can be expressed as follows:

$$[R] = [R_\phi][R_{\theta_x}][R_{\theta_y}] \tag{2.23}$$

As an orthogonal matrix, $[R]$ has an inverse equal to its transpose. Therefore, the $\hat{J}$ unit vector in Equation 2.20 is obtained from the second equation of the following three

$$\begin{Bmatrix} \hat{I} \\ \hat{J} \\ \hat{K} \end{Bmatrix} = [R]^T \begin{Bmatrix} \hat{i} \\ \hat{j} \\ \hat{k} \end{Bmatrix} \tag{2.24}$$

to obtain the following expression for $\hat{J}$:

$$\hat{J} = (\sin\phi\cos\theta_x)\hat{i} + (\cos\phi\cos\theta_x)\hat{j} - (\sin\theta_x)\hat{k} \tag{2.25}$$

Since $\hat{i}' = \hat{i}''$, inverting the 3rd of Equations 2.21 yields the following:

$$\hat{i}' = \hat{i}\cos\phi - \hat{j}\sin\phi \tag{2.26}$$

Substituting Equations 2.25 and 2.26 into Equation 2.20 produces the following result:

$$\vec{\omega} = (\dot\theta_y \sin\phi \cos\theta_x + \dot\phi_x \cos\phi)\hat{i} + (\dot\theta_y \cos\phi \cos\theta_x - \dot\theta_x \sin\phi)\hat{j}$$
$$+ (-\dot\theta_y \sin\theta_x + \dot\omega)\hat{k} \tag{2.27}$$

Equation 2.27 provides the proper components for ω_x, ω_y, and ω_z to insert into Equation 2.19 for the disk's rotational kinetic energy, T_{rot}, as follows:

$$T_{\text{rot}} = \tfrac{1}{2}I_T\left(\omega_x^2 + \omega_y^2\right) + \tfrac{1}{2}I_P\omega_z^2 = \tfrac{1}{2}[I_T(\dot\theta_y \sin\phi \cos\theta_x + \dot\theta_x \cos\phi)^2$$
$$+ I_T(\dot\theta_y \cos\phi \cos\theta_x - \dot\theta_x \sin\phi)^2 + I_P(-\dot\theta_y \sin\theta_x + \omega)^2] \tag{2.28}$$

Simplifications utilizing $\cos\theta_x \cong 1$, $\sin\theta_x \cong \theta_x$, and $\sin^2\theta_x \ll \sin\theta_x$ then yield the following expression for the disk's rotational kinetic energy:

$$T_{\text{rot}} = \tfrac{1}{2}\left[I_T\left(\dot\theta_x^2 + \dot\theta_y^2\right) + I_P\left(\omega^2 - 2\omega\dot\theta_y\theta_x\right)\right] \quad (2.29)$$

A potential point of confusion is avoided here if one realizes that θ_x and θ_y are both very small and are applied in the Euler angle sequence ahead of ϕ, which is not small ($\phi = \omega t$). Thus, $\dot\theta_x$ and $\dot\theta_y$ are directed along axes that are basically aligned with the nonrotating inertial $x-y$ coordinates, not those spinning with the disk. As with the Lagrange approach (i), the disk's total kinetic energy is expressible as the sum of the mass-center kinetic energy plus the rotational kinetic energy as follows:

$$T_{\text{disk}} = T_{\text{cg}} + T_{\text{rot}} \quad (2.30)$$

The total system kinetic energy is thus expressible for this Lagrange approach by the following equation:

$$T = \tfrac{1}{2}\left[m_1\left(\dot x_1^2 + \dot y_1^2\right) + m_2\left(\dot x_2^2 + \dot y_2^2\right) + I_T\left(\dot\theta_x^2 + \dot\theta_y^2\right) + I_P\left(\omega^2 - 2\omega\dot\theta_y\theta_x\right) \\ + m_3\left(\dot x_3^2 + \dot y_3^2\right)\right] \quad (2.31)$$

Equation 2.31 differs from its Lagrange approach (i) counterpart, Equation 2.12, only by its I_P term that contains the disk's gyroscopic effect.

The potential energy formulation and bearing dynamic force expressions used here are identical to those in Lagrange approach (i), Equations 2.11 and 2.13, respectively. However, here the gyroscopic effect is contained within the kinetic energy function in Equation 2.31. Therefore, Equations 2.17 used in the Lagrange approach (i) for gyroscopic moment components upon the disk are not applicable here. Implementing the clearly indicated math steps implicit in Equations 2.4, this approach yields the same eight equations given by Equations 2.18.

2.2.3.3 Direct F = ma Approach

In this approach, the sum of x-forces and the sum of y-forces on m_1, m_2, and m_3 equated to their respective $m\ddot q$ terms yields six of the eight motion equations. The sum of x-moments and the sum of y-moments on the disk equated to their respective $I_T\ddot\theta$ terms yields the other two motion equations. This can be summarized as follows.

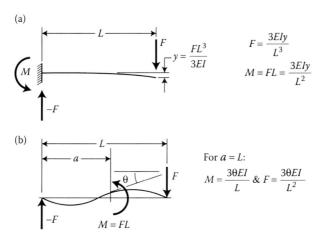

FIGURE 2.6 Beam deflection formulas.

Bearing forces and gyroscopic moment are taken directly from Equations 2.13 and 2.17, respectively. Thus, only the beam-deflection reaction forces and moments need developing here, and these can be derived using superposition of the two cases given in Figure 2.6. All reaction force and moment components due to x and y translations with θ_x and θ_y both zero are obtained using the *cantilever beam* end-loaded case given in Figure 2.6a. Likewise, all reaction force and moment components due to θ_x and θ_y with x and y translations both zero are obtained using the *simply supported beam* with an end moment, that is, case with $a = L$ in Figure 2.6b. Superimposing these two cases provides all the beam reaction force and moment components due to all eight displacements and these are summarized as follows:

Beam-Deflection Reaction	Force and Moment Components	
$f_{1x} = \dfrac{3EI}{L^3}(-x_1 + x_2 - \theta_y L)$	$M_{2x} = \dfrac{3EI}{L^3}\left(y_1 L - 2\theta_x L^2 - y_3 L\right)$	
$f_{1y} = \dfrac{3EI}{L^3}(-y_1 + y_2 + \theta_x L)$	$M_{2y} = \dfrac{3EI}{L^3}\left(-x_1 L - 2\theta_y L^2 + x_3 L\right)$	(2.32)
$f_{2x} = \dfrac{3EI}{L^3}(x_1 - 2x_2 + x_3)$	$f_{3x} = \dfrac{3EI}{L^3}(x_2 - x_3 + \theta_y L)$	
$f_{2y} = \dfrac{3EI}{L^3}(y_1 - 2y_2 + y_3)$	$f_{3y} = \dfrac{3EI}{L^3}(y_2 - y_3 - \theta_x L)$	

The eight equations of motion are constructed from $F = ma$ and $M = I\ddot{\theta}$ utilizing Equations 2.13 for bearing forces, Equations 2.17 for gyroscopic moments, and Equations 2.32 for beam-bending force and moment reactions, as follows:

$$m_1\ddot{x}_1 = f_{1x} + f_x^{(1)} \qquad I_T\ddot{\theta}_x = M_{2x} + M_{gyro,x}$$
$$m_1\ddot{y}_1 = f_{1y} + f_y^{(1)} \qquad I_T\ddot{\theta}_y = M_{2y} + M_{gyro,y}$$
$$m_2\ddot{x}_2 = f_{2x} \qquad m_3\ddot{x}_3 = f_{3x} + f_x^{(2)}$$
$$m_2\ddot{y}_2 = f_{2y} \qquad m_3\ddot{y}_3 = f_{3y} + f_y^{(2)} \qquad (2.33)$$

Substituting the appropriate expressions from Equations 2.13, 2.17, and 2.32 into Equations 2.33 yields the 8-DOF model's equations of motion given in Equations 2.18.

Equations 2.18 have been derived here in three somewhat different approaches. However, all three approaches are based on Newton's second law and thus must yield the same result.

The right-hand side of Equations 2.18, $\{R\}$, is strictly for time-dependent forcing functions and viewed as being externally applied on the system. No specific examples of $\{R\}$ were needed to develop the three derivations of Equations 2.18, but two important cases are now delineated: (i) *eigenvalue extraction* and (ii) *steady-state unbalance response*. For eigenvalue extraction, such as performed in searching for operating zones where dynamic instability (self-excited vibration) is predicted, $\{R\} = 0$ can be used since $\{R\}$ does not enter into that mathematical process (see Section 1.3, subheading "Dynamic Instability: The Complex Eigenvalue Problem"). For an unbalance response example, the combination of so-called *static unbalance* and *dynamic unbalance* are simultaneously applied on the 8-DOF model's disk, as shown in Figure 2.7. An unbalance is modeled by its equivalent centrifugal force.

Here, the static unbalance mass is chosen as the angular reference point (*key phaser*) on the rotor and ϕ (90° for illustrated case in Figure 2.7) is the phase angle between m_s and the rotating moment produced by the two 180° out-of-phase m_d dynamic unbalance masses. Equations 2.18 then have the

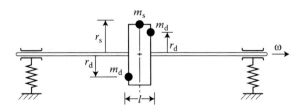

FIGURE 2.7 Combination of static and dynamic rotor disk unbalance.

right-hand side $\{R\}$ shown in the following equations:

$$[M]\{\ddot{q}\} + [C]\{\dot{q}\} + [K]\{q\} = \omega^2 \begin{Bmatrix} 0 \\ 0 \\ m_s r_s \cos \omega t \\ m_s r_s \sin \omega t \\ m_d r_d l \cos(\omega t + \phi) \\ m_d r_d l \sin(\omega t + \phi) \\ 0 \\ 0 \end{Bmatrix} \quad (2.34)$$

The four zeros in $\{R\}$ reflect *no unbalances* at the two bearing stations.

2.3 Formulations for RDA Software

The vibration fundamentals covered in Chapter 1 and the foregoing sections of this chapter provide ample background to follow the development of the governing formulations for the RDA code. RDA is a user-friendly PC-based user-interactive software package that is structured on the finite-element method. It was developed in the Rotor Dynamics Laboratory at Case Western Reserve University to handle the complete complement of *linear LRV analyses*, and it is supplied with this book. In this section, the focus is on formulation, solution, and computation aspects of the RDA code. In Part 2 of this book (Chapters 4, 5, and 6), the focus shifts to the use of RDA in problem solving.

2.3.1 Basic Rotor Finite Element

Development of the RDA model starts with the *basic rotor finite-element building block*, which is comprised of two disks (or any M, I_T, I_P) connected by a beam of uniform circular-cross-section (shaft), as shown in Figure 2.8.

For the rotor finite element shown in Figure 2.8, the following two lists summarize its elementary parameters.

Shaft element properties:

Mass, $\quad M^{(s)} = \dfrac{\gamma_s \pi (d_o^2 - d_i^2) L}{4g}$

Transverse inertia at c.g., $\quad I_T^{(s)} = \dfrac{1}{12} M^{(s)} \left[3\left(\dfrac{d_o^2 + d_i^2}{4}\right) + L^2 \right]$

Polar inertia, $\quad I_P^{(s)} = \dfrac{1}{2} M^{(s)} \left(\dfrac{d_o^2 + d_i^2}{4}\right) \quad (2.35a)$

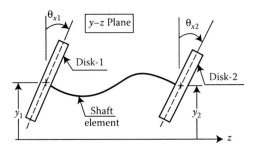

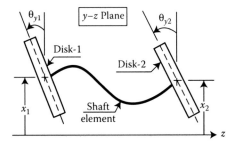

FIGURE 2.8 Basic rotor finite-element building block.

Area moment of inertia, $I = \dfrac{\pi(d_o^4 - d_i^4)}{64}$

Modulus of elasticity, E

where d_o is the shaft outside diameter (OD), d_i is the shaft inside diameter (ID) (optional concentric hole), L is the shaft length, γ_s is the shaft weight density, and g is the gravity constant.

The formulas for concentrated disk masses are essentially the same as those for the shaft element, and thus listed here as follows.

Concentrated disk mass properties:
Any axially symmetric mass specified by its M, I_T, and I_P can be used, for example, couplings, impellers, gears, and so on

Mass, $M^{(d)} = \dfrac{\gamma_d \pi \left(D_o^2 - D_i^2\right) l}{4g}$

Transverse inertia at c.g., $I_T^{(d)} = \dfrac{1}{12} M^{(d)} \left[3\left(\dfrac{D_o^2 + D_i^2}{4}\right) + l^2 \right]$ (2.35b)

Polar inertia, $I_P^{(d)} = \dfrac{1}{2} M^{(d)} \left(\dfrac{D_o^2 + D_i^2}{4}\right)$

where D_o is the disk OD, D_i is the disk ID (concentric hole), l is the disk axial thickness, and γ_d is the disk weight density.

As shown in Figure 2.8, each mass station has four DOFs, that is, $x, y, \theta_x,$ and θ_y. Thus, with θ_x and θ_y coordinates included at every mass station, beam-bending transverse rotary inertia, an effect of increased importance for higher frequency bending modes, is included. In addition therefore, either an optional concentrated *point mass* or concentrated *disk mass* (or other axially symmetric mass) can be added at each mass station after the complete rotor matrices are assembled from all the individual shaft element matrices.

The programmed steps in building the RDA equations of motion for arbitrary model configurations are essentially the encoding of the [M], [C], and [K] matrices, as well as the right-hand side column of the applied forces, {R}. These matrices are essentially the discrete model's equations of motion. Using the basic rotor finite-element *building block* shown in Figure 2.8, the total system stiffness and damping matrices are single-option paths, in contrast to the mass matrix that has three options, *lumped mass, distributed mass,* and *consistent mass* discretizations. For most rotor vibration models, the *consistent* and *distributed* mass formulations provide significantly better model resolution accuracy (i.e., converge with fewer finite elements or DOFs) than the *lumped* mass formulation. Furthermore, based on the author's experience, the consistent mass model seems to be marginally better for rotors than the distributed mass model. RDA is coded to allow the user to select any of these three mass models. While the consistent mass model is usually the preferred option, it is occasionally informative to be able to easily switch between these three mass models to parametrically study model convergence characteristics. That is, to study if a selected number of rotor elements is adequate for the needs of a particular analysis. The three mass-matrix options are covered here first, followed by the stiffness and gyroscopic matrices.

2.3.2 Shaft Element Lumped Mass Matrix

In this approach, it is assumed that the shaft element's mass is lumped at the element's two end points according to static weight-equilibrating forces at the element end points. For the uniform diameter shaft element programmed into RDA this means lumping half the shaft element's mass at each of the mass stations at the two ends of the element. Implicit in this approximation is a step change in the shaft element's lateral (radial) acceleration at its axial midpoint. In other words, the actual continuous axial variation in radial acceleration is approximated by a series of small discrete step changes. Similarly, half the beam element's transverse moment of inertia is transferred to each of its two ends points using the *parallel-axis theorem*, shown as follows:

$$I_{T_i} = \frac{1}{12}\left(\frac{M^{(s)}}{2}\right)\left[3\left(\frac{d_o^2 + d_i^2}{4}\right) + \left(\frac{L}{2}\right)^2\right] + \frac{M^{(s)}}{2}\left(\frac{L}{4}\right)^2 \qquad (2.36)$$

With the coordinate vector ordering $\{x_1, y_1, \theta_{x1}, \theta_{y1}, x_2, y_2, \theta_{x2}, \theta_{y2}\}$ employed, the shaft element *lumped mass* matrix is then as follows:

$$[M]_i^l = \begin{bmatrix} \frac{1}{2}M_i^{(s)} & 0 & 0 & 0 & 0 & 0 & 0 & 0 \\ 0 & \frac{1}{2}M_i^{(s)} & 0 & 0 & 0 & 0 & 0 & 0 \\ 0 & 0 & I_{T_i} & 0 & 0 & 0 & 0 & 0 \\ 0 & 0 & 0 & I_{T_i} & 0 & 0 & 0 & 0 \\ 0 & 0 & 0 & 0 & \frac{1}{2}M_i^{(s)} & 0 & 0 & 0 \\ 0 & 0 & 0 & 0 & 0 & \frac{1}{2}M_i^{(s)} & 0 & 0 \\ 0 & 0 & 0 & 0 & 0 & 0 & I_{T_i} & 0 \\ 0 & 0 & 0 & 0 & 0 & 0 & 0 & I_{T_i} \end{bmatrix} \qquad (2.37)$$

2.3.3 Shaft Element Distributed Mass Matrix

The underlying assumption for the distributed mass formulation is that the shaft element's lateral acceleration varies linearly in the axial direction, a logical first-order improvement over the axial step-change approximation implicit in the lumped mass formulation. An axial linear variation of lateral acceleration requires that the element's lateral velocity also varies linearly in the axial direction. The derivation here considers two adjacent mass stations, as shown in Figure 2.9, to formulate the linear variation of lateral velocity.

The linear variation of x-velocity is expressed as follows:

$$\dot{x} = \dot{x}_i + \frac{1}{L_i}(\dot{x}_{i+1} - \dot{x}_i)z \qquad (2.38)$$

The x-direction derivation is shown here, but the y-direction derivation is identical. The x-translation kinetic energy of a shaft element can thus be

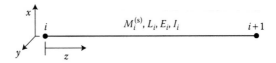

FIGURE 2.9 Two consecutive rotor-model mass stations.

Lateral Rotor Vibration Analysis Models

expressed as follows (similar for y-translation kinetic energy):

$$T_i^{(x)} = \frac{1}{2}\frac{M_i^{(s)}}{L_i}\int_0^{L_i} \dot{x}^2\, dz \tag{2.39}$$

Substituting Equation 2.38 into Equation 2.39 and integrating yields the portion of the total system's kinetic energy function that is needed to extract the shaft element's lateral acceleration terms associated with the x_i and x_{i+1} Lagrange equations of motion. This leads to the following two results:

$$\begin{aligned}\frac{d}{dt}\left(\frac{\partial T}{\partial \dot{x}_i}\right) &= \frac{1}{3}M_i^{(s)}\ddot{x}_i + \frac{1}{6}M_i^{(s)}\ddot{x}_{i+1} \\ \frac{d}{dt}\left(\frac{\partial T}{\partial \dot{x}_{i+1}}\right) &= \frac{1}{6}M_i^{(s)}\ddot{x}_i + \frac{1}{3}M_i^{(s)}\ddot{x}_{i+1}\end{aligned} \tag{2.40}$$

Since the beam element transverse rotary inertia effect is secondary to its lateral mass acceleration effect, the inclusion of shaft element transverse rotary inertia is included here, as already shown for the lumped mass formulation, Equations 2.36. That is, beam element transverse moment of inertia is not "distributed" in the manner just derived for the lateral mass acceleration components. With the coordinate vector ordering $\{x_1, y_1, \theta_{x1}, \theta_{y1}, x_2, y_2, \theta_{x2}, \theta_{y2}\}$ employed, the shaft element's *distributed mass* matrix thus obtained is as follows:

$$[M]_i^d = \begin{bmatrix} \frac{1}{3}M_i^{(s)} & 0 & 0 & 0 & \frac{1}{6}M_i^{(s)} & 0 & 0 & 0 \\ 0 & \frac{1}{3}M_i^{(s)} & 0 & 0 & 0 & \frac{1}{6}M_i^{(s)} & 0 & 0 \\ 0 & 0 & I_{T_i} & 0 & 0 & 0 & 0 & 0 \\ 0 & 0 & 0 & I_{T_i} & 0 & 0 & 0 & 0 \\ \frac{1}{6}M_i^{(s)} & 0 & 0 & 0 & \frac{1}{3}M_i^{(s)} & 0 & 0 & 0 \\ 0 & \frac{1}{6}M_i^{(s)} & 0 & 0 & 0 & \frac{1}{3}M_i^{(s)} & 0 & 0 \\ 0 & 0 & 0 & 0 & 0 & 0 & I_{T_i} & 0 \\ 0 & 0 & 0 & 0 & 0 & 0 & 0 & I_{T_i} \end{bmatrix} \tag{2.41}$$

2.3.4 Shaft Element Consistent Mass Matrix

When the spatial distribution of acceleration (and therefore velocity) in a finite element is formulated with the same shape function as static deflection, it is referred to as the consistent mass approach. The shaft element in Figure 2.8 is postulated to be a uniform cross-section beam in bending. Thus, its static beam deflection can be expressed as cubic functions in the

$x-z$ and $y-z$ planes, as follows (z referenced to left mass station, as shown in Figure 2.9):

$$\Psi(z) = az^3 + bz^2 + cz + d \tag{2.42}$$

A general state of shaft element deflection in a plane ($x-z$ or $y-z$) can be expressed as a linear superposition of four cases, each having a unity displacement for one of the four generalized coordinates in the plane with zero displacement for the other three coordinates in the plane. These four cases are specified by the following tabulated sets of boundary conditions.

Correspondence between Ψ_j, Ψ'_j and Rotor Element Coordinates

$x-z$ plane →	x_1	θ_{y1}	x_2	θ_{y2}
$y-z$ plane →	y_1	$-\theta_{x1}$	y_2	$-\theta_{x2}$
Case 1:	$\Psi_1(0) = 1$	$\Psi'_1(0) = 0$	$\Psi_1(L) = 0$	$\Psi'_1(L) = 0$
Case 2:	$\Psi_2(0) = 0$	$\Psi'_2(0) = 1$	$\Psi_2(L) = 0$	$\Psi'_2(L) = 0$
Case 3:	$\Psi_3(0) = 0$	$\Psi'_3(0) = 0$	$\Psi_3(L) = 1$	$\Psi'_3(L) = 0$
Case 4:	$\Psi_4(0) = 0$	$\Psi'_4(0) = 0$	$\Psi_4(L) = 0$	$\Psi'_4(L) = 1$

Substituting each of the four above boundary condition sets into Equation 2.42 and solving in each case for the four coefficients in Equation 2.42 yields the following four *deflection shape functions*:

$$\Psi_1(z) = 1 - 3\left(\frac{z}{L}\right)^2 + 2\left(\frac{z}{L}\right)^3, \quad \Psi_2(z) = z - 2\frac{z^2}{L} + \frac{z^3}{L^2}$$

$$\Psi_3(z) = 3\left(\frac{z}{L}\right)^2 - 2\left(\frac{z}{L}\right)^3, \quad \Psi_4 = \frac{z^2}{L}\left(\frac{z}{L} - 1\right) \tag{2.43}$$

The general state of shaft element deflection can be expressed as follows:

$$x = x_1 \Psi_1(z) + \theta_{y1} \Psi_2(z) + x_2 \Psi_3(z) + \theta_{y2} \Psi_4(z)$$
$$y = y_1 \Psi_1(z) - \theta_{x1} \Psi_2(z) + y_2 \Psi_3(z) - \theta_{x2} \Psi_4(z) \tag{2.44}$$

Thus, the general state of shaft element velocity can be expressed as follows:

$$\dot{x} = \dot{x}_1 \Psi_1(z) + \dot{\theta}_{y1} \Psi_2(z) + \dot{x}_2 \Psi_3(z) + \dot{\theta}_{y2} \Psi_4(z)$$
$$\dot{y} = \dot{y}_1 \Psi_1(z) - \dot{\theta}_{x1} \Psi_2(z) + \dot{y}_2 \Psi_3(z) - \dot{\theta}_{x2} \Psi_4(z) \tag{2.45}$$

Lateral Rotor Vibration Analysis Models

The total shaft element kinetic energy is derived by substituting Equations 2.45 into the following equation:

$$T_i = \frac{1}{2}\frac{M_i^{(s)}}{L_i}\int_0^{L_i}(\dot{x}^2 + \dot{y}^2)\,dz \qquad (2.46)$$

The element consistent mass matrix is obtained by substituting the integrated result from Equation 2.46 into the acceleration portion for each of the eight Lagrange equations for the shaft element, as follows:

$$\frac{d}{dt}\left(\frac{\partial T_i}{\partial \dot{q}_r}\right) \equiv [M_{rs}]_i^c\{\ddot{q}_s\}, \quad r = 1, 2, \ldots, 8 \qquad (2.47)$$

With $\{\ddot{q}_s\} = \{\ddot{x}_1, \ddot{y}_1, \ddot{\theta}_{x1}, \ddot{\theta}_{y1}, \ddot{x}_2, \ddot{y}_2, \ddot{\theta}_{x2}, \ddot{\theta}_{y2}\}$, the shaft element *consistent mass* matrix thus obtained is as follows:

$$[M]_i^c = \frac{M_i^{(s)}}{420}\begin{bmatrix} 156 & 0 & 0 & 22L_i & 54 & 0 & 0 & -13L_i \\ 0 & 156 & -22L_i & 0 & 0 & 54 & 13L_i & 0 \\ 0 & -22L_i & 4L_i^2 & 0 & 0 & 0 & 0 & 0 \\ 22L_i & 0 & 0 & 4L_i^2 & 13L_i & 0 & 0 & -3L_i^2 \\ 54 & 0 & 0 & 13L_i & 156 & 0 & 0 & -22L_i \\ 0 & 54 & 0 & 0 & 0 & 156 & 22L_i & 0 \\ 0 & 13L_i & 0 & 0 & 0 & 22L_i & 4L_i^2 & 0 \\ -13L_i & 0 & 0 & -3L_i^2 & -22L_i & 0 & 0 & 4L_i^2 \end{bmatrix}$$

(2.48)

2.3.5 Shaft Element Stiffness Matrix

Borrowing from Equation 2.5, the potential energy for the shaft element in bending can be expressed as follows:

$$V_i = \frac{1}{2}E_i I_i \int_0^{L_i}[(x'')^2 + (y'')^2]\,dz \qquad (2.49)$$

Substituting Equations 2.44 into Equation 2.49 provides the shaft element V_i as a function of the element's eight generalized coordinates, similar to the detailed development of Equation 2.11 for the 8-DOF "Simple Nontrivial Model." The element stiffness matrix is obtained by substituting the integrated result from Equation 2.49 into the potential energy term for each of the eight Lagrange equations for the shaft element, as follows:

$$\frac{\partial V_i}{\partial q_r} \equiv [K_{rs}]_i \{q_s\}, \quad r = 1, 2, \ldots, 8 \quad (2.50)$$

With $\{q_s\} = \{x_1, y_1, \theta_{x1}, \theta_{y1}, x_2, y_2, \theta_{x2}, \theta_{y2}\}$, the element *stiffness* matrix thus obtained is as follows:

$$[K]_i = \frac{2E_i I_i}{L_i^3} \begin{bmatrix} 6 & 0 & 0 & 3L_i & -6 & 0 & 0 & 3L_i \\ 0 & 6 & -3L_i & 0 & 0 & -6 & -3L_i & 0 \\ 0 & -3L_i & 2L_i^2 & 0 & 0 & 3L_i & L_i^2 & 0 \\ 3L_i & 0 & 0 & 2L_i^2 & -3L_i & 0 & 0 & L_i^2 \\ -6 & 0 & 0 & -3L_i & 6 & 0 & 0 & -3L_i \\ 0 & -6 & 3L_i & 0 & 0 & 6 & 3L_i & 0 \\ 0 & -3L_i & L_i^2 & 0 & 0 & 3L_i & 2L_i^2 & 0 \\ 3L_i & 0 & 0 & L_i^2 & -3L_i & 0 & 0 & 2L_i^2 \end{bmatrix}$$

(2.51)

2.3.6 Shaft Element Gyroscopic Matrix

Half the shaft element's polar moment of inertia, $I_P^{(s)}$, is transferred to each of its two ends points. Utilizing Equation 2.17, the shaft element's gyroscopic matrix is accordingly given by the following:

$$[G]_i^s = \begin{bmatrix} 0 & 0 & 0 & 0 & 0 & 0 & 0 & 0 \\ 0 & 0 & 0 & 0 & 0 & 0 & 0 & 0 \\ 0 & 0 & 0 & \omega I_{P_i} & 0 & 0 & 0 & 0 \\ 0 & 0 & -\omega I_{P_i} & 0 & 0 & 0 & 0 & 0 \\ 0 & 0 & 0 & 0 & 0 & 0 & 0 & 0 \\ 0 & 0 & 0 & 0 & 0 & 0 & 0 & 0 \\ 0 & 0 & 0 & 0 & 0 & 0 & 0 & \omega I_{P_i} \\ 0 & 0 & 0 & 0 & 0 & 0 & -\omega I_{P_i} & 0 \end{bmatrix} \quad (2.52)$$

$$I_{P_i} \equiv \frac{1}{2} I_P^{(s)} = \frac{1}{4} M^{(s)} \left(\frac{d_o^2 + d_i^2}{4} \right)$$

2.3.7 Addition of Nonstructural Mass and Inertia to Rotor Element

Nonstructural mass is added mass and inertia, lumped at mass stations, that does not contribute to element flexibility. The rotor element in Figure 2.8 shows a concentrated disk at each end. A concentrated disk

Lateral Rotor Vibration Analysis Models

($M^{(d)}$, $I_P^{(d)}$, and $I_T^{(d)}$) may be added at any rotor mass station. For a purely concentrated *nonstructural point mass*, $I_P^{(d)} = I_T^{(d)} = 0$. Since construction of the complete matrices for the rotor alone (next topic) overlays the element matrices at their connection stations, nonstructural mass and inertia is added to the left mass station of each element prior to that overlay of element matrices, as reflected in the following equations. The exception is the far right rotor station, where nonstructure mass is added to the right station.

Complete element mass matrix $\equiv [M]_i = \left[[M]_i^l \text{ or } [M]_i^d \text{ or } [M]_i^c \right]$

$$\begin{bmatrix} M_{\text{left}}^{(d)} & 0 & 0 & 0 & 0 & 0 & 0 & 0 \\ 0 & M_{\text{left}}^{(d)} & 0 & 0 & 0 & 0 & 0 & 0 \\ 0 & 0 & I_{T,\text{left}}^{(d)} & 0 & 0 & 0 & 0 & 0 \\ 0 & 0 & 0 & I_{T,\text{left}}^{(d)} & 0 & 0 & 0 & 0 \\ 0 & 0 & 0 & 0 & 0 & 0 & 0 & 0 \\ 0 & 0 & 0 & 0 & 0 & 0 & 0 & 0 \\ 0 & 0 & 0 & 0 & 0 & 0 & 0 & 0 \\ 0 & 0 & 0 & 0 & 0 & 0 & 0 & 0 \end{bmatrix} \quad (2.53)$$

Complete element gyroscopic matrix $\equiv [G]_i^s$

$$\begin{bmatrix} 0 & 0 & 0 & 0 & 0 & 0 & 0 & 0 \\ 0 & 0 & 0 & 0 & 0 & 0 & 0 & 0 \\ 0 & 0 & 0 & \omega I_{P,\text{left}}^{(d)} & 0 & 0 & 0 & 0 \\ 0 & 0 & -\omega I_{P,\text{left}}^{(d)} & 0 & 0 & 0 & 0 & 0 \\ 0 & 0 & 0 & 0 & 0 & 0 & 0 & 0 \\ 0 & 0 & 0 & 0 & 0 & 0 & 0 & 0 \\ 0 & 0 & 0 & 0 & 0 & 0 & 0 & 0 \\ 0 & 0 & 0 & 0 & 0 & 0 & 0 & 0 \end{bmatrix} \quad (2.54)$$

2.3.8 Matrices for Complete Free–Free Rotor

The $[M]$, $[C]$, and $[K]$ matrices for the complete *free–free* rotor (i.e., *free of connections* to ground and *free of external forces*) are assembled by linking all the corresponding individual rotor-element matrices. The right mass station of each rotor element is overlaid on to the left mass station of its immediate right neighbor. Thus, the total number of rotor mass stations (N_{ST}) is equal to the total number of rotor elements (N_{EL}) plus one. The total number of rotor DOFs is 4 times N_{ST}.

$$N_{ST} = N_{EL} + 1$$
$$N_{RDOF} = 4N_{ST} \qquad (2.55)$$

Accordingly, the rotor matrices are expressible as follows:

$$[M]_R = \begin{bmatrix} [M'_1] & & & & \\ & [M'_2] & & & \\ & & [M'_3] & & \\ & & & \ddots & \\ & & & & [M'_{N_{EL}}] \end{bmatrix}_{N_{RDOF} \times N'_{RDOF}} \qquad (2.56)$$

$$[C]_R = \begin{bmatrix} [G'_1] & & & & \\ & [G'_2] & & & \\ & & [G'_3] & & \\ & & & \ddots & \\ & & & & [G'_{N_{EL}}] \end{bmatrix}_{N_{RDOF} \times N'_{RDOF}} \qquad (2.57)$$

$$[K]_R = \begin{bmatrix} [K'_1] & & & & \\ & [K'_2] & & & \\ & & [K'_3] & & \\ & & & \ddots & \\ & & & & [K'_{N_{EL}}] \end{bmatrix}_{N_{RDOF} \times N'_{RDOF}} \qquad (2.58)$$

Note that the free–free rotor damping matrix contains only the shaft gyroscopic terms. As further explained in Section 2.4, although the gyroscopic effect is imbedded in [C] it is not really "damping" in the energy dissipation sense. It is an inertia effect and therefore energy conservative.

2.3.9 Radial-Bearing and Bearing-Support Models

The RDA code is configured so that inputs for a radial-bearing stiffness and damping model, such as illustrated in Figure 2.2, may be applied at any rotor model mass station. In a complete model, *at least two radial bearings are needed* to provide stiffness connections between the rotor and the inertial reference frame (ground), because the rotor possesses two static equilibrium conditions for the x–z plane and two for the y–z plane. That is, it is necessary to have rotor-to-ground stiffness connections for at least two x-coordinates and two y-coordinates for there to exist a static equilibrium state to exist, to which the computed linear-model vibrations are referenced. The strictly mathematical statement of this is that the total model's stiffness matrix [K] must be nonsingular, which it would not be if at least the minimum required number of rotor-to-ground stiffness connections was not incorporated. The obvious practical way of viewing this is that a minimum of two radial bearings are required to confine a rotor to its prescribed rotational center line within the machine; otherwise, "look out!"

Not surprisingly, the most typical rotor–bearing configuration has two radial bearings, but large steam turbine-generator sets may have 10 or more journal bearings on one continuous flexible rotor. For most LRV computer models, one typically uses 10–20 rotor-model mass stations between adjacent bearings. Bearing rotor dynamic properties present probably the biggest challenge in undertaking LRV analyses. This is because the bearing "inputs" (stiffness and damping coefficients), while very important to the accuracy of computed results, inherently have a high degree of uncertainty. Chapter 5 of this book is devoted entirely to bearing and seal rotor dynamic inputs. Although bearings and seals are different machine elements, both are included in LRV analysis models in the same manner, that is, as radial connections between the rotor and the inertial reference frame. In contrast to bearings, seals often need fluid-inertia effects to be incorporated into the rotor-to-ground connection model, as detailed in Chapter 5. Here, the focus is on how the bearing rotor-to-ground stiffness and damping connections are incorporated into the matrices for the complete equations of motion.

The x and y components of the total radial force ($\vec{F}$) exerted upon the rotor from a bearing can be separated into static-equilibrium and dynamic-deviation parts, as follows:

$$F_x = -W_x + f_x \quad \text{and} \quad F_y = -W_y + f_y \qquad (2.59)$$

where W_x and W_y are the x and y components, respectively, of the static load ($\vec{W}$) exerted *upon the bearing*, whereas f_x and f_y are the x and y components, respectively, of the dynamic deviation of total bearing force exerted *upon the rotor*. This is illustrated by the vector diagram in Figure 2.10.

Fluid-film journal bearings provide the most typical example upon which the inclusion of radial bearing dynamic compliance into linear LRV analyses can be explained. As fully developed in Chapter 5, the dynamic-deviation interactive force between a bearing and its rotating journal can be described as a continuous function of journal-to-bearing

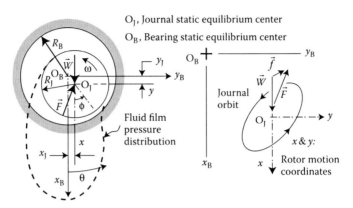

FIGURE 2.10 Force vectors and rotor-to-bearing vibration orbit at a bearing.

position and velocity components. This assumes that lubricant viscosity effects are dominant and lubricant fluid inertia effects are negligible; otherwise journal-to-bearing acceleration component effects should also be included. A continuous function that also has continuous derivatives to any order can be expanded in a Taylor series. Therefore, relative to the static equilibrium state, the x and y components of the dynamic-deviation of bearing force *upon the rotor* can be expressed as follows (under static equilibrium, $\vec{F} + \vec{W} = 0$):

$$F_x + W_x = f_x = \frac{\partial F_x}{\partial x} x + \frac{\partial F_x}{\partial \dot{x}} \dot{x} + \frac{\partial F_x}{\partial y} y + \frac{\partial F_x}{\partial \dot{y}} \dot{y} + \text{(higher order terms)}^{\,0}$$

$$F_y + W_y = f_y = \frac{\partial F_y}{\partial x} x + \frac{\partial F_y}{\partial \dot{x}} \dot{x} + \frac{\partial F_y}{\partial y} y + \frac{\partial F_y}{\partial \dot{y}} \dot{y} + \text{(higher order terms)}^{\,0} \quad (2.60)$$

$F_x = F_x(x, y, \dot{x}, \dot{y})$ and $F_y = F_y(x, y, \dot{x}, \dot{y})$. x and y are referenced relative to the static equilibrium position, as shown in Figure 2.10. It is postulated that vibration signal amplitudes $(x, y, \dot{x}, \dot{y})$ are sufficiently small for the "higher order terms" in Equations 2.60 to be vanishingly small compared to the linear terms. Thus, only the linear terms are retained. Essentially, this is how all linear vibration models are justified, because all real systems have some nonlinearity.

Fortunately, the assumption of linearity leads to adequate answers in most vibration engineering analyses, and simplifies considerably the tasks of making calculations and understanding what is calculated. Some specialized large-amplitude rotor vibration problems justify treating nonlinear effects, for example, large rotor unbalance such as that from turbine blade loss, shock and seismic base-motion excitations, rotor rub-impact phenomena, and self-excited vibration limit cycles. These topics are described at the end of this chapter in Section 2.5. With the "higher order terms" dropped,

Lateral Rotor Vibration Analysis Models

it is convenient to put Equations 2.60 into the following matrix form:

$$\begin{Bmatrix} f_x \\ f_y \end{Bmatrix} = - \begin{bmatrix} k_{xx} & k_{xy} \\ k_{yx} & k_{yy} \end{bmatrix} \begin{Bmatrix} x \\ y \end{Bmatrix} - \begin{bmatrix} c_{xx} & c_{xy} \\ c_{yx} & c_{yy} \end{bmatrix} \begin{Bmatrix} \dot{x} \\ \dot{y} \end{Bmatrix} \qquad (2.61)$$

where $k_{ij} \equiv -(\partial F_i / \partial x_j)$ and $c_{ij} \equiv -(\partial F_i / \partial \dot{x}_j)$ are the eight bearing stiffness and damping coefficients. In general, these coefficient matrices for bearings and seals cannot be simultaneously diagonalized in a single x–y coordinate system, in contrast to the model shown in Figure 2.2. In fact, the bearing and seal stiffness coefficient matrices are often nonsymmetric and their damping coefficient matrices may also be nonsymmetric when certain fluid dynamical factors are significant (e.g., fluid inertia). Such nonsymmetries are somewhat of an anomaly within the broader field of linear vibration analysis, but are quite the usual circumstance in rotor dynamics. These nonsymmetries mathematically embody important physical aspects of rotor dynamical systems that are explained in some depth in Section 2.4. The minus signs in Equations 2.61 stem from definitions of the stiffness and damping coefficients that are based on two implicit assumptions: (i) a spring-like stiffness restoring force resisting radial displacement from the equilibrium position and (ii) a damping drag force resisting radial-plane velocity. This is identical to the sign convention shown at the beginning of Chapter 1 for the 1-DOF spring–mass–damper system.

The most commonly used option in bearing LRV models is to "connect" the rotor to ground directly through the bearing stiffness and damping coefficients, and this is quite appropriate when very stiff bearing support structures are involved. In that case, bearing coefficients embody exclusively the bearing's own dynamic characteristics. Conversely, when the bearing support structure's flexibility is not negligible, then the bearing coefficients should either be modified to incorporate the support structure's compliance, or additional DOFs should be added to the complete system to include appropriate modeling for the support structure. RDA is configured with two options, (i) bearing coefficients connect rotor directly to ground and (ii) bearing coefficients connect to an intermediate mass which then connects to ground through its own x and y stiffness and damping coefficients. This second option, referred to here as the *2-DOF bearing pedestal* model, adds two DOFs to the complete system for each bearing on which it is used. Figure 2.2, previously introduced to illustrate a simple point-mass model, alternately provides an adequate schematic illustration of the 2-DOF bearing pedestal model. One may visualize a rotating journal inside a concentric hole of the mass illustrated in Figure 2.2. Both RDA bearing-support options are now shown.

2.3.9.1 Bearing Coefficients Connect Rotor Directly to Ground

As observed in Section 2.2, Equations 2.18, for the 8-DOF model's equations of motion, handling of this option is quite simple. That is, each of the bearing stiffness and damping coefficients are just added to their respective rotor mass station's x or y components within the total rotor $[K]_R$ and $[C]_R$ matrices. The total system stiffness matrix is thus described as follows. $[K] = [K]_R + [K]_B$, where $[K]_B \equiv N_{RDOF} \times N_{RDOF}$ matrix containing all the bearing stiffness coefficients in their proper locations. This is shown as follows for the embedding of a bearing within $[K]$:

$$[K]\{q\} = \begin{bmatrix} \ddots & & \\ & \begin{bmatrix} \dfrac{12 E_i I_i}{L_i^3} + k_{xx}^{(n)} & k_{xy}^{(n)} \\ k_{yx}^{(n)} & \dfrac{12 E_i I_i}{L_i^3} + k_{yy}^{(n)} \end{bmatrix} & \\ & & \ddots \end{bmatrix}_{N_{RDOF} \times N_{RDOF}} \begin{Bmatrix} \vdots \\ x \\ y \\ \vdots \end{Bmatrix} \quad (2.62)$$

Similarly, $[C] = [C]_R + [C]_B$, where $[C]_B \equiv N_{RDOF} \times N_{RDOF}$ matrix containing all the bearing damping coefficients in their proper locations.

$$[C]\{\dot{q}\} = \begin{bmatrix} \ddots & & \\ & \begin{bmatrix} c_{xx}^{(n)} & c_{xy}^{(n)} \\ c_{yx}^{(n)} & c_{yy}^{(n)} \end{bmatrix} & \\ & & \ddots \end{bmatrix}_{N_{RDOF} \times N_{RDOF}} \begin{Bmatrix} \vdots \\ \dot{x} \\ \dot{y} \\ \vdots \end{Bmatrix} \quad (2.63)$$

where n is the bearing no. $= 1, 2, \ldots, N_B$.

$$[M] = [M]_R \quad (2.64)$$

For this option, the total number of DOFs is $N_{DOF} = N_{RDOF}$. The 8-coefficient bearing model does not include any acceleration effects, thus $[M] = [M]_R$. At least two bearings must have nonzero principle values for their $[k_{ij}^{(n)}]$, for the total model stiffness matrix $[K]$ to be nonsingular, which is a requirement fully explained at the beginning of this subsection.

2.3.9.2 Bearing Coefficients Connect to an Intermediate Mass

The total system $[M], [C]$, and $[K]$ matrices with no bearing pedestals in the model, Equations 2.62 through 2.64, are split after the coordinates of

Lateral Rotor Vibration Analysis Models

each station where a 2-DOF bearing pedestal is located, to insert the additional two rows and two columns containing the corresponding matrix coefficients for the two additional DOFs. This is easy to demonstrate by showing the following expressions for the example of adding a 2-DOF bearing pedestal model only to a bearing at rotor station no. 1:

$$[M] = \begin{bmatrix} [4 \times 4] & \begin{bmatrix} 0 & 0 \\ 0 & 0 \\ 0 & 0 \\ 0 & 0 \end{bmatrix} & [4 \times N_{\text{RDOF}}] \\ \begin{bmatrix} 0 & 0 & 0 & 0 \\ 0 & 0 & 0 & 0 \end{bmatrix} & \begin{bmatrix} M_{B,x}^{(1)} & 0 \\ 0 & M_{B,y}^{(1)} \end{bmatrix} & \begin{bmatrix} 0 & 0 & 0 & 0 \\ 0 & 0 & 0 & 0 \end{bmatrix} \\ [N_{\text{RDOF}} \times 4] & \begin{bmatrix} 0 & 0 \\ 0 & 0 \\ 0 & 0 \\ 0 & 0 \end{bmatrix} & [N_{\text{RDOF}} \times N_{\text{RDOF}}] \end{bmatrix}$$

(2.65)

Pedestal-expanded $[C]$ and $[K]$ matrices must be formulated to account for the bearing $[k_{ij}^{(n)}]$ and $[c_{ij}^{(n)}]$ stiffness and damping coefficients being driven by the differences between rotor and bearing pedestal displacement and velocity components, respectively. The $[2 \times 4]$ and $[4 \times 2]$ off-diagonal coefficient arrays shown within the following two equations accomplish that

$$[C] = \begin{bmatrix} [4 \times 4] & \begin{bmatrix} -c_{xx}^{(1)} & -c_{yx}^{(1)} \\ -c_{xy}^{(1)} & -c_{yy}^{(1)} \\ 0 & 0 \\ 0 & 0 \end{bmatrix} & [4 \times N_{\text{RDOF}}] \\ \begin{bmatrix} -c_{xx}^{(1)} & -c_{xy}^{(1)} & 0 & 0 \\ -c_{yx}^{(1)} & -c_{yy}^{(1)} & 0 & 0 \end{bmatrix} & \begin{bmatrix} c_{xx}^{(1)} + C_{B,xx}^{(1)} & c_{xy}^{(1)} + C_{B,xy}^{(1)} \\ c_{yx}^{(1)} + C_{B,yx}^{(1)} & c_{yy}^{(1)} + C_{B,yy}^{(1)} \end{bmatrix} & \begin{bmatrix} 0 & 0 & 0 & 0 \\ 0 & 0 & 0 & 0 \end{bmatrix} \\ [N_{\text{RDOF}} \times 4] & \begin{bmatrix} 0 & 0 \\ 0 & 0 \\ 0 & 0 \\ 0 & 0 \end{bmatrix} & [N_{\text{RDOF}} \times N_{\text{RDOF}}] \end{bmatrix}$$

(2.66)

$$[K] = \begin{bmatrix} \begin{bmatrix} 4 \times 4 \end{bmatrix} & \begin{bmatrix} -k_{xx}^{(1)} & -k_{yx}^{(1)} \\ -k_{xy}^{(1)} & -k_{yy}^{(1)} \\ 0 & 0 \\ 0 & 0 \end{bmatrix} & \begin{bmatrix} 4 \times N_{RDOF} \end{bmatrix} \\ \begin{bmatrix} -k_{xx}^{(1)} & -k_{xy}^{(1)} & 0 & 0 \\ -k_{yx}^{(1)} & -k_{yy}^{(1)} & 0 & 0 \end{bmatrix} & \begin{bmatrix} k_{xx}^{(1)} + K_{B,xx}^{(1)} & k_{xy}^{(1)} + K_{B,xy}^{(1)} \\ k_{yx}^{(1)} + K_{B,yx}^{(1)} & k_{yy}^{(1)} + K_{B,yy}^{(1)} \end{bmatrix} & \begin{bmatrix} 0 & 0 & 0 & 0 \\ 0 & 0 & 0 & 0 \end{bmatrix} \\ \begin{bmatrix} N_{RDOF} \times 4 \end{bmatrix} & \begin{bmatrix} 0 & 0 \\ 0 & 0 \\ 0 & 0 \\ 0 & 0 \end{bmatrix} & \begin{bmatrix} N_{RDOF} \times N_{RDOF} \end{bmatrix} \end{bmatrix}$$

(2.67)

For this example, $\{q\} = \{x_1, y_1, \theta_{1x}, \theta_{1y}, x_{B,1x}, y_{B,1y}, x_2, y_2, \theta_{2x}, \theta_{2y}, \ldots\}$ is the generalized coordinate vector. Note the additional two coordinates that are added at the end of station 1 rotor coordinates.

$M_{B,x}^{(n)}$ and $M_{B,y}^{(n)}$ are the nth bearing pedestal's x and y modal masses, respectively. $[C_{B,ij}^{(n)}]_{2\times2}$ and $[K_{B,ij}^{(n)}]_{2\times2}$ are the nth bearing pedestal's damping and stiffness connection-to-ground coefficients, respectively. The total number of system DOFs is equal to the rotor DOF (N_{RDOF}) plus 2 times the number of bearing pedestals (N_P) employed in the model, where $N_P \leq N_B$.

$$N_{DOF} = N_{RDOF} + 2N_P \tag{2.68}$$

2.3.10 Completed RDA Model Equations of Motion

The complete RDA N_{DOF} equations of motion can now be written in the compact matrix form introduced in Equation 1.15. All the analysis options available within the RDA code have one of two $\{f(t)\}$ right-hand sides as follows: $\{f(t)\} = \{0\}$ for eigenvalue analyses (e.g., instability searches), and at rotor stations with unbalance inputs, for steady-state unbalance response.

$$\begin{Bmatrix} \vdots \\ f_x \\ f_y \\ \vdots \end{Bmatrix} = \omega^2 \begin{Bmatrix} \vdots \\ m_{ub} r_{ub} \cos(\omega t + \phi_{ub}) \\ m_{ub} r_{ub} \sin(\omega t + \phi_{ub}) \\ \vdots \end{Bmatrix} \tag{2.69}$$

2.4 Insights into Linear LRVs

Successful rotating machinery developments need reliable analyses to predict vibration performance. Predictive analyses can also be an invaluable tool in successful troubleshooting of vibration problems in existing machinery. Present computerized rotor vibration analyses provide many software options in this regard, such as the RDA code supplied with this book. Equally important, but frequently overlooked and not well understood, are the basic physical insights, which can easily be obscured in the presence of enormous computational power. Basic physical insights are essential for one to understand, explain, and apply what advanced analyses predict. This section relates important physical characteristics for LRV to the mathematical structure of the governing equations of motion. The centerpiece here is the decomposition of the equation-of-motion matrices into their symmetric and skew-symmetric parts, and the relation of these parts to the conservative and nonconservative forces of rotor dynamical systems.

It has been recognized for quite sometime that, aside from journal bearings, other fluid annuli such as sealing clearances and even complete turbo-machinery stages produce rotor dynamically significant interactive rotor–stator forces. These forces must be adequately characterized and included in many rotor vibration analyses if reliable prediction and understanding of machinery vibration is to be realized. The most complete rotor–stator interactive linear radial force model currently in wide use is shown in the following equation, which can be referred to Figure 2.10 and its associated nomenclature.

$$\begin{Bmatrix} f_x \\ f_y \end{Bmatrix} = - \begin{bmatrix} k_{xx} & k_{xy} \\ k_{yx} & k_{xx} \end{bmatrix} \begin{Bmatrix} x \\ y \end{Bmatrix} - \begin{bmatrix} c_{xx} & c_{xy} \\ c_{yx} & c_{yy} \end{bmatrix} \begin{Bmatrix} \dot{x} \\ \dot{y} \end{Bmatrix} - \begin{bmatrix} m_{xx} & m_{xy} \\ m_{yx} & m_{yy} \end{bmatrix} \begin{Bmatrix} \ddot{x} \\ \ddot{y} \end{Bmatrix}$$

(2.70)

$k_{ij} \equiv -(\partial F_i/\partial x_j)$, $c_{ij} \equiv -(\partial F_i/\partial \dot{x}_j)$, and $m_{ij} \equiv -(\partial F_i/\partial \ddot{x}_j)$ are defined at static equilibrium and have an orthogonal transformation property of the Cartesian second-rank tensor, that is, they are second-rank tensors just like stress. Chapter 5 provides a more in-depth treatment of how these stiffness, damping, and virtual mass (inertia) coefficients are determined. At this point, suffice it to say that both first-principle-based computations as well as some highly challenging experimental approaches are utilized to quantify these rotor dynamic coefficients, because they are crucial to meaningful rotor vibration analyses.

2.4.1 Systems with Nonsymmetric Matrices

The decomposition of any $n \times n$ matrix $[A]$ into its *symmetric* ("s") and *skew-symmetric* ("ss") parts is an elementary technique of Matrix Algebra, expressed as follows:

$$[A_{ij}] = \tfrac{1}{2}[A_{ij} + A_{ji}] + \tfrac{1}{2}[A_{ij} - A_{ji}] \equiv [A_{ij}^s] + [A_{ij}^{ss}]$$

where $[A_{ij}^s] \equiv \tfrac{1}{2}[A_{ij} + A_{ji}]$ and $[A_{ij}^{ss}] \equiv \tfrac{1}{2}[A_{ij} - A_{ji}]$ (2.71)

giving $[A_{ij}^s] = [A_{ij}^s]^T$ and $[A_{ij}^{ss}] = -[A_{ij}^{ss}]^T$

As shown in Equations 2.71, the symmetric part of [A] is equal to its own transpose ("T," i.e., interchange of rows and columns) whereas its skew-symmetric part is equal to minus its own transpose. This matrix decomposition technique can therefore be applied to any of the square matrices associated with the equations of motion for LRV. Clearly, if an $n \times n$ matrix is symmetric to begin with, then its skew-symmetric part is zero and this matrix decomposition does not accomplish anything. Although most linearized vibration models have symmetric [M], [C], and [K] matrices, LRV models typically have some nonsymmetries. There are compelling physical reasons to justify that the 2×2 interaction-force gradient coefficient matrices $[k_{ij}]$ and $[c_{ij}]$ defined in Equation 2.70 can be nonsymmetric, and conversely that the 2×2 array $[m_{ij}]$ should be symmetric. Furthermore, as already shown for spinning rotors in Equations 2.17, the gyroscopic moment effect manifests itself in the motion equations as a skew-symmetric additive to the [C] matrix, for example, Equations 2.18, 2.52, and 2.54. In a series of papers some years ago, listed in the Bibliography at the end of this chapter, the author related the somewhat unique nonsymmetric structure of rotor–bearing dynamics equation-of-motion matrices to certain physical characteristics of these systems. The main points of those papers are treated in the remainder of this section.

The complete linear LRV equations of motion can be compactly expressed in standard matrix form as follows:

$$[M]\{\ddot{q}\} + [C]\{\dot{q}\} + [K]\{q\} = \{f(t)\} \quad (2.72)$$

First, the matrices in this equation are decomposed into their symmetric and skew-symmetric parts as follows:

$$[K] = [K^s] + [K^{ss}], \quad [C] = [C^s] + [C^{ss}], \quad [M] = [M^s] + [M^{ss}] \quad (2.73)$$

where the decompositions in Equations 2.73 are defined by Equations 2.71. The fundamental demonstration is to show that these decompositions amount to a separation of dynamical effects into *energy conservative* and *energy nonconservative* parts. That $[K^s]$ is conservative, $[C^s]$ is nonconservative, and $[M^s]$ is conservative can automatically be accepted, being the standard symmetric stiffness, damping, and mass matrices, respectively.

Lateral Rotor Vibration Analysis Models

$[C^{ss}]$ is handled here first since there is a similarity in the treatments of $[K^{ss}]$ and $[M^{ss}]$.

Attention is first on some 2×2 submatrix within the $[C^{ss}]$ matrix that contains $[c_{ij}^{ss}]$, the skew-symmetric part of $[c_{ij}]$ for a radial bearing, seal, or other fluid-containing confine between the rotor and nonrotating member. The incremental work dw (i.e., force times incremental displacement) done on the rotor by the $[c_{ij}^{ss}]$ terms at any point on any orbital path (refer to journal center orbital trajectory shown in Figure 2.10) is expressible as follows:

$$dw = -\left[c_{ij}^{ss}\right]\begin{Bmatrix}\dot{x}\\ \dot{y}\end{Bmatrix}\{dx\ dy\} \qquad (2.74)$$

where $[c_{ij}^{ss}] = \begin{bmatrix} 0 & c_{xy}^{ss} \\ -c_{xy}^{ss} & 0 \end{bmatrix}$.

Performing the indicated multiplications in Equation 2.74 and substituting $dx = \dot{x}\,dt$ and $dy = \dot{y}\,dt$ yields the following result:

$$dw = -c_{xy}^{ss}(\dot{x}\dot{y} - \dot{y}\dot{x})\,dt \equiv 0 \qquad (2.75)$$

This result simply reflects that the force vector here is always perpendicular to its associated velocity vector, and thus no work (or power) is transmitted. Similarly, focusing on some 2×2 submatrix within the $[C^{ss}]$ matrix that contains a pair of gyroscopic moment terms, as provided in Equations 2.17, the identical proof applies to the gyroscopic moment effects, shown as follows:

$$dw = -\begin{bmatrix} 0 & \omega I_P \\ -\omega I_P & 0 \end{bmatrix}\begin{Bmatrix}\dot{\theta}_x\\ \dot{\theta}_y\end{Bmatrix}\{d\theta_x\ d\theta_y\}$$

$$= -\begin{bmatrix} 0 & \omega I_P \\ -\omega I_P & 0 \end{bmatrix}\begin{Bmatrix}\dot{\theta}_x\\ \dot{\theta}_y\end{Bmatrix}\{\dot{\theta}_x\ \dot{\theta}_y\}\,dt \equiv 0 \qquad (2.76)$$

The gyroscopic moment vector is perpendicular to its associated angular velocity vector, and thus no work (or power) is transmitted. The skew-symmetric part of the total system $[C]$ matrix thus embodies only *conservative force* fields and is therefore not really damping in the energy dissipation or addition sense, in contrast to the symmetric part of $[C]$ which embodies only nonconservative forces.

Turning attention to the skew-symmetric part of $[K]$, consider some 2×2 submatrix within the $[K^{ss}]$ matrix that contains $[k_{ij}^{ss}]$, the skew-symmetric part of $[k_{ij}]$ for a radial bearing, seal, or other fluid-containing confine between rotor and nonrotating member. The incremental work done by the

$[k_{ij}^{ss}]$ terms on any point on any orbital trajectory is expressible as follows:

$$dw = -\begin{bmatrix} 0 & k_{xy}^{ss} \\ -k_{xy}^{ss} & 0 \end{bmatrix} \begin{Bmatrix} x \\ y \end{Bmatrix} \{dx\ dy\} = -k_{xy}^{ss} y\ dx + k_{xy}^{ss} x\ dy \equiv f_x\ dx + f_y\ dy \tag{2.77}$$

$$\therefore \frac{\partial f_x}{\partial y} = -k_{xy}^{ss} \quad \text{and} \quad \frac{\partial f_y}{\partial x} = k_{xy}^{ss}$$

Obviously, $(\partial f_x/\partial y) \neq (\partial f_y/\partial x)$, that is, dw here is not an *exact differential*; hence, the $[k_{ij}^{ss}]$ energy transferred over any portion of a trajectory between two points "A" and "B" is *path dependent*, and thus the force field is *nonconservative*. The skew-symmetric part of the total system $[K]$ matrix thus embodies only *nonconservative force* fields and is therefore not really stiffness in the energy conservative sense, in contrast to the symmetric part of $[K]$ which embodies only conservative forces. An additional interesting insight is obtained here by formulating the net energy-per-cycle exchange from the $[k_{ij}^{ss}]$ terms (see Figure 2.11).

$$E_{\text{cyc}} = \oint dw = -k_{xy}^{ss} \oint (y\ dx - x\ dy) \tag{2.78}$$

Splitting the integral in Equation 2.78 into two line integrals between points "A" and "B," and integrating the dy terms "by parts" yield the following result:

$$E_{\text{cyc}} = 2k_{xy}^{ss} \int_{x_A}^{x_B} (y_2 - y_1)\ dx \tag{2.79}$$

The integral in Equation 2.79 is clearly the orbit area. Typically, $k_{xy}^{ss} \geq 0$ for journal bearings, seals, and other rotor–stator fluid annuli, even complete centrifugal pump stages. Thus, the k_{xy}^{ss} effect represents negative damping for *forward* (corotational) orbits and positive damping for *backward* (counter-rotational) orbits. Only for orbits where the integral in

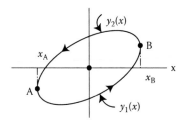

FIGURE 2.11 Any periodic orbit of rotor relative to nonrotating member.

Lateral Rotor Vibration Analysis Models

Equation 2.79 is zero will the net exchange of energy per cycle be zero. One such example is a straight-line cyclic orbit. Another example is a "figure 8" orbit comprised of a positive area and a negative area of equal magnitudes.

The complete nonconservative radial interaction force vector $\{P\}$ on the rotor at a journal bearing, for example, is thus embodied only in the symmetric part $[c_{ij}^S]$ and the skew-symmetric part $[k_{ij}^{SS}]$, and expressible as follows (actually, $c_{xx}^S = c_{xx}$ and $c_{yy}^S = c_{yy}$):

$$\begin{Bmatrix} P_x \\ P_y \end{Bmatrix} = -\begin{bmatrix} c_{xx}^S & c_{xy}^S \\ c_{xy}^S & c_{xx}^S \end{bmatrix} \begin{Bmatrix} \dot{x} \\ \dot{y} \end{Bmatrix} - \begin{bmatrix} 0 & k_{xy}^{SS} \\ -k_{xy}^{SS} & 0 \end{bmatrix} \begin{Bmatrix} x \\ y \end{Bmatrix} \quad (2.80)$$

The parametric equations, $x = X\sin(\Omega t + \phi_x)$ with $y = Y\sin(\Omega t + \phi_y)$, are used here to specify a harmonic rotor orbit for the purpose of formulating the energy imparted to the rotor per cycle of harmonic motion, as follows:

$$E_{cyc} = \oint (P_x\, dx + P_y\, dy) = \int_0^{2\pi/\Omega} (P_x \dot{x}\, dt + P_y \dot{y}\, dt)$$

$$= -\pi \left\{ \Omega \left[c_{xx}^S X^2 + 2c_{xy}^S XY \cos(\phi_x - \phi_y) + c_{yy}^S Y^2 \right] \right.$$

$$\left. - 2k_{xy}^{SS} XY \sin(\phi_x - \phi_y) \right\} \quad (2.81)$$

By casting in the x–y orientation of the principal coordinates of $[c_{ij}^S]$, the c_{xy}^S term in Equation 2.81 disappears, yielding the following result, which is optimum for an explanation of rotor dynamical instability self-excited vibration:

$$E_{cyc} = -\pi \left[\Omega \left(c_{xx}^P X^2 + c_{yy}^P Y^2 \right) - 2k_{xy}^{SS} XY \sin(\phi_x - \phi_y) \right] \quad (2.82)$$

Since $[k_{ij}^{SS}]$ is an *isotropic tensor*, its coefficients are invariant to orthogonal transformation, that is, do not change in transformation to the principal coordinates of $[c_{ij}^S]$. Furthermore, Ω, c_{xx}^P, c_{yy}^P, k_{xy}^{SS}, X and Y are all positive in the normal circumstance. For corotational orbits the difference in phase angles satisfies $\sin(\phi_x - \phi_y) > 0$, and conversely for counterrotational orbits $\sin(\phi_x - \phi_y) < 0$. For a straight-line orbit, which is neither forward nor backward whirl, $\phi_x = \phi_y$ so $\sin(\phi_x - \phi_y) = 0$, yielding zero destabilizing energy input to the rotor from the k_{xy}^{SS} effect. From Equation 2.82, one thus sees the presence of positive and negative damping effects for any forward whirling motion. Typically, as rotor speed increases, the k_{xy}^{SS} effect becomes progressively stronger in comparison with the c_{ij}^S (squeeze-film damping) effect. At the *instability threshold speed*, the two effects exactly balance on an energy-per-cycle basis, and Ω is the natural frequency of the rotor–bearing resonant mode which is on the threshold of "self-excitation." From Equation 2.82 it therefore becomes clear as to why this type of instability always produces a self-excited orbital vibration with *forward whirl* (corotational orbit), since the k_{xy}^{ss} term actually adds positive damping to a backward whirl. It also becomes clear as to why the instability mechanism usually excites the *lowest-frequency* forward-whirl mode, because the energy dissipated per cycle by the velocity-proportional drag force is also proportional to Ω, but the energy input per cycle from the k_{xy}^{ss} destabilizing effect is not proportional to Ω. In other words, the faster an orbit is traversed, the greater the energy dissipation per cycle by the drag force. However, the energy input per cycle from the k_{xy}^{ss} destabilizing effect is only proportional to the orbit area, not to how fast the orbit is traversed. Consequently, as rotor speed is increased, the first mode to be "attacked" by instability is usually the *lowest-frequency forward-whirl mode*.

Harmonic motion is also employed to investigate the m_{xy}^{ss} effect. The net energy per cycle imparted to the rotor by such a skew-symmetric additive to the mass matrix is accordingly formulated similar to Equation 2.78, as follows:

$$E_{cyc} = -m_{xy}^{ss} \oint (\ddot{y}\,dx - \ddot{x}\,dy) = \Omega^2 m_{xy}^{ss} \oint (y\,dx - x\,dy) \quad (2.83)$$

The factor $(-\Omega^2)$ comes from twice differentiating the sinusoidal functions for x and y to obtain $\ddot{x}$ and $\ddot{y}$, respectively. With reference to Figure 2.11, utilizing the same steps in Equation 2.83 as in going from Equation 2.78 to Equation 2.79, the following result is obtained:

$$E_{cyc} = -2\Omega^2 m_{xy}^{ss} \int_{x_A}^{x_B} (y_2 - y_1)\,dx \quad (2.84)$$

It is clear from Equation 2.84 that an m_{xy}^{ss} effect would be nonconservative, similar to the k_{xy}^{ss} effect, but differing by the multiplier $(-\Omega^2)$. For $m_{xy}^{ss} > 0$, such a skew-symmetric additive to an otherwise symmetric mass matrix would therefore "attack" one of the highest frequency backward-whirl modes of a rotor–bearing system and drive it into a self-excited vibration. Even if m_{xy}^{ss} were very small (positive or negative), the Ω^2 multiplier would seek a high-enough-frequency natural mode in the actual continuous-media rotor system spectrum to overpower any velocity-proportional drag-force damping effect, which has only an Ω multiplier. No such very high-frequency backward-whirl ($m_{xy}^{ss} > 0$) or forward-whirl ($m_{xy}^{ss} < 0$) instability has ever been documented for any type of machinery. Thus, it must be concluded that $m_{xy}^{ss} = 0$ is consistent with physical reality. In other words, the mass matrix should be symmetric to be

Lateral Rotor Vibration Analysis Models

consistent with real machinery. An important directive of this conclusion is the following: For laboratory experimental results from bearings, seals, or other fluid-containing confines between rotor and nonrotating member, schemes for fitting measured data to linear models like Equation 2.70 should *constrain* $[m_{ij}]$ *to symmetry*.

Even with symmetry imposed on $[m_{ij}]$, the model in Equation 2.70 still has 11 coefficients (instead of 12), which must be obtained either from quite involved computational fluid mechanics analyses or from quite specialized and expensive experimental efforts, as more fully described in Chapter 5. Thus, any justifiable simplification to Equation 2.70 model that reduces the number of its coefficients is highly desirable. For conventional oil-film journal bearings, the justified simplification is to discount the lubricant's fluid inertia effects, which automatically reduces the number of coefficients to *eight*. For seals and other rotor–stator fluid confines that behave more like rotationally symmetric flows than do bearings, the *isotropic* model is employed as described in Section 2.4.3.

2.4.2 Explanation of Gyroscopic Effect

During the author's 1986 spring series of lectures on *Turbomachinery Rotor Vibration Problems*, at the Swiss Federal Institute (ETH-Zurich), one student commented as here paraphrased. "Professor, I have studied gyroscopic effects in dynamics theory and have seen laboratory demonstrations of it, but I still do not really understand it." He was asking for a *layman's explanation*. The student's comment warranted more than just a short answer. So I promised to come to my next lecture prepared with an illustrated explanation, Figure 2.12. Even best known twentieth century vibrations engineer, MIT Professor J. P. Den Hartog (1901–1989), wavered on this phenomenon in his book *Mechanical Vibrations*, McGraw-Hill, 1940.

To understand the pivot forces needed to support the simultaneous spin and transverse precession angular velocities imposed upon the disk illustrated in Figure 2.12, one need only understand the pivot forces $(F, -F)$ required just for the two mass points (m_1, m_2) shown in Figure 2.12. The disk's whole continuum of mass particles carry out the same effect to varying degrees, collectively yielding the required moment couple $(F_T, -F_T)$ upon the disk as per Newton's Second Law $(mR^2 \omega_s \omega_p \hat{k}/2 = \Sigma \vec{M} = 2RF_T\hat{k})$. By viewing the mass points' (m_1, m_2) respective trajectories, Figure 2.12 clearly shows why the instantaneous axial components of acceleration $(\vec{a}_1, \vec{a}_2 = -\vec{a}_1)$ of the two mass points are in opposite directions, thus creating the need for the applied moment about the z-axis perpendicular to the precession and spin axes.

As rigorously derived in Sections 2.2.3 (Equations 2.18) and 2.3.6 (Equation 2.52) of this chapter, the presence of a polar moment-of-inertia on a rotor model adds a skew-symmetric additive to the damping matrix $[C]$,

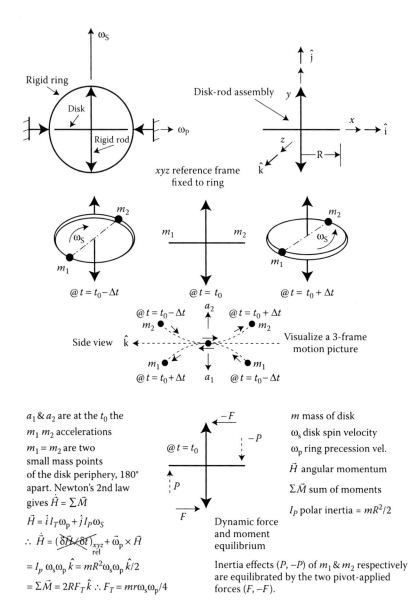

FIGURE 2.12 Illustrated explanation of the gyroscopic effect.

even though it embodies a conservative effect. However, gyroscopic effects produce a spectral *bifurcation of rotor natural frequencies* along *forward* and *backward* whirl rotor vibration orbits. This has been recognized for a long time, as described by Professor Den Hartog (1940) in his book. The same type of spectral bifurcation is also created for any physical effect that imbeds itself in the model as a skew-symmetric additive to the [C] matrix. Primary important examples of this are the rotor–stator fluid interaction forces in bearings and seals when fluid inertia is not negligible. In low Reynolds number fluid annuli, like typical oil-film journal bearings, fluid inertia effects are usually neglected. However, for many other rotor–stator cylindrical fluid annuli, such as high Reynolds number journal bearings, seals, and the motor radial gap (filled with water) of canned-motor pumps, fluid inertia is a dominant influence (see Section 6.3.1 of Chapter 6 and Figure 6.6). The convective inertia terms in the Navier–Stokes fluid dynamics equations embody a skew-symmetric additive to the [C] matrix. This can be readily understood if one visualizes that the rotor–stator convective fluid inertia effect must be rotational-direction biased. That is, its radial force influence will not be the same on backward-whirl orbits as on forward-whirl orbits, given the directional bias of the inherent fluid flow field. This is in contrast to the temporal fluid inertia effect (embodied in the [M] matrix), which is the same for either orbit direction and also remains even if the rotor speed is zero. The temporal fluid inertia effect is the fluid annulus counterpart of a disk's transverse moment-of-inertia, also embodied in the [M] matrix and also having its effect remain even if the rotor speed is zero.

A canned-motor pump model case is described by Adams and Padovan (1981), which shows the combined gyroscopic-like effects of the canned-motor fluid annulus and impeller wear-ring seal. They show two spectral branches (modes) that emerge from one point as the influence of the skew-symmetric additive to the [C] matrix is increased from zero to 100% full effect. Surprisingly, the two mode shapes they show for the full 100% effect are quite similar in shape, even though they differ in frequency by nearly a 2:1 ratio, with the backward-whirl branch being the lower-frequency mode and the forward-whirl branch being the higher-frequency mode.

2.4.3 Isotropic Model

The underlying assumptions for the isotropic model are that (i) the rotating and nonrotating members forming an annular fluid-filled gap are concentric; (ii) the annular gap has geometric variations, if any, in the axial direction only; and (iii) the inlet flow boundary conditions are rotationally symmetric. As a consequence, it is assumed that the rotor orbital vibrations impose only small dynamic perturbations upon an otherwise rotationally symmetric primary steady flow field within the annular gap. Rotational symmetry requires that the k_{ij}, c_{ij}, and m_{ij} coefficients in Equation 2.70 be invariant to orthogonal transformation, that is, have the same values in all orientations of the radial plane $x-y$ coordinate system. It is relevant to mention here that k_{ij}, c_{ij}, and m_{ij} are coefficients of single-point second-rank tensors, just like stress and rigid-body mass moment of inertia, which is not typically so in the broader class of linear vibration model matrices. Thus, for the case of rotationally symmetric flow, these tensors are *isotropic*. This justifies that Equation 2.70 can be simplified to the following form for the *isotropic model*:

$$\begin{Bmatrix} f_x \\ f_y \end{Bmatrix} = -\begin{bmatrix} k^s & k^{ss} \\ -k^{ss} & k^s \end{bmatrix} \begin{Bmatrix} x \\ y \end{Bmatrix} - \begin{bmatrix} c^s & c^{ss} \\ -c^{ss} & c^s \end{bmatrix} \begin{Bmatrix} \dot{x} \\ \dot{y} \end{Bmatrix}$$
$$- \begin{bmatrix} m^s & m^{ss} \\ -m^{ss} & m^s \end{bmatrix} \begin{Bmatrix} \ddot{x} \\ \ddot{y} \end{Bmatrix} \qquad (2.85)$$

Clearly, the isotropic assumption by itself reduces the number of coefficients to *six*. However, the constraint of symmetry on m_{ij} developed in the previous subsection means that $m^{ss} = 0$; so in fact only *five* coefficients are required for the isotropic model. The major limitation of the isotropic model is that it does not accommodate nonzero rotor-to-stator static eccentricities or other rotational asymmetries between rotor and stator. Thus, this model would be physically inconsistent for journal bearings since they derive their static load capacity from significant static eccentricity ratios. However, it is widely applied for seals and other rotor–stator fluid confines, but not for journal bearings.

The isotropic model lends itself to an insightful visualization of how the linear interaction force model separates its single force vector into the distinct parts delineated by the model. Such an illustration using the full anisotropic model of Equation 2.70 would be too complicated an illustration to be as insight provoking as Figure 2.13, which is based on the isotropic model. The force vector directions shown in Figure 2.13 are for the six coefficients of the isotopic model, all assumed to be positive, and all brackets are omitted from the indicated matrix multiplications. Although it has already been established that $m^{ss} = 0$, Figure 2.13 shows a component that would be present if $m^{ss} > 0$, to illustrate its nonconservative nature.

Figure 2.13 visually embodies all the major points provided thus far in this section, and more. First, note that the all the k_{ij} and m_{ij} force parts are in their same directions for both forward and backward whirl, whereas the c_{ij} force parts are all direction reversed between the forward- and backward-whirl cases. The case of circular whirl is easiest to visualize, with all force parts being either tangent or perpendicular to the path at the instantaneous position. Note that the symmetric stiffness part provides a *centering* force for $k^s > 0$ and would thus represent a decentering force for $k^s < 0$. The c^s force part is opposite the instantaneous orbit velocity in both orbit direction cases and thus always provides a *drag force* for $c^s > 0$. The $c^{ss} > 0$ force part provides a centering force for forward whirl and a decentering force for backward whirl, thus imposing a *gyroscopic-like* effect that tends to bifurcate the system natural frequency spectrum along higher-frequency forward-whirl branches and lower-frequency backward-whirl branches. The $k^{ss} > 0$ force part, being tangent to the path and in the velocity direction for forward whirl, thus provides an energy (power) input to forward-whirl rotor orbital motion and is thus a destabilizing influence for forward-whirl modes, as previously described. Similarly, the $m^{ss} > 0$ force part, which should actually be omitted from models, would impose a destabilizing energy input to backward-whirl rotor orbits.

Parts c and d of Figure 2.13 show force part delineation for the more general case of elliptical orbits. A general harmonic orbit is an ellipse; thus circular orbits result only when $X = Y$ with $\phi_x - \phi_y = \pi/2$ radians or 90°. For elliptical orbits, the relationship of each force part to a physical effect is the same as just described for circular orbits. However, the picture is slightly more complicated to visualize. Both force parts c^s and c^{ss} are still tangent and perpendicular to the path, respectively. However, all the stiffness and inertia parts (k^s, k^{ss}, m^s, and m^{ss}) are either colinear or perpendicular to the instantaneous position vector of the journal center relative to the static equilibrium point, as shown. Only where the trajectory crosses the major and minor axes of the orbital ellipse are all force parts either tangent or perpendicular to the path.

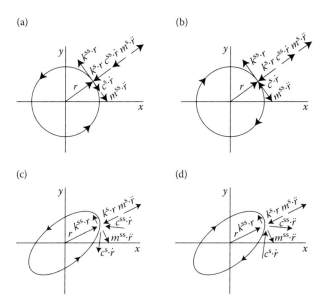

FIGURE 2.13 Force components delineated by the symmetric/skew-symmetric decomposition of the k_{ij}, c_{ij}, and m_{ij} coefficient matrices for the *isotropic* model: (a) circular orbit, forward whirl, (b) circular orbit, backward whirl, (c) elliptical orbit, forward whirl, and (d) elliptical orbit, backward whirl.

2.4.4 Physically Consistent Models

Allowing a bearing $[c_{ij}]$ matrix to be nonsymmetric without including a companion symmetric $[m_{ij}]$ matrix can falsify the predicted natural frequency spectrum of the rotor–bearing system because such would constitute a physically inconsistent or incomplete model for inertia of the fluid within an annular gap. This would be comparable to treating a concentrated rotor disk by including its polar moment of inertia but excluding its transverse moment of inertia. It is equally valid to argue in the same way regarding the bearing $[k_{ij}]$ matrix. That is, a nonsymmetric $[k_{ij}]$ without its companion symmetric $[c_{ij}]$ in the model would provide a destabilizing influence without the companion stabilizing influence to counter it, that is, a physically inconsistent nonconservative characteristic. These two arguments here, combined with the earlier argument that the $[m_{ij}]$ matrix must be symmetric to be consistent with physical reality, suggest the following axiom: *The coefficient matrix of the highest order term for an interactive rotor force should be symmetric to avoid physical inconsistencies in the model.* Hence if only $[k_{ij}]$ coefficients are included in a model, so as to evaluate undamped natural frequencies, only the symmetric part of the $[k_{ij}]$ coefficient matrices should be included. Likewise, if the $[m_{ij}]$ (symmetric) inertia effects are excluded for an interactive rotor force, as is typical for oil-film journal bearings, then its $[c_{ij}]$ coefficient matrix should include only its symmetric part in the model. In fact, as shown in Chapter 5, computational determinations for journal-bearing stiffness and damping coefficients based on the Reynolds lubrication equation yield symmetric damping coefficients, as should be expected, since the Reynolds equation is based on purely viscous flow with all fluid inertia effects omitted. That is, any skew-symmetric part of a $[c_{ij}]$ coefficient matrix must represent an inertia effect since it embodies a conservative force field.

2.4.5 Combined Radial and Misalignment Motions

A shortcoming of Equation 2.70 rotor–stator radial interaction force model is its lack of an account of angular misalignment motions between the rotor and stator centerlines. Figure 2.14 illustrates the case of simultaneous radial and misalignment motions.

For bearings and seals, misalignment motion effects naturally become more important, the larger the length-to-diameter ratio. There are always practical limitations on just how close to "perfection" any engineering analysis model can be. Researchers in the field are still working to obtain more accurate and diversified coefficient inputs for the Equation 2.70 model. For angular misalignment motion effects to be included in the model, the $[k_{ij}]$, $[c_{ij}]$, and $[m_{ij}]$ coefficient matrices each require to be 4×4 instead of only 2×2, because the local generalized coordinates then include

Lateral Rotor Vibration Analysis Models

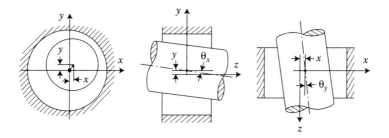

FIGURE 2.14 Radial bearing/seal radial and misalignment coordinates.

$\{x, y, \theta_x, \theta_y\}$ instead of only $\{x, y\}$. Consequently, the number of coefficients would increase by a factor of *four*, as shown in Equation 2.86 for such a model. Along practical lines of argument, optimum designs hopefully have minimal static and dynamic misalignment effects. While the definitive pronouncement on such effects may not have yet been rendered, other uncertainties such as from the manufacturing tolerances affecting journal-bearing clearance are more significant and prevalent (Chapter 5).

$$\begin{Bmatrix} f_x \\ f_y \\ M_x \\ M_y \end{Bmatrix} = - \begin{bmatrix} k_{xx} & k_{xy} & k_{x\theta_x} & k_{x\theta_y} \\ k_{yx} & k_{yy} & k_{y\theta_x} & k_{y\theta_y} \\ k_{\theta_x x} & k_{\theta_x y} & k_{\theta_x \theta_x} & k_{\theta_x \theta_y} \\ k_{\theta_y x} & k_{\theta_y y} & k_{\theta_y \theta_x} & k_{\theta_y \theta_y} \end{bmatrix} \begin{Bmatrix} x \\ y \\ \theta_x \\ \theta_y \end{Bmatrix}$$
$$- \begin{bmatrix} c_{xx} & c_{xy} & c_{x\theta_x} & c_{x\theta_y} \\ c_{yx} & c_{yy} & c_{y\theta_x} & c_{y\theta_y} \\ c_{\theta_x x} & c_{\theta_x y} & c_{\theta_x \theta_x} & c_{\theta_x \theta_y} \\ c_{\theta_y x} & c_{\theta_y y} & c_{\theta_y \theta_x} & c_{\theta_y \theta_y} \end{bmatrix} \begin{Bmatrix} \dot{x} \\ \dot{y} \\ \dot{\theta}_x \\ \dot{\theta}_y \end{Bmatrix}$$
$$- \begin{bmatrix} m_{xx} & m_{xy} & m_{x\theta_x} & m_{x\theta_y} \\ m_{yx} & m_{yy} & m_{y\theta_x} & m_{y\theta_y} \\ m_{\theta_x x} & m_{\theta_x y} & m_{\theta_x \theta_x} & m_{\theta_x \theta_y} \\ m_{\theta_y x} & m_{\theta_y y} & m_{\theta_y \theta_x} & m_{\theta_y \theta_y} \end{bmatrix} \begin{Bmatrix} \ddot{x} \\ \ddot{y} \\ \ddot{\theta}_x \\ \ddot{\theta}_y \end{Bmatrix} \quad (2.86)$$

The model in Equation 2.86 has both a radial force vector and a radial moment vector, thus spawning *48 coefficients*, as shown. The author feels this is sufficient reason to move to a different next topic at this point!

2.5 Nonlinear Effects in Rotor Dynamical Systems

The vast majority of LRV analyses justifiably utilize linear models. However, for postulated operating conditions that yield large vibration amplitudes, linear models do not give realistic predictions of rotor vibrations and the attendant dynamic forces because of the significant dynamic nonlinearities controlling the phenomena of such operating conditions. Virtually any condition that causes a significantly high vibration level will invariably be accompanied by significant dynamic nonlinearity. When the journal vibration orbit fills up a substantial portion of a bearing or seal radial clearance, the corresponding interactive rotor–stator force is no longer well approximated by the truncated Taylor series linear model introduced in Equations 2.60 along with Figure 2.10.

2.5.1 Large Amplitude Vibration Sources that Yield Nonlinear Effects

Well-recognized operating conditions, albeit out of the ordinary, that cause large rotor-to-bearing vibration orbits include the following:

- *Very large rotor unbalance*, for example, sudden detachment loss of large turbine or fan blades at running speed.
- Rotor–bearing *self-excited* orbital vibration *limit cycles*.
- Explosive detonation (*shock*) near underwater naval vessels.
- Unbalance-driven resonance at an *inadequately damped critical speed*.
- Resonance build-up resulting from *earthquakes*.

When such large vibration-causing phenomena occur, the following additional rotor dynamic nonlinear phenomenon is likely to be produced in the process:

- Rotor-to-stator *rub-impacting*.

Rub-impact phenomena can also result from other initiating factors such as misalignment and differential thermal growths and/or distortions. In such cases, the resulting influence of a rub-impact condition may or may not by itself lead to high vibration levels, but it will likely inject a significant nonlinear dynamic effect into the system.

Where risk assessments warrant, the added cost of performing appropriate nonlinear rotor dynamic analyses to properly evaluate potential failure modes associated with such unusually large vibration events is a prudent investment. However, such analyses are more likely to be performed only after a catastrophic failure occurs, to "do battle" in the resulting "contest" in order to determine who was at fault and consequently who must pay. The author spearheaded some of the early efforts in this problem area in the 1970s while at Westinghouse Electric's Corporate R&D Center near Pittsburgh. A primary paper (1980) by the author stemming from that work is included in the Bibliography at the end of this chapter.

Lateral Rotor Vibration Analysis Models 85

In all but a few classic 1-DOF nonlinear models, making computational predictions of dynamic response when one or more nonlinear effects are incorporated into the model requires that the equations of motion be numerically integrated *marching forward in time*. This means that the parameter "time" in the motion equations is subdivided into many very small but finite "slices" and within each one of these time slices, the force model associated with a particular nonlinear effect is linearized or at least held constant. This is quite similar to drawing a curved line by joining many short straight-line segments, that is, as the length of each straight-line segment gets smaller and smaller, their visual effect becomes the curved line. Various numerical integration schemes are available for this purpose, and with the advent of high-speed digital computers, such analyses first became feasible in the 1960s, but were quite expensive. Subsequently, with the evolution of modern PCs and workstations, at least the computational costs of such analyses are now negligible.

2.5.2 Journal-Bearing Nonlinearity with Large Rotor Unbalance

Fluid-film journal bearings are a prominent component where dynamic nonlinearity can play a controlling role in rotor vibration when the journal-to-bearing orbital vibration amplitude becomes a substantial portion of the bearing clearance circle. When this is the case, the linear model introduced in Equations 2.60 fails to provide realistic rotor dynamic predictions, as previously explained at the beginning of this section. As detailed in Chapter 5, computation of the fluid-film separating force that keeps the journal from contacting the bearing starts by solving the *lubricant pressure distribution* within the separating film. The film's pressure distribution is computed by solving the PDE known as the *Reynolds lubrication equation*, the solution for which other types of CPU-intensive numerical computations are required (e.g., finite difference, finite element). Performing a numerical *time marching* integration of the motion equations for a rotor supported by fluid-film journal bearings requires that the fluid-film bearing forces to be recomputed at each time step of the *time marching* computation. Thus, the fluid-film pressure distributions at each journal bearing must be recomputed at each time step. Therefore, depending upon the level of approximation used in solving the Reynolds equation, it can be quite CPU intensive to perform a *time marching* integration of the motion equations for a rotor supported by fluid-film journal bearings. For an instantaneous journal-to-bearing $\{x, y, \dot{x}, \dot{y}\}$, the x and y components of fluid-film force upon the journal are computed by integrating the instantaneous x and y projections of the film pressure distribution upon the journal surface, as expressed by the following equation:

$$\begin{Bmatrix} f_x - W_x \\ f_y - W_y \end{Bmatrix} = -R \int_{-L/2}^{L/2} \int_0^{2\pi} p(\theta, z, t) \begin{Bmatrix} \cos\theta \\ \sin\theta \end{Bmatrix} d\theta \, dz \qquad (2.87)$$

Referring to Figure 2.10, W_x and W_y are the x and y components, respectively, of the static load vector $\vec{W}$ acting upon the bearing. Thus, $-W_x$ and $-W_y$ are the corresponding static reaction load components acting upon the journal. Just as in linear LRV models, it is convenient in nonlinear analyses to formulate the equations of motion relative to the static equilibrium state. In so formulating the nonlinear LRV motion equations, the journal static loads ($-W_x$ and $-W_y$) are moved to the right-hand side of Equation 2.87, leaving $\{f_x, f_y\}$ as the instantaneous nonequilibrium dynamic force upon the journal.

The photographs in Figure 2.15 show some of the aftermath from two 1970s catastrophic failures of large steam turbo-generators. Both these failures occurred without warning and totally destroyed the machines. The author is familiar with other similar massive failures. Miraculously, in none of the several such failures to which the author has been familiarized have any serious personal injuries or loss of life occurred, although the potential for such personal mishap is surely quite possible in such events. The two early 1970s failures led to the author's work in developing computerized analyses to research the vibration response when *very large rotor mass unbalance is imposed on a multibearing flexible rotor*. For in-depth treatment of computational methods and results for nonlinear LRV, the group of ten papers on nonlinear rotor dynamics topics, listed in the Bibliography at the end of this chapter, are suggested. Some of the author's reported results are presented here.

The rotor illustrated in Figure 2.16 is one of the two identical low-pressure (LP) turbine rotors of a 700 MW 3600 rpm steam turbo-generator unit. It was used to computationally research the nonlinear vibrations resulting from unusually large mass unbalance. Using methods presented in Sections 1.3 and 2.3, the free–free rotor's undamped natural frequencies and corresponding planar mode shapes were determined from a finite-element model. All static and dynamic forces acting upon the rotor are applied on the free–free model as "external forces" including nonlinear forces, for example, bearing static and dynamic loads, unbalances, gyroscopic moments, weight, and so on. This approach is detailed in Adams (1980) and supplemented in the other associated papers referenced. The computation essentially entails solving the rotor response as a transient motion, numerically integrating forward in time for a sufficiently large number of shaft revolutions until a steady-state or motion envelope is determined.

Steady-state large unbalance results for the rotor in Figure 2.16 are shown in Figure 2.17a for the rotor supported in standard fixed-arc journal bearings, and in Figure 2.17b for the rotor supported in pivoted-pad journal

Lateral Rotor Vibration Analysis Models

(a)

(b)

(c)

FIGURE 2.15 (See color insert following page 262.) Photos from the two 1970s catastrophic failures of large 600 MW steam turbine-generator sets. Using nonlinear rotor dynamic response computations, failures could be potentially traced to the large unbalance from loss of one or more large LP turbine blades at running speed, coupled with behavior of fixed-arc journal bearings during large unbalance. (a) LP steam turbine outer casing. (b) Brushless exciter shaft. (c) Generator shaft. (d) LP steam turbine last stage.

(d)

FIGURE 2.15 Continued.

bearings. In both of these cases, it is assumed that one-half of a complete last-stage turbine blade detaches at 3600 rpm. This is equivalent to a 100,000 pound corotational 3600 cpm rotating load imposed at the last-stage blade row where the lost blade piece is postulated to separate from the rotor. As a point of magnitude reference, this LP turbine rotor weighs approximately 85,000 pounds. The Figure 2.17 results show four orbit-like plots as follows:

- Journal-to-bearing orbit normalized by radial clearance
- Total bearing motion (see bearing pedestal model, Section 2.3.9.2)
- Total journal motion
- Total fluid-film force transmitted to bearing

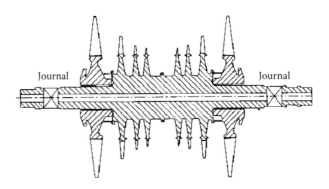

FIGURE 2.16 LP rotor portion of a 3600 rpm 700 MW steam turbine.

Lateral Rotor Vibration Analysis Models

The normalized journal-to-bearing orbit is simply the journal motion minus the bearing motion divided by the bearing radial clearance. For the cylindrical journal bearing of the Figure 2.17a results, this clearance envelope is thus a circle of unity radius. In contrast, for the pivoted four-pad journal bearing of the Figure 2.17b results, the clearance envelope is a square of unity side. A prerequisite to presenting a detailed explanation of these results are the companion steady-state vibration and dynamic force amplitude results presented in Figure 2.18 for unbalance conditions from zero to 100,000 pounds imposed at the same last-stage blade row of the same nonlinear model.

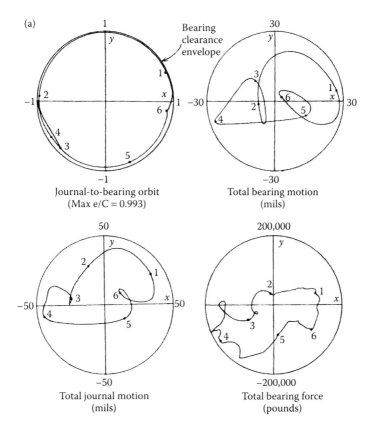

FIGURE 2.17 (a) Steady-state periodic response at bearing nearest the unbalance with force magnitude of 100,000 pounds, rotor supported on two identical fixed-arc journal bearings modeled after the actual rotor's two journal bearings. Timing marks at each one-half revolution, that is, 3 rev shown. (b) Steady-state periodic response at bearing nearest unbalance with force magnitude of 100,000 pounds, rotor supported on two identical four-pad pivoted-pad bearings with the gravity load directed between the bottom two pads. Bearings have same film diameter, length, and clearance as the actual fixed-arc bearings. Timings mark each one-half revolution, that is, 3 rev shown.

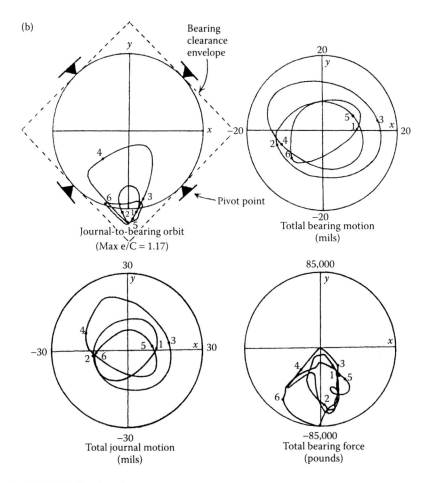

FIGURE 2.17 Continued.

An informative transition between 30,000 and 40,000 pounds unbalance is shown in Figure 2.18, from essentially a linear behavior, through a classic *nonlinear jump phenomenon*, and into a quite nonlinear dynamic motion detailed by the Figure 2.17 results for a 100,000 pound unbalance force. The explanation for the results in Figures 2.17 and 2.18 can be secured to the well-established knowledge of fixed-arc and pivoted-pad journal bearings concerning instability self-excited vibrations. The $x-y$ signals displayed in Figures 2.17a and b contain sequentially numbered timing marks for each one-half rotor revolution time interval at 3600 rpm. The observed steady-state motions therefore require three revolutions to complete one vibration cycle for both cases shown in Figure 2.17. Thus, these steady-state motions both fall into the category of a *period-3* motion since they both contain

Lateral Rotor Vibration Analysis Models

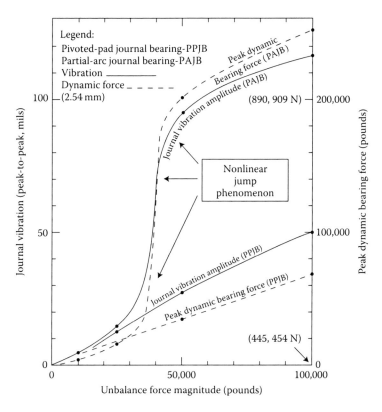

FIGURE 2.18 Comparison between partial-arc and pivoted-pad journal-bearing vibration control capabilities under large unbalance operating conditions of an LP steam turbine rotor at 3600 rpm; steady-state journal motion and transmitted peak dynamic bearing force over a range of unbalance magnitudes (data points mark computed simulation cases).

a 1/3 *subharmonic* frequency component along with a once-per-rev (synchronous) component. But these two cases are clearly in stark contrast to each other.

With the partial-arc bearings, Figure 2.17a, steady-state motion is dominated by the 1/3 subharmonic component and the journal motion virtually fills up the entire bearing clearance circle. However, in the second case which employs a tilting-pad bearing model, the 1/3 subharmonic component is somewhat less than the synchronous component and the journal motion is confined to the lower portion of the bearing clearance envelope. As is clear from Figure 2.18, with partial-arc bearings, motion undergoes a nonlinear jump phenomenon as unbalance magnitude is increased. With pivoted-pad bearings, a nonlinear jump phenomenon is not obtained. This contrast is even clearer when the motions are transformed into the frequency domain, as provide by FFT in Figure 2.19. The journal-to-bearing trajectories in these two cases provide the instantaneous lubricant minimum film thickness. For the partial-arc case, a smallest computed transient minimum film thickness of 0.1 mil (0.0001 in.) was obtained, thus surely indicating that hard journal-on-bearing rubbing would occur and as a consequence seriously degrade the bearings' catastrophe containment abilities. For the pivoted-pad case, a smallest computed transient minimum film thickness of 2 mils (0.002 in.) was obtained, thus indicating a much higher probability of maintaining bearing (film) integrity throughout such a large vibration event, especially considering that the pivoted pads are also inherently self-aligning.

The comparative results collectively shown by Figures 2.17 through 2.19 show a phenomenon that is probably possible for most such machines that operate only marginally below the threshold speed for the bearing-induced self-excited rotor vibration, commonly called *oil whip*. That is, with *fixed-arc journal bearings* and a large mass unbalance above some critical level (between 30,000 and 40,000 pounds for the simulated case here), a very large subharmonic resonance is a strong possibility.

In linear systems, steady-state response to harmonic excitation forces can only contain the frequencies of the sinusoidal driving forces, as can be rigorously shown from the basic mathematics of differential equations. However, in nonlinear systems, the response to sinusoidal driving forces has many more possibilities, including periodic motion (possibly with subharmonics and/or superharmonics), quasiperiodic motion (two or more noninteger-related harmonics), and chaos motion. Here, as the rotor mass unbalance is progressively increased, the journal-bearing forces become

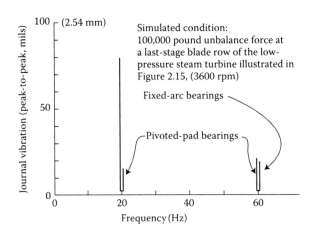

FIGURE 2.19 Fast Fourier transform of peak-to-peak journal vibration displacement amplitudes.

progressively more nonlinear, thus increasing the opportunity for dynamic behavior which deviates in some way from the limited behavior allowed for linear systems. Such LP turbines typically have a fundamental corotational mode which is in the frequency vicinity of one-third the 3600 rpm rotational frequency. This mode typically has adequate damping to routinely be *passed through* as a critical speed at approximately 1100–1300 rpm. However, up at 3600 rpm the speed-dependent destabilizing effect (k_{xy}^{ss}) of the fixed-arc bearings upon this mode places the rotor–bearing system only marginally below the instability threshold speed, as dissected in Section 2.4. Therefore, what is indicated by the results in Figure 2.17a is that the progressively increased bearing nonlinearity allows some energy to "flow" into the lightly damped 1/3 subharmonic, whose amplitude then adds to the overall vibration level and thus adds to the degree of bearing nonlinearity, thus increasing further the propensity for energy to flow into the 1/3 subharmonic, and so on. This synergistic mechanism manifests itself as the *nonlinear jump* in vibration and dynamic force shown in Figure 2.18. In other words, it is consistent with other well-known dynamic features of rotor–bearing systems.

Because of the emergence of strong nonlinearity in such a sequence of events, an exact integer match (e.g., 3:1 in this case) between the forcing frequency and the linearized subharmonic mode is not needed for the above-described scenario to occur.

Pivoted-pad journal bearings have long been recognized as not producing the destabilizing influence of fixed-arc journal bearings. The four-pad bearing modeled in these simulations has a symmetric stiffness coefficient matrix, consistent with its recognized inherent stability. Therefore, the case with pivoted-pad bearings gives results that are consistent with the prior explanation for the high amplitude subharmonic resonance exhibited with fixed-arc bearings. That is, *the inherent characteristic of the pivoted-pad-type journal bearing that makes it far more stable than fixed-arc bearings also makes it far less susceptible to potentially catastrophic levels of subharmonic resonance under large unbalance conditions*. If the static bearing load vector is subtracted from the total bearing force, the dynamic bearing force transmissibility is approximately 4 for the Figure 2.17a results and 1 for the Figure 2.17b results. Thus, the pivoted-pad bearing's superiority in this context is again manifest, in a 4:1 reduction in dynamic forces transmitted to the bearing support structure, *the last line of defense*.

The topic covered in the subsequent Section 2.5.4 would appear to be related to this type of large unbalance-excited subharmonic resonance nonlinear jump phenomenon, which occurs at an exact integer fraction of the unbalance forcing frequency. However, the two phenomena are not exactly the same thing, since the journal-bearing hysteresis-loop phenomenon is self-excited and has its own frequency, being initiated only by a large bump or ground motion disturbance.

2.5.3 Unloaded Tilting-Pad Self-Excited Vibration in Journal Bearings

Symptoms of this problem first arose with routine bearing inspections during scheduled outages of large power plant machinery. Some large steam turbo-generator units employing tilting-pad journal bearings exhibited fatigue crack damage to the leading edge of statically unloaded top pads of large four-pad tilting-pad journal bearings, illustrated in Figure 2.20. Around the same time, the author was investigating rotor dynamical characteristics of a pressurized water reactor (PWR) vertical centerline canned-motor pump rotor supported in tilting-pad journal bearings, utilizing nonlinear time-transient rotor vibration simulations. In that work, an unanticipated discovery occurred that found a previously unrecognized dynamical phenomenon of tilting-pad journal bearings, namely unloaded-pad self-excited flutter. The computational tools previously employed to obtain the large unbalance results presented in Section 2.5.2 of this chapter were employed by the author for nonlinear rotor dynamical simulations of a canned-motor pump rotor supported in tilting pad journal bearings. Other mission objectives drove these canned-motor pump simulations. But an unexpected self-excited pad-flutter motion of a statically unloaded pad, at a frequency slightly below one-half the rotational frequency, was observed from the simulations.

Not focal to the mission at hand, the discovered pad flutter was not further researched at that time. Some years later, the author reopened research on this self-excited pad-flutter phenomenon.

Adams and Payandeh (1983) present their nonlinear dynamical simulations for a single-pad 2-DOF model with simultaneous radial and pitching pad motions. Figure 2.21 illustrates this model, and provides a four-frame one-period time sequence of a typical case of unloaded pad flutter. They

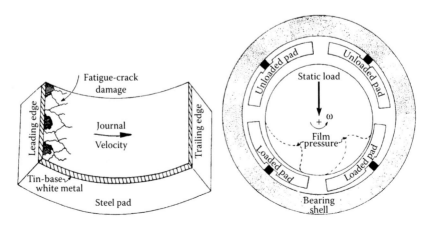

FIGURE 2.20 Four-pad tilting-pad bearing with unloaded pads.

Lateral Rotor Vibration Analysis Models 95

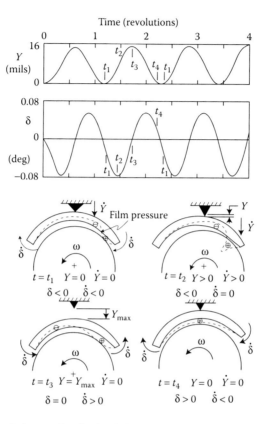

FIGURE 2.21 Simulation results of unloaded pad self-excited vibration.

present a tabulation of the results of several different pad configurations and operating parameters that provide a broad coverage of design and operating conditions that will or will not promote pad flutter. Their work provides some general observations on the pad flutter phenomenon. Namely, when a pad's operating pivot clearance is larger than the concentric clearance, a stable static equilibrium pad position may not exist, and if not, self-excited subsynchronous pad vibration will occur. The self-excited motion continuously seeks to find an instantaneous film pressure distribution that produces a null force and moment. The base frequency of this vibration is somewhat below 0.5 times the rotational speed, just like classical rotor–bearing instability. In fact, if one observes the stationary journal centerline from a reference frame fixed in the fluttering pad, the journal appears to be undergoing self-excited whirling. So this pad flutter phenomenon is really a camouflaged version of the classical journal-bearing instability phenomenon.

2.5.4 Journal-Bearing Hysteresis Loop

The hysteresis loop associated with the journal bearing caused dynamic instability self-excited vibration mechanism called *oil whip* was for a long time an interesting topic for the academics. But it did not attract the close scrutiny of rotating machinery development engineers. However, in the seismically active region of Japan, a team headed by Professor Y. Hori at the University of Tokyo brought the practical importance of the journal-bearing hysteresis loop to the wider engineering community. In the paper by Hori and Kato (1990), the distinct possibility of an earthquake-initiated high-amplitude sustained self-excited rotor vibration is addressed. That work helped initiate subsequent research by the author and his team, reported in the paper by Adams et al. (1996). A generic illustration of their journal-bearing hysteresis loop and computational model are shown in Figure 2.22.

Figure 2.22 encapsulates the imbedding of the classical oil-whip phenomenon within an expanded view that shows two stable vibration solutions at speeds below the oil-whip threshold speed ω_{th} (Hopf biburcation) and one unstable solution, which is a boundary between the two

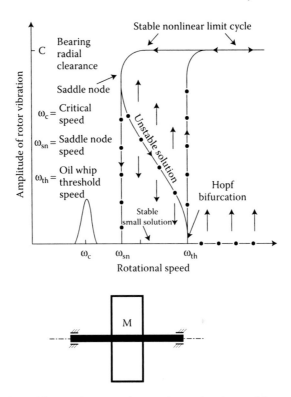

FIGURE 2.22 Journal-bearing hysteresis-loop and rotor–bearing model.

stable solutions. Adams et al. (1996) demonstrate this through computational simulations of a rigid rotor in two identical journal bearings, which they treat as a 2-DOF point mass rotor supported in one hydrodynamic fluid-film journal bearing. They computationally constructed hysteresis loop examples covering a wide range of parameter combinations, especially static load range from nearly zero to high loads. They confirm their simulation constructed hysteresis loop examples using a specially configured laboratory test rig. At vanishing small static bearing loads, a hysteresis loop does not occur and the oil-whip threshold speed ω_{th} is near twice the critical speed. At progressively higher bearing static loads, the hysteresis loops progressively open up wider, with the whip threshold speed ω_{th} occurring at progressively higher speeds, as is well known. But the lower speed limit ω_{sn} of the hysteresis loop gets progressively lower, asymptotically approaching 1.72 times the saddle node speed ω_{sn}. The quite practical importance of this is that given a substantially large dynamic disturbance, a large amplitude oil-whip limit cycle vibration can occur at a rotor speed less than twice the critical (resonance) speed. So while increased bearing static load raises the expected oil-whip threshold speed, it also lowers the speed above which a large amplitude oil whip limit cycle vibration can occur.

One can see a similarity between the hysteresis loop and the nonlinear jump phenomenon discussed in Section 2.5.2. However, the two phenomena are not exactly the same thing, since the journal-bearing hysteresis loop phenomenon is self-excited and has its own frequency, being initiated only by a large bump or transient ground motion disturbance.

The unstable solution shown in Figure 2.22 is qualitatively superimposed to show its relationship to the hysteresis loop. Khonsari and Chang (1993) show the existence of a closed position boundary encircling the static equilibrium point, which delineates between initial positions that die out and initial positions that grow to a stable large amplitude limit cycle orbit. The author believes they have encountered the unstable solution that exists within the framework of the hysteresis loop. Although they do not apparently recognize this, their contribution to the subject is nonetheless valuable. The complete initial condition boundary provided by the unstable solution would have to be described in the appropriated dimensioned phase space, of which the orbital position space is a subset.

2.5.5 Shaft-on-Bearing Impacting

Impacting is a quite nonlinear dynamic phenomenon. In Chapter 9 Section 9.8.3, *rotor–stator rub-impacting* is treated from the point of view as a cause of excessive rotor vibration, and its identifying symptoms are treated. In order to computationally model rotor–bearing impact conditions, there is the need for an *impact restitution coefficient*, a necessary input

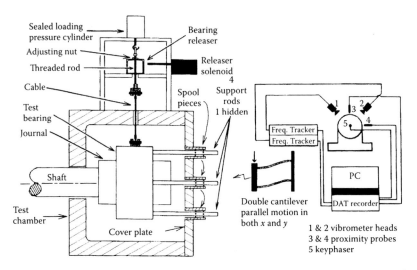

FIGURE 2.23 Test for rotor–bearing restitution coefficient measurement.

for impact dynamics analyses. In general, the coefficient-of-restitution is based on experimentally determined information, a strictly theoretical modeling approach being at the limits of what modern solid mechanics analysis tools can provide. Adams et al. (2000) configured a quite elaborate experimental setup employing orthogonal x–y laser vibrometers to directly measure velocities through controlled bearing-shaft impacts. That test apparatus is illustrated in Figure 2.23. Test results obtained with this apparatus are presented in Figure 2.24, and cover wide ranges of journal speed and impact velocity.

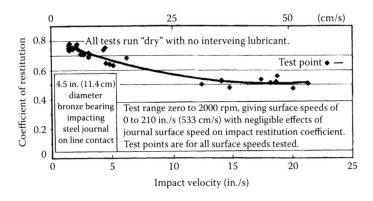

FIGURE 2.24 Bearing-on-journal impact restitution coefficient.

Lateral Rotor Vibration Analysis Models   99

As is clear from Figure 2.24, test results obtained with this apparatus exhibit a fairly close grouping of test points, which is significant considering the nontriviality of capturing impact velocities, even with modern sensors and data reduction methods. These results indicate virtually no influence of journal sliding velocity, which probably reflects the quite small relative impact time during bearing–journal contact. These results approach a maximum restitution coefficient of 0.8 as impact velocity is relatively small, and asymptotically approach a value of 0.5 as impact velocity increases.

2.5.6 Chaos in Rotor Dynamical Systems

The necessary, but not sufficient, ingredient for chaos in vibrations is nonlinearity. As part of its research on new methods for diagnostics in rotating machinery vibration, the author's research team has done extensive numerical simulations and some experimental investigations to discover if chaos tracking tools can offer any new information to facilitate machinery monitoring and diagnostics. The main body of this research is reported by Adams and Abu-Mahfouz (1994), who investigated three different types of systems to search for routes to chaos through nonlinear rotor dynamical phenomena. These are (1) rotor–bearing rub-impact, (2) cylindrical bearing supported journal, and (3) tilting-pad bearing-supported journal. In Chapter 9 Section 9.7, *chaos analysis tools*, several examples of the research into this topic are presented to show the potential utility of chaos measures in order to diagnostically identify a variety of internal machinery distress sources. The reader is referred to Section 9.7 of Chapter 9 for the explanation of simulation results from the three system categories.

When a system has zero damping, the type of chaotic motion that can occur is referred to as *Hamiltonian*. Where there is finite damping, *dissipative chaos* can occur. For the purposes of studying routes to chaos in rotor dynamical systems, dissipative chaos is the more useful since real systems always have energy dissipation mechanisms present. Dissipative chaos is analyzed by employing various established methods. Stroboscopic records (i.e., Poincaré maps) and their fractal dimensions are the most commonly used tools. Significant progress in the general modern study of chaotic dynamics has been pioneered by researchers in applied mathematics, and its disclosures are thus often imbedded in the more abstract language of mathematics. In contrast, the research reported by Adams and Abu-Mahfouz has the goal of exploring and explaining ways in which chaos tracking techniques can be used to potentially further the field of rotating machinery vibration *monitoring and diagnostics*. When the mathematicians' jargon is filtered from the topic of chaos, as Adams and Abu-Mahfouz have done, it is quite easy for engineers to understand the essence of the topic and thereby conceptualize applications potentially useful for engineering purposes.

2.5.7 Nonlinear Damping Masks Oil Whip and Steam Whirl

While employed in the 1970s at the Westinghouse Electric Corporate Research Labs near Pittsburgh, the author first became aware of a quite interesting nonlinear damping phenomenon that can mask oil whip and steam whirl. As is well known and treated in Sections 2.5.1 and 2.5.2, if rotor vibration levels grow progressively higher, the journal-bearing fluid-film force characteristic becomes progressively more nonlinear.

2.5.7.1 Oil Whip Masked

A hypothetical case serves well the fundamental explanation. Assume that a large steam turbine-generator machine has an oil-whip threshold speed that over years has "in secrecy" gradually moved lower, approaching the 3600 rpm operating speed. This is not unusual because of support structure shifting and various accrued wear that can gradually reduce the effective residual damping of the potentially unstable mode associated with the oil-whip threshold. Further assume that the rotor of this machine has gradually undergone a deterioration of its state of rotor mass balance. It has therefore undergone progressively increased vibration levels and has become "rough running" and thus in need of rebalancing measures. So at a convenient point, like over a weekend or other power-lowered demand periods, the unit is brought down, cooled, and balance correction weight(s) appropriately attached.

On completion of this rebalancing, the machine is put through its standard several-hour start-up sequence, approaching the 3600 rpm operating speed. But before quite reaching the 3600 rpm speed, a high-level rotor vibration comes in at a frequency somewhat below half the rotor spin speed. The level of this subsynchronous vibration is significantly above allowable operating vibration levels; therefore, the machine is immediately brought down so that the plant engineers can determine what it is that they have just encountered. What they have just encountered is oil whip. By rebalancing the rotor and thereby significantly reducing the unbalance vibration levels, the bearing dynamic nonlinearities accompanying the rough running condition are largely removed. And thereby the extra damping of the potentially unstable oil-whip vibration mode resulting from bearing nonlinearities is also largely removed. Without this extra damping, the self-excited oil-whip instability vibration is able to "kick in." So by reducing the state of rotor unbalance-driven synchronous vibration, the attendant reduction in nonlinear damping allows the oil whip to finally "come out of the closet," after years "in hiding" above but close to

Lateral Rotor Vibration Analysis Models

the 3600 rpm operating speed. This is akin to trading a *headache* for *appendicitis*. This unit will have to be fixed using a *cost-effective option*, which the author strongly suggests *is not adding the unbalance back* on to the rotor (see Chapter 11 case studies).

2.5.7.2 Steam Whirl Masked

In addition to the above oil-whip scenario, a similar scenario involving steam whirl also occurs in large high-pressure steam turbines. As in the oil-whip case described, steam whirl can occur immediately following a rotor rebalancing to reduce synchronous unbalance forced vibration of a rough running machine. The unit's rotor is balanced over its speed range and subsequently brought up to operating synchronous speed and electrically connected to the grid. The unit is then powered up at its allowed power ramp-up (e.g., 5 MW/min). But as it approaches say 85% of the rated capacity, a high level of subsynchronous rotor vibration (also somewhat below half the rotor spin frequency) kicks in, primarily in the high-pressure turbine rotor. The nature of this vibration is quite similar to oil whip, but it is not oil whip. Until the problem is fixed, the plant is restricted to operate this unit at loads below 85% of the unit's rated capacity. That is a significant loss in generating revenue and yields a degradation of heat rate (higher fuel cost via lower efficiency). The unit owners are informed by their rotating machinery vibration consultant that the unit is load limited by the long recognized *steam-whirl* phenomenon. This unit will have to be fixed using one of the options available. Again, adding back the removed rotor unbalance is definitely not the recommended fix to allow the unit to be operated at its full-rated power capacity (see Chapter 11 case studies).

2.5.8 Nonlinear Bearing Dynamics Explains Compressor Bearing Failure

The piston and connecting rod subassembly shown in Figure 2.25 is from a single-piston reciprocating compressor designed for use in both a home refrigerator and a window air conditioner produced by a major home appliances manufacturer. Several months after some design modifications to this compressor were released into these two refrigerant products, the refrigerator application started to show 4 times the rate of warranty compressor failures as the window air conditioning application. This resulted in an annual multimillion dollar loss on the refrigerator product from warranty replacements and it thereby quickly got onto the upper management's "radar screen." Close study of several of the failed compressors revealed that it was the wrist pin bearing that was failing. The wrist pin bearing is the sleeve bearing press fitted into the piston end of the connecting rod and surrounds the wrist pin, which is press fitted into the piston as shown in Figure 2.25.

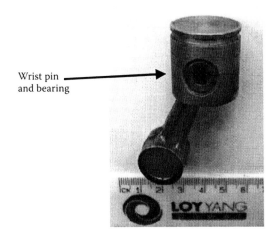

FIGURE 2.25 (**See color insert following page 262.**) Piston and connecting rod of a small reciprocating compressor.

That the refrigerator compressor failure rate was 4 times that of the air conditioner compressor mystified the manufacturer's top compressor engineers, because the wrist pin peak load in the air conditioner was approximately 25% higher than in the refrigerator. The wrist pin bearing radial load versus crank angle is illustrated for both applications in Figure 2.26. In an attempt to uncover the root cause for the relatively large warranty failure rate in the refrigerator application, many different analyses and tests were conducted, sort of a "fishing expedition."

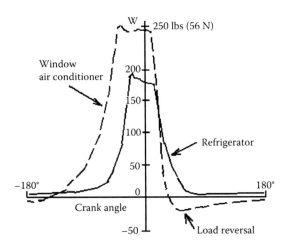

FIGURE 2.26 Wrist pin bearing load (W) curves versus crank angle.

One of the many analyses pursued was computation of the wrist pin bearing's minimum film thickness within the 0–360° crank cycle. The author was assigned this task. Unlike the calculation of journal-bearing minimum oil-film thickness under bearing static load, the nonlinear dynamical orbit of the wrist pin relative to the bearing must be modeled to predict the transient minimum oil film thickness. This is a well-established computational approach for reciprocating compressors and internal combustion piston engines. Both the air conditioner and refrigerator time-varying load were entered into the author's inputs simulation, which employed a nonlinear time-transient marching algorithm. The computed wrist pin orbits, not the minimum film thickness, unexpectedly provided the answer to the bearing failure root cause.

Two simulated wrist pin orbits from this analysis are illustrated in Figure 2.27. They clearly show the root cause of the refrigerator compressor's higher warranty failure rate. The load curves illustrated in Figure 2.26 show that one loading function goes slightly negative and one does not, a feature that was not previously noted when focus was on the maximum peak loads. That is, in the air conditioner application, just a slight amount of load reversal causes the wrist pin to substantially separate away from the oil-feed hole that channels oil from the rod bearing through a connecting hole in the rod. In contrast, in the refrigerator application, there is no load reversal and thus the wrist pin does not lift off the oil-feed hole as its oscillatory trajectory clearly shows. Subsequent endurance tests completely confirmed what the computed nonlinear dynamical orbits imply. That is, the refrigerator wrist pin continuously rubs on the bearing over the oil hole and thereby does not separate to allow oil in for the next

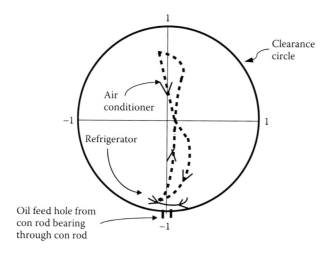

FIGURE 2.27 Wrist pin orbital trajectories from nonlinear simulations.

squeeze film action of each loading cycle. With the root cause uncovered by the author's nonlinear analysis, modifications were then implemented to insure that the refrigerator compressor load curve included some load reversal. The high compressor failure rate ceased, once those units still "in the pipeline" cleared the retailers' stocks.

2.6 Summary

Chapter 2 is the "backbone" of this book. The fundamental formulations and basic physical insight foundations for LRV are presented. The focus is primarily on the construction of linear analysis models. However, the last section on nonlinear effects should sensitize the troubleshooter to the fact that all real systems have some nonlinearity. Therefore, one should not expect even the best linear models to portray all the vibration features that might be obtained from vibration measurements on actual machinery or even on laboratory test rigs.

Bibliography

Rotating machinery vibration cuts across many different industries and equipment types, making it difficult for any single individual to provide a perfectly balanced treatment of the complete subject. Several comprehensive English texts on rotor dynamics are listed here, and each one reflects the experience focus of its author(s), as this book surely must. Collectively, they provide what is tantamount to an "encyclopedia" on Rotor Dynamics, citing a large portion of the major published papers and reports in the field. The intent of the author of this book is not to provide an encyclopedia on the subject, but to provide the tools to solve real-world problems and to prepare readers to use the "encyclopedia" when deemed necessary.

Textbooks

Adams, M. L., *Rotating Machinery Vibration—From Analysis to Troubleshooting*, 1st edn, Marcel Dekker, New York, 2001, pp. 354.
Bently, D. E. and Hatch, C. T., *Fundamentals of Rotating Machinery Diagnostics*, Bently Pressurized Bearing Corporation, Minden, NV, 1999.
Childs, D., *Turbomachinery Rotordynamics—Phenonmena, Modeling, and Analysis*, Wiley, New York, 1993, pp. 476.

Dimarogonas, A. D. and Paipetis, S. A., *Analytical Methods in Rotor Dynamics*, Applied Science Publishers, London, 1983, pp. 217.
Dimentberg, F. M., *Flexural Vibrations of Rotating Shafts*, Butterworths, London, 1961, pp. 243.
Gasck, R., Nordmann, R., and Pfutzner, H., *Rotordynamik* (in German), Springer, Berlin, 2006, pp. 705.
Gunter, E. J., *Dynamic Stability of Rotor-Bearing Systems*, NASA SP-113, Washington, 1966, pp. 228.
Lalanne, M. and Ferraris, G., *Rotordynamics Prediction in Engineering*, 2nd edn, Wiley, New York, 1997, pp. 254.
Kramer, E., *Dynamics of Rotors and Foundations*, Springer, Berlin, 1993, pp. 383.
Muszynska, A., *Rotor Dynamics*, CRC Taylor & Francis Group, London, 2005, pp. 1120.
Tondl, A., *Some Problems of Rotor Dynamics*, Chapman & Hall, London, 1965.
Vance, J. M., *Rotordynamics of Turbomachinery*, Wiley, New York, 1988, pp. 388.

Selected Papers Concerning Rotor Dynamics Insights

Adams, M. L. and Padovan, J., "Insights into linearized rotor dynamics," *Journal of Sound & Vibration*, 76(1):129–142, 1981.
Adams, M. L., "A note on rotor-bearing stability," *Journal of Sound & Vibration*, 86(3):435–438, 1983.
Adams, M. L., "Insights into linearized rotor dynamics, Part 2," *Journal of Sound & Vibration*, 112(1):97–110, 1987.
Yu, H. & Adams, M. L., "The linear model for rotor-dynamic properties of journal bearings and seals with combined radial and misalignment motions," *Journal of Sound & Vibration*, 131(3):367–378, 1989.

Selected Papers on Nonlinear Rotor Dynamics

Adams, M. L., "Non-linear dynamics of flexible multi-bearing rotors," *Journal of Sound & Vibration*, 71(1):129–144, 1980.
Adams, M. L., Padovan, J., and Fertis, D., "Engine dynamic analysis with general nonlinear finite-element codes, Part 1: Overall approach and development of bearing damper element," *ASME, Journal of Engineering for Power*, 104(3):586–593, 1982.
Adams, M. L. and Payandeh, S., "Self-excited vibration of statically unloaded pads in tilting-pad journal bearings," *ASME, Journal of Lubrication Technology*, 105(3):377–384, 1983.
Adams, M. L. and McCloskey, T. H., *Large Unbalance Vibration in Steam Turbine-Generator Sets*, Third IMechE International Conference on Vibrations in Rotating Machinery, York, England, 1984.

Adams, M. L. and Abu-Mahfouz, I., *Exploratory Research on Chaos Concepts as Diagnostic Tools*, Proceedings of the IFTOMM Fourth International Conference on Rotor Dynamics, Chicago, Illinois, September 6–9, 1994.

Adams, M. L., Adams, M. L., and Guo, J. S., *Simulations & Experiments of the Non-Linear Hysteresis Loop for Rotor-Bearing Instability*, Proceedings, Sixth IMechE International Conference on Vibration in Rotating Machinery, Oxford University, September 1996.

Adams, M. L., Afshari, F., and Adams, M. L., *An Experiment to Measure the Restitution Coefficient for Rotor-Stator Impacts*, Seventh IMechE International Conference on Vibrations in Rotating Machinery, Nottingham University, September 2000.

Hori, Y., *Anti-Earthquake Considerations in Rotor Dynamics*, Key note address paper, Fourth IMechE International Conference on Vibrations in Rotating Machinery, Edinburgh, September 1988.

Hori, Y. and Kato, T., "Earthquake induced instability of a rotor supported by oil film bearing," *ASME Journal of Vibration and Acoustics*, 112:160–165, 1990.

Konsari, M. M. and Chang, Y. T., "Stability boundary of nonlinear orbits within clearance circle of journal bearings," *ASME Journal of Vibration and Acoustics*, 115:303–307, 1993.

PROBLEM EXERCISES

1. The simple rotor dynamics model here illustrated is for a small turbomachinery rotor that runs at 6000 rpm. The mid-span 14 kg disk has an unbalance of 0.0028 kg-m. Neglecting the mass of the steel shaft and neglecting bearing flexibility, determine the amplitude of disk radial orbital vibration and amplitude of rotating dynamic force at each bearing for a 25 mm diameter shaft.

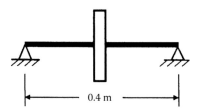

2. For the rotor model in Problem 1 estimate the error of neglecting the shaft mass m_s. The method for this is to use the simple energy approach to derive the equation of motion. That is, since the system is conservative (no damping) the time rate of change of system energy is zero. For the 1-DOF spring-mass model, this yields the following path to the equation of motion and natural frequency.

$$\frac{d}{dt}\left(\frac{1}{2}m\dot{x}^2 + \frac{1}{2}kx^2\right) = 0 \therefore m\dot{x}\ddot{x} + kx\dot{x} = 0 \text{ giving}$$

$$m\ddot{x} + kx = 0 \quad \text{and} \quad \omega_n = \sqrt{k/m}.$$

To estimate the effect of shaft mass, integrate its kinetic energy over (0, L) using the static deflection shape under a center load, scaled to the center deflection x and shaft mass density per unit length. Then add this shaft kinetic energy to the disks kinetic energy to obtain total kinetic energy. This results in the amount of shaft mass to be added to the disk mass as $17m_s/35$, yielding the following.

$$\omega_n = \sqrt{\frac{k}{(m + 17m_s/35)}} \quad \text{where } k = 48EI/L^3$$

The reason this is not theoretically an exact answer is because the static deflection shape under a center load is slightly different than the dynamic deflection shape.

3. The model here illustrated is for small rotor with a 140 kg disk mounted as shown on a 75 mm diameter steel shaft. The bearing flexibility vertically is sufficiently small to be treated as rigid. But bearing flexibility in the horizontal direction is not negligible, each having a horizontal stiffness k of 50 million N/m. Neglecting shaft mass and gyroscopic effect, determine the critical speeds.

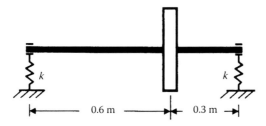

4. Shown is a rotor model for a simply supported steel shaft carrying two disks, left disk at 45 kg and right disk at 70 kg. $L_1 = 150$ mm and $L = 600$ mm. Using the flexibility matrix approach (see Problem 7 in Chapter 1), develop the equations of motion and determine the critical speeds. Assume that both bearing flexibility and disk gyroscopic effects are negligible.

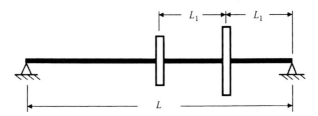

5. For the three-bearing shaft supporting two disks, use the influence coefficient matrix approach (see Problem 7 in Chapter 1)

to formulate the equations of motion, and determine the critical speeds and corresponding mode shapes for the configuration.

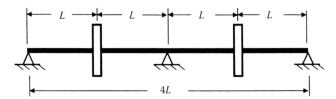

6. Formulate the lumped-mass equations of motion for an 8-DOF flexible rotor-bearing configuration similar to that shown in Figure 2.4, but with the two shafting pieces of different length L and different EI. Put the equations in matrix form. (a) Use a Lagrange approach, (b) use the direct $F = ma$ approach.
7. Formulate the equations of motion for an 8-DOF flexible rotor-bearing configuration with the disk mounted at the free end of an overhung portion of the shaft, not between the two bearings. Let the shaft piece between the two bearings and the overhung shaft piece have arbitrarily different lengths and different EI. Put the equations in matrix form. (a) Use a Lagrange approach, (b) use direct $F = ma$ approach.
8. Formulate the lumped-mass equations of motion for a 12-DOF flexible rotor-bearing configuration with two equal disks symmetrically located between two end bearings on three equal shaft pieces each having a length of one-third the total shaft length bearing span. Put the equations in matrix form. (a) Use a Lagrange approach, (b) use the direct $F = ma$ approach.
9. Formulate the lumped-mass equations of motion for a 12-DOF flexible rotor-bearing configuration with two different disks non-symmetrically located between two end bearings with three different shaft pieces. Put the equations in matrix form. (a) Use a Lagrange approach, (b) use the direct $F = ma$ approach.
10. Formulate the lumped-mass equations of motion for a 12-DOF flexible rotor-bearing configuration with two equal disks, one disk positioned at the mid-span location between the two bearings and the second disk positioned at the free end of an overhung portion of the shaft, not between the two bearings. Let the bearing shaft span and overhung shaft portion be of arbitrarily different lengths and different EI. Put the equations in matrix form. (a) Use a Lagrange approach, (b) use direct $F = ma$ approach.
11. Formulate the lumped-mass equations of motion for a 12-DOF flexible rotor-bearing configuration with two equal disks, each positioned at a free end of two overhung portion of the shaft, not between the two bearings. Let the bearing shaft span be different than the two overhung shaft pieces which are identical. Put the equations in matrix form. (a) Use a Lagrange approach, (b) use direct $F = ma$ approach.

12. *Project on nonlinear rotor vibration hysteresis loop.* Using the modeling procedures cited in Adams, Adams, and Guo (1996) formulate and implement a computer code of a time-marching 2-DOF model for a point-mass rotor supported by a single 360° fluid-film journal bearing. Model the fluid-film bearing force by re-solving the Reynolds PDE lubrication equation at each marching time step. With the debugged code, simulate the cases reported in Adams, Adams, and Guo (1996). Then computationally explore the unbalance induced nonlinear jump phenomenon described in Section 2.5.2 of this chapter.
13. *Project on nonlinear self-excited vibration simulation of unloaded bearing tilting pads.* Employing the 2-DOF pad model of Adams and Payandeh (1983) reproduce the simulation results reported by them.

3
Torsional Rotor Vibration Analysis Models

3.1 Introduction

Torsional rotor vibration (TRV) is angular vibratory twisting of a rotor about its centerline superimposed on its angular spin velocity. TRV analysis is not needed for many types of rotating machinery, particularly those machines with a single uncoupled rotor. Many single-drive-line rotors are stiff enough in torsion so that torsional natural frequencies are sufficiently high to avoid forced resonance by the time-varying torque components transmitted in the rotor. A notable exception is the quite long rigidly coupled rotors in modern large steam turbine generator sets, examined later in this chapter. When single rotors are coupled together, the possibility is greater for excitation of coupled-system torsional natural frequency modes. Coupling of single rotors in this context can also be through standard so-called *flexible couplings* connecting coaxial rotors and/or through *gears*. In most coupled drive trains, it is the characteristics of the couplings, gear trains, and electric motors or generators that instigate TRV problems.

It is typical for dynamic coupling between the *torsional* and *lateral* rotor vibration (LRV) characteristics to be discounted. The generally accepted thinking is that while potentially coexisting to significant degrees in the same rotor(s), TRV and LRV do not significantly interact in most machinery types. There are a few exceptions to this as noted at the end of Section 2.1.

As previously summarized in Table 2.1, LRV is always an important consideration in virtually all types of rotating machinery. Conversely, TRV is often not an important consideration in many machinery types, especially machines with single uncoupled rotors as previously noted. Consequently, TRV has not received nearly as much attention in engineering publications as LRV. Of the rotor dynamics books listed in the Chapter 2 Bibliography, most do not cover TRV. This is a measure of the extent to which rotor dynamics technologists have focused on the admittedly much better funded topics within the LRV category. For rotating machinery products where TRV considerations are now part of standard design analyses, dramatic past failures were often involved in making TRV get its deserved recognition.

Unlike typical LRV modes, TRV modes are usually quite lightly damped. For example, the significant squeeze-film damping inherently provided to LRV modes by fluid-film journal bearings and/or squeeze-film dampers does not help damp TRV modes because the TRV modes are nearly uncoupled from the LRV modes in most cases. With very little damping, excitation of a TRV mode can readily lead to a serious machine failure without warning. Because TRV modes are usually uncoupled from LRV modes, TRV modes can be continuously or intermittently undergoing large amplitude forced resonance without the machine exhibiting any readily monitored or outward signs of distress or "shaking." That is, there is no sign of distress until the shaft suddenly fails from a through-propagated fatigue-initiated crack in consequence of vibration-caused material fatigue. When a machine specified to have say a 40-year design life experiences such a failure after say 6 months in service, one strongly competing conclusion is the following: *some discounted phenomenon* (like TRV) *had in fact become significant to the product*. Some notable examples of TRV "earning" its deserved recognition as an important design consideration involves *synchronous motors* with slow speed-ups of large rotary inertia drive lines and early *frequency-inverter variable-speed induction electric motor* drives. In both of these cases, pulsating motor torque is the source of excitation.

As described in the book by Vance (1988), the start-up of a large power *synchronous electric motor* produces a pulsating torque with a frequency that changes from twice line frequency at start, down to zero at synchronous operating speed. The peak-to-peak magnitude of the torque pulsation varies with speed and motor design, but is often larger than the average torque. Any TRV mode with a natural frequency in the zero to twice-line-frequency range is therefore potentially vulnerable to large amplitude excitation during the start-up transient. In a *worst-case scenario*, a number of large single rotors are coupled (tandem, parallel, or other), yielding some coupled-system low natural frequencies, and a large total rotary inertia for a relatively long start-up exposure time to forced resonances. In an application where such a machine must undergo a number of start–stop cycles each day, a shaft or other driveline failure within 6 months of service is a likely outcome. Synchronous motor-powered drive trains are just one important example where the rotor system must be analyzed at the design stage to avoid such TRV-initiated failures. Special couplings that provide TRV damping, or act as torsional low-pass filters, and other design approaches are used. Successful use of such approaches requires careful analyses, thus the major thrust of this chapter is to present the formulation of TRV models for such analyses.

3.2 Rotor-Based Spinning Reference Frames

To properly visualize TRV, one must consider that the relatively small torsion-twisting angular velocities of TRV are superimposed on the considerably larger rotor spin velocity. That is, the TRV angular displacements, velocities, and accelerations are referenced to a *rotating* (noninertial) *reference frame* that rotates at the spin velocity. However, TRV equations of motion are generally derived as though the coordinate system is not rotating. The reason this produces proper motion equations warrants a fundamental explanation. As developed in Chapter 2, the rate-of-change of a rigid body's angular momentum vector, prescribed in a coordinate system rotating at $\vec{\Omega}$, is given by Equation 2.15. The same form of equation applies to time differentiation of any vector prescribed in a rotating reference frame. The instantaneous total angular velocity ($\vec{\dot{\theta}}_i^T$) at a rotor mass station is the sum of the instantaneous TRV velocity ($\vec{\dot{\theta}}_i$) and the instantaneous rotor spin velocity ($\vec{\omega}$), shown as follows:

$$\vec{\dot{\theta}}_i^T = \vec{\dot{\theta}}_i + \vec{\omega} \qquad (3.1)$$

Thus, the inertial angular acceleration at a rotor station is as follows:

$$\vec{\ddot{\theta}}_i^T = \vec{\ddot{\theta}}_i + \vec{\dot{\omega}} = (\vec{\ddot{\theta}}_i)_\omega + \overset{\nearrow 0}{\vec{\omega} \times \vec{\dot{\theta}}_i} + \vec{\dot{\omega}} \qquad (3.2)$$

The spin and TRV velocity vectors are coaxial; thus their cross product is zero, as indicated in Equation 3.2. Furthermore, for most TRV analysis purposes, rotor spin acceleration ($\dot{\omega}$) is taken as zero, that is, $\omega \cong$ constant, so $\vec{\ddot{\theta}}_i^T = \vec{\ddot{\theta}}_i$. Inertial angular acceleration vectors for TRV can then be given as follows:

$$\vec{\ddot{\theta}}_i = (\vec{\ddot{\theta}}_i)_\omega \qquad (3.3)$$

That TRV equations of motion are derived as though the rotor is not spinning about its axis is thus shown to be valid.

3.3 Single Uncoupled Rotor

Although TRV analysis is not needed for many single-rotor drive lines, there are notable exceptions like large steam turbine generator sets. Furthermore, the single uncoupled rotor model is the basic analysis model building block for the general category of coupled rotors. It is thus logical to begin TRV model development at the single rotor level. Chapter 1 provides the essential vibration concepts and methods to follow the presentations in this chapter. Also, the developments on LRV analysis models presented in Chapter 2 have many similarities to the TRV model developments presented in this chapter. In particular, TRV equations of motion are systematically assembled in matrix form based on standard finite-element procedures, combining both structural (i.e., flexible) and nonstructural (i.e., lumped) mass contributions.

As explained in Sections 1.1 and 1.3, undamped models are accurate for the prediction of natural frequencies in most mechanical systems. Since most TRV systems are quite lightly damped, these prior arguments are especially valid for TRV models. Thus, the focus here is on developing undamped models. The TRV rotor model in Figure 3.1 shows a number of rigid disks, each with a flexible torsional shaft connection to its immediate neighbors, but no connections to ground. As shown, this model therefore has one so-called *rigid-body mode* about the rotor axis and its twisting modes must each conserve angular momentum about the rotor axis. For configurations where the driver (e.g., electric motor, turbine) and/or the driven component (e.g., pump, compressor, fan) provide only relatively low torsional stiffness connections to ground, a *free–free* TRV model, such as in Figure 3.1, may be appropriate. However, in some applications the interactive torsional stiffness of the driver and/or the driven components cannot be neglected in making accurate predictions for natural frequencies and mode shapes. For example, with a variable speed electric motor drive having a high-gain feedback speed control, the effective rotor-to-ground torsional stiffness effect at the motor may be quite significant.

Vance (1988) describes the relation of feedback speed control to TRV. Stability analysis of the speed controller should couple the TRV differential equations of motion to the differential equations of state for the controller (see the last topic in Section 1.3 of Chapter 1). Such a stability analysis will

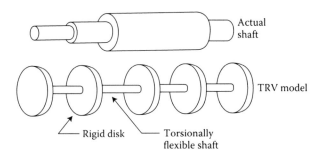

FIGURE 3.1 Multielement TRV model for a single-shaft rotor.

determine whether or not the controller gain must be reduced to avoid TRV-controller system dynamic instabilities that could potentially excite one or more of the rotor system's many lightly damped torsional vibration modes. Although long recognized, this is now a timely point to emphasize because of the significantly increasing trend to use variable speed motors with feedback speed control. This trend is fostered by the introduction of many new types of variable speed electric motors utilizing microprocessor speed controllers.

Assemblage of the mass and stiffness matrices for a single rotor such as in Figure 3.1 follows the same approach used in Chapter 2 for LRV analysis models. The basic TRV finite element used here has the identical geometric features as its LRV counterpart shown in Figure 2.8. However, the TRV model is postulated with torsional "twistability," not the beam-bending flexibility of the LRV model. The basic TRV finite element is shown in Figure 3.2 and is comprised of a uniform diameter shaft element connecting two rigid disks.

3.3.1 Lumped and Distributed Mass Matrices

As shown in Figure 3.2, the basic TRV finite element has only two DOFs and is thus simpler than its 8-DOF LRV counterpart shown in Figure 2.8. Both the *lumped mass* and *distributed mass* shaft element matrices are presented here. The *consistent mass* matrix for the 2-DOF element in Figure 3.2 is the same as its *distributed mass* matrix. This is because the shaft element's torsional deflection is a linear (shape) function of axial position between its two end point stations, and thus is *consistent* with the piecewise linear variation of acceleration implicit in the *distributed mass* formulation.

3.3.1.1 Lumped Mass Matrix

In this approach, it is assumed that for each uniform-diameter shaft element, half its polar moment of inertia, $I^{(s)}$, is lumped at each of the

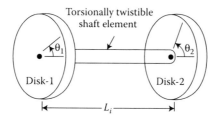

FIGURE 3.2 Rotor torsional finite-element 2-DOF building block.

element's two end points (stations). Implicit in this approximation is an incremental step change in angular acceleration for each shaft element at its axial midpoint. That is, the continuous axial variation in angular acceleration is approximated by a series of small discrete step changes. A concentrated (nonstructural) polar moment of inertia, $I^{(d)}$, may be optionally added at any rotor station as appropriate to model gears, couplings, impellers, turbine disks, pulleys, flywheels, thrust-bearing collars, nonstiffening motor and generator rotor components, and so on. The complete *single-rotor* ("sr") *lumped* ("l") *mass matrix* is thus a diagonal matrix, given as follows:

$$[M]_{sr}^l =$$

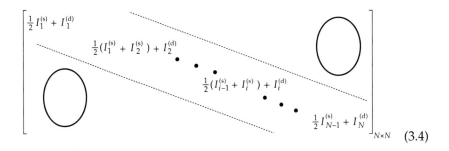

$$(3.4)$$

N = No. of rotor stations = No. of DOFs = No. of elements + 1.
Subscript on $I^{(s)}$ = Element Number, Subscript on = $I^{(d)}$ Station Number.

3.3.1.2 Distributed Mass Matrix

As similarly explained in Section 2.3 for LRV models, the underlying assumption here is that the angular acceleration of each shaft element about it axis varies linearly over its own length. Therefore, model resolution accuracy is better with the *distributed mass* formulation than with the *lumped mass* formulation. The better the model resolution accuracy, the fewer the number of finite elements (or DOFs) needed to accurately characterize the relevant modes of the actual continuous media system using a discrete model. Consistent with the assumption that angular acceleration varies linearly between rotor stations, the angular velocity then also must vary linearly between rotor stations. The instantaneous TRV kinetic energy stored in the ith single shaft element can be formulated from the integration of kinetic energy distributed over the ith element's length, similar to Equation 2.39 for LRV radial velocity components,

Torsional Rotor Vibration Analysis Models

as follows:

$$T_i^{(s)} = \frac{1}{2}\frac{I_i^{(s)}}{L_i}\int_0^{L_i}(\dot{\theta}+\omega)^2\,dz \qquad (3.5)$$

Substituting a linearly varying $\dot{\theta}$ and $\omega \equiv$ constant into Equation 3.5 yields the ith shaft element's torsional kinetic energy terms associated with the θ_i and θ_{i+1} Lagrange equations for the ith and $(i\text{th}+1)$ rotor stations. This yields the following results, consistent with Equation 3.2 (i.e., $\omega \equiv$ constant, $\therefore \dot{\omega}=0$).

$$\frac{d}{dt}\left(\frac{\partial T}{\partial \dot{\theta}_i}\right) = \tfrac{1}{3}I_i^{(s)}\ddot{\theta}_i + \tfrac{1}{6}I_i^{(s)}\ddot{\theta}_{i+1}$$
$$\frac{d}{dt}\left(\frac{\partial T}{\partial \dot{\theta}_{i+1}}\right) = \tfrac{1}{6}I_i^{(s)}\ddot{\theta}_i + \tfrac{1}{3}I_i^{(s)}\ddot{\theta}_{i+1} \qquad (3.6)$$

The complete *single-rotor distributed mass matrix* is thus a tridiagonal matrix, as follows. Note the optional $I^{(d)}$ at each station, just as in Equation 3.4.

Polar moment-of-inertia formulas for *shaft elements* and *concentrated disks* are the same as given at the beginning of Section 2.3 for LRV models.

3.3.2 Stiffness Matrix

The TRV stiffness matrix $[K]_{ff}$ for a *free–free* single rotor, such as shown in Figure 3.1, is quite simple to formulate. It is the torsional equivalent of the type of translational system shown in Figure 1.8. That is, each rotor mass station has elastic coupling only to its immediate neighbors. Therefore, the single-rotor TRV stiffness matrix, shown as follows, is tridiagonal just as shown for the system in Figure 1.8.

$[M]_{sr}^d =$

$$\begin{bmatrix} \tfrac{1}{3}I_1^{(s)}+I_1^{(d)} & \tfrac{1}{6}I_1^{(s)} & & & & & 0 \\ \tfrac{1}{6}I_1^{(s)} & \tfrac{1}{3}(I_1^{(s)}+I_2^{(s)})+I_2^{(d)} & \tfrac{1}{6}I_2^{(s)} & & & & \\ & & \tfrac{1}{6}I_{i-1}^{(s)} & \tfrac{1}{3}(I_{i-1}^{(s)}+I_i^{(s)})+I_i^{(d)} & \tfrac{1}{6}I_i^{(s)} & & \\ & & & & & \tfrac{1}{6}I_{N-1}^{(s)} & \tfrac{1}{3}I_{N-1}^{(s)}+I_N^{(d)} \\ 0 & & & & & & \end{bmatrix}_{N\times N}$$

$$(3.7)$$

$[K]_{ff} =$

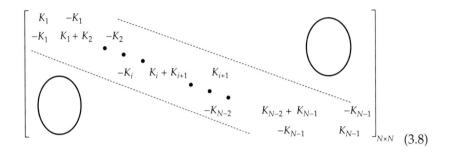

$$\begin{bmatrix} K_1 & -K_1 & & & & & & \\ -K_1 & K_1+K_2 & -K_2 & & & & & \\ & \ddots & \ddots & \ddots & & & & \\ & & -K_i & K_i+K_{i+1} & K_{i+1} & & & \\ & & & \ddots & \ddots & \ddots & & \\ & & & & -K_{N-2} & K_{N-2}+K_{N-1} & -K_{N-1} \\ & & & & & -K_{N-1} & K_{N-1} \end{bmatrix}_{N \times N}$$ (3.8)

Subscript on K = Element Number.

Shaft element torsional stiffness:

$$K = \frac{\pi(d_o^4 - d_i^4)G}{32L}$$

where d_o is the outside diameter, d_i is the inside diameter (optional concentric hole), L is the element length, and G is the element material modulus of rigidity.

The term "free–free" refers to a model that is both *free* of external forces or torques and *free* of connections to the inertial frame of reference. The free–free rotor stiffness matrix given in Equation 3.8 is a singular matrix because it contains no torsional stiffness connections to the inertial frame of reference. Therefore, the complete model obtained by combining the stiffness matrix of Equation 3.8 with either the *lumped* or *distributed* mass matrix in Equations 3.4 and 3.5, respectively, is a system with one rigid-body mode (refer to Table 1.1, Case 4). All the generalized coordinates, velocities, and accelerations here (i.e., θ_i, $\dot{\theta}_i$, $\ddot{\theta}_i$) are coaxial; thus the model is one-dimensional even though it is multi-DOF. This is why it has only one rigid-body mode and why only one rigid or stiffness connection to the inertial frame of reference, added at a single mass station, is required to make the stiffness matrix nonsingular. With the addition of one or more connections to ground, the model will not have a rigid-body mode. By comparison, general LRV models (Chapter 2) have four dimensions of motion $(x, y, \theta_x, \text{and } \theta_y)$. LRV models thus can have as many as four rigid-body modes (two in the x–z plane and two in the y–z plane) if the rotor is completely unconnected to ground. Thus, for an LRV stiffness matrix to be nonsingular, there must be a minimum of two stiffness connections in the x–z plane plus two in the y–z plane, that is, like having a minimum of two radial bearings. Discussion of this aspect of LRV models is also given in Section 2.3.9.

The general single-rotor TRV stiffness matrix $[K]_{sr}$ is the sum of the *free–free* stiffness matrix $[K]_{ff}$ and a diagonal matrix $[K]_c$ containing the optional one or more stiffness connections to the inertial frame, as follows. A rigid connection to the inertial frame has no DOF at the connection point.

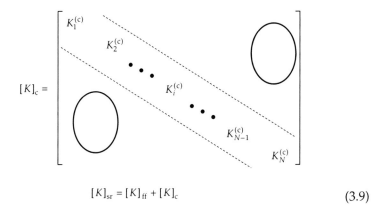

$$[K]_{sr} = [K]_{ff} + [K]_c \qquad (3.9)$$

Equations of motion for the undamped single-rotor model are then as follows:

$$[M]_{sr}\{\ddot{\theta}\} + [K]_{sr}\{\theta\} = \{m(t)\} \qquad (3.10)$$

Here, $\{m(t)\}$ contains any *externally applied* time-dependent torque components, such as to model synchronous generators of turbo-generators during severe electrical disturbances like *high-speed reclosure* (HSR) of circuit breakers after fault clearing of transmission lines leaving power stations. Of course, to compute the undamped natural frequencies and corresponding mode shapes, only the mass and stiffness matrices are utilized.

3.4 Coupled Rotors

The single-rotor mass and stiffness matrices developed in the previous section form the basic model building blocks for TRV coupled-rotor models. One of the many advantages of assembling the equations of motion in matrix form is the ease with which modeled substructures can be joined to assemble the complete equations of motion of a multisubstructure system.

Coaxial same-speed coupled-rotor configurations are the simplest TRV coupled-rotor models to assemble, and are treated here first. Model construction for a broader category of coupled-rotor TRV systems has additional inherent complexities stemming from the following three features:

- Coupled rotors may have *speed ratios* other than 1:1.
- Torsional coupling may be either *rigid* (e.g., gears) or *flexible* (e.g., belt).
- System may be *branched* instead of *unbranched*.

An understanding of these complexities can be obtained by following the formulation details of their TRV model constructions, which are presented subsequently in this section. The handling of these complexities is simplified by the fact that correct TRV equations of motion can be derived as though the rotors are not spinning, as shown in Section 3.2, Equation 3.3, that is, modeled as though the coupled-rotor machine is not running.

3.4.1 Coaxial Same-Speed Coupled Rotors

This is a quite common configuration category and the most typical case involves two single rotors joined by a so-called *flexible coupling*. Assembling the mass and stiffness matrices for this case is quite simple, as shown by the following equations. The total mass matrix can be expressed as follows:

$$[M] = \begin{bmatrix} [M_1]_{sr} & [0] \\ [0] & [M_2]_{sr} \end{bmatrix} \quad (3.11)$$

Usually, a flexible coupling can be adequately modeled by two concentrated polar moment-of-inertias connected by a torsional spring stiffness. The two concentrated coupling inertias $I_1^{(c)}$ and $I_2^{(c)}$ are added as concentrated inertias to the last diagonal element of $[M_1]_{sr}$ and the first diagonal element of $[M_2]_{sr}$, respectively. To assemble the total stiffness matrix, the equivalent torsional spring stiffness $K^{(c)}$ of the coupling is used to join the respective single-rotor stiffness matrices of the two rotors, as follows:

$$[K] = \begin{bmatrix} [K_1]_{sr} & [0] \\ [0] & [K_2]_{sr} \end{bmatrix} + [K_c]_{2\times 2} = \begin{bmatrix} \begin{bmatrix} K_{ij}^{(1)} & & \\ & K^{(c)} & -K^{(c)} \\ & -K^{(c)} & K^{(c)} \\ & & & K_{ij}^{(2)} \end{bmatrix}_{2\times 2} \end{bmatrix} \quad (3.12)$$

The complete equations of motion for two coaxially coupled rotors are then expressible in the same matrix format as Equation 3.10, that is, $[M]\{\ddot{\theta}\} + [K]\{\theta\} = \{M(t)\}$. For three or more simply connected same-speed flexible-coupled rotors, the above process is taken to its natural extension.

3.4.2 Unbranched Systems with Rigid and Flexible Connections

For rotors coupled by *gears*, the appropriate model for TRV coupling could be *flexible* or *rigid*, depending on the particulars of a given application. When the gear teeth contact and gear wheel combined equivalent torsional stiffness is much greater than other torsional stiffnesses in the system, it is best to model the geared connection as *rigid* so as to avoid any computational inaccuracies stemming from large disparities in connecting stiffnesses. The configuration shown in Figure 3.3 contains both a geared connection and a pulley–belt connection of a three-shaft assembly. Although the shown shafts are mutually parallel, this is not a restriction for the TRV models developed here.

The configuration shown in Figure 3.3 is categorized as an *unbranched* system. The full impact of this designation is fully clarified in the next subsection that treats *branched* systems. Whether a TRV system is *branched* or *unbranched* made a lot of difference in computer programming complexity when older solution algorithms (e.g., *transfer matrix* method) were used. With the modern finite-element-based matrix approaches used exclusively in this book, additional programming complexities with branched systems are not nearly so significant as with the older algorithmic methods that were better matched to the relatively limited memory of early generation computers. A coupled-rotor TRV system is defined here as *unbranched* when each of the coupled rotors in the drive train is connected to the *next* or *previous* rotor only at its two end stations (i.e., first or last), is not connected to more than one rotor at either end station, and is not connected to the same rotor at both end stations. When this is the case, the stiffness matrix for the coupled system is *tridiagonal*, just like the individual rotor stiffness matrices. The simplest example of this is the stiffness matrix for *coaxial same-speed coupled rotors*, Equation 3.12, which is *tridiagonal*. The Figure 3.3 system also clearly fits the definition of an *unbranched* TRV system, and is

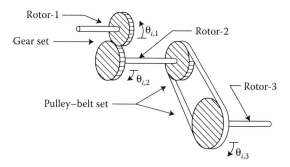

FIGURE 3.3 Unbranched three-rotor system with a gear set and a pulley–belt set.

here used to show the formulation for TRV equation-of-motion matrices of *arbitrary-speed-ratio rotors with rigid and flexible unbranched connections.*

3.4.2.1 Rigid Connections

The gear set of the system in Figure 3.3 will be assumed to be torsionally much stiffer than other torsional flexibilities of the system, and thus taken as perfectly *rigid*. The TRV angular displacements of the two gears are then constrained to have the same ratio as the nominal speed ratio of the two-gear set. Thus, one equation of motion must be eliminated either from rotor-1 (last station) or rotor-2 (first station). Here the equation of motion for the first station of rotor-2 is absorbed into the equation of motion for last station of rotor-1. The concentrated inertia of the rotor-2 gear is thus transferred to the rotor-1 station with the mating gear. Defining n_{21} as the speed ratio of rotor-2 to rotor-1, and $\theta_{i,j}$ as ith angular coordinate of the jth rotor, the TRV angular coordinate of the rotor-2 gear is expressed in terms of the rotor-1 gear's coordinate, as follows. Note the opposite positive sense for $\theta_{i,2}$, Figure 3.3.

$$\theta_{1,2} = n_{21}\theta_{N_1,1} \tag{3.13}$$

where N_1 = number of stations on rotor-1 = station number rotor-1's last station.

The TRV kinetic energy of the two rigidly coupled gears is thus expressible as follows:

$$T_{12}^{\text{gears}} = \tfrac{1}{2}I_{N_1,1}^{(d)}\dot{\theta}_{N_1,1}^2 + \tfrac{1}{2}I_{1,2}^{(d)}\dot{\theta}_{1,2}^2 = \tfrac{1}{2}\left(I_{N_1,1}^{(d)} + n_{21}^2 I_{1,2}^{(d)}\right)\dot{\theta}_{N_1,1}^2 \tag{3.14}$$

where $I_{i,j}^{(d)} \equiv$ nonstructural concentrated inertia for the ith station of the jth rotor.

The combined TRV nonstructural inertia of the two gears is thus lumped in the motion equation for station N_1 of rotor-1 as follows:

$$\frac{d}{dt}\left(\frac{\partial T_{12}^{\text{gears}}}{\partial \dot{\theta}_{N_1,1}}\right) = \left(I_{N_1,1}^{(d)} + n_{21}^2 I_{1,2}^{(d)}\right)\ddot{\theta}_{N_1,1} \tag{3.15}$$

As previously detailed in Section 3.3, shaft element structural mass is included by using either the *lumped mass* or the *distributed mass* approach. For the *lumped mass* approach, the shaft element connecting the rotor-2 first and second stations has half its inertia lumped at the last station of rotor-1 (with the n_{21}^2 multiplier) and half its inertia lumped at the rotor-2 second station. For the *distributed mass* approach (in TRV, same as the *consistent mass* approach), the shaft element kinetic energy is integrated along the first shaft element of rotor-2, as is similarly shown in Equations 3.5 and 3.6.

That is, postulating a linear variation of angular velocity along the shaft element, and substituting from Equation 3.13 for $\theta_{1,2}$, yields the following equation for the TRV kinetic energy of shaft element-1 of rotor-2.

$$T_{1,2}^{(s)} = \frac{I_{1,2}^{(s)}}{6}\left(n_{21}^2 \dot{\theta}_{N_1,1}^2 + n_{21}\dot{\theta}_{N_1,1}\dot{\theta}_{2,2} + \dot{\theta}_{2,2}^2\right) \tag{3.16}$$

The following equation-of-motion *distributed mass* inertia contributions of this shaft element to the stations that bound it are accordingly obtained:

$$\frac{d}{dt}\left(\frac{\partial T_{1,2}^{(s)}}{\partial \dot{\theta}_{N_1,1}}\right) = \tfrac{1}{3}n_{21}^2 I_{1,2}^{(s)}\ddot{\theta}_{N_1,1} + \tfrac{1}{6}n_{21}I_{1,2}^{(s)}\ddot{\theta}_{2,2}$$

$$\frac{d}{dt}\left(\frac{\partial T_{1,2}^{(s)}}{\partial \dot{\theta}_{2,2}}\right) = \tfrac{1}{6}n_{21}I_{1,2}^{(s)}\ddot{\theta}_{N_1,1} + \tfrac{1}{3}I_{1,2}^{(s)}\ddot{\theta}_{2,2} \tag{3.17}$$

where $I_{i,j}^{(s)} \equiv$ structural inertia for the ith shaft element of the jth rotor.

Postulating a *rigid connection* between the two gears in Figure 3.3 eliminates one DOF (i.e., the first station of rotor-2). The corresponding detailed formulations needed to merge the rotor-1 and rotor-2 mass matrices are contained in Equations 3.13 through 3.17. Merging the rotor-1 and rotor-2 stiffness matrices must also incorporate the same elimination of one DOF. Specifically, shaft element-1 of rotor-2 becomes a direct torsional stiffness between the last station of rotor-1 and the second station of rotor-2. This stiffness connection is almost as though these two stations were adjacent to each other on the same rotor, except for the speed-ratio effect. The easiest way to formulate the details for merging rotor-1 and rotor-2 stiffness matrices is to use the potential energy term of the Lagrange formulation for the equations of motion, as follows (see Equation 2.50):

$$V_{1,2} = \tfrac{1}{2}K_{1,2}(\theta_{1,2} - \theta_{2,2})^2 \tag{3.18}$$

where $V_{i,j} \equiv$ TRV potential energy stored in ith shaft element of the jth rotor and $K_{i,j} \equiv$ Torsional stiffness of the ith shaft element of the jth rotor.

Substituting from Equation 3.13 for $\theta_{1,2}$ into Equation 3.18 thus leads to the following terms for merging rotor-1 and rotor-2 stiffness matrices:

$$\frac{\partial V_{1,2}}{\partial \theta_{N_1,1}} = K_{1,2}(n_{21}^2 \theta_{N_1,1} - n_{21}\theta_{2,2})$$

$$\frac{\partial V_{1,2}}{\partial \theta_{2,2}} = K_{1,2}(-n_{21}\theta_{N_1,1} + \theta_{2,2}) \tag{3.19}$$

Before implementing the terms for connecting rotor-1 to rotor-2, the detailed formulations for connecting rotor-2 to rotor-3 are first developed so that the mass and stiffness matrices for the complete Figure 3.3 system can be assembled.

3.4.2.2 Flexible Connections

The pulley–belt set in Figure 3.3 connecting rotor-2 to rotor-3 is assumed to be a *flexible connection* and thus no DOF is eliminated, contrary to the *rigid connection* case. A flexible connection does not entail modifications to the mass matrix of either of the two flexibly connected rotors. Only the stiffness of the belt must be added to the formulation to model the flexible connection. It is assumed that both straight spans of the belt connecting the two pulleys are in tension, and thus both spans are assumed to have the same tensile stiffness, k_b, and their TRV stiffness effects are additive like two springs in parallel. The easiest way to formulate the merging rotor-2 and rotor-3 stiffness matrices is to use the potential energy term of the Lagrange formulation, as shown in Equation 2.50. To model *gear-set flexibility*, replace $2k_b$ with pitch-line k_g and define R_j as jth pitch radius, not jth pulley radius.

$$V_b = \tfrac{1}{2}(2k_b)(\theta_{N_2,2}R_2 - \theta_{1,3}R_3)^2$$
$$= k_b(\theta_{N_2,2}^2 R_2^2 - 2\theta_{N_2,2}\theta_{1,3}R_2 R_3 + \theta_{1,3}^2 R_3^2) \tag{3.20}$$

$$\frac{\partial V_b}{\partial \theta_{N_2,2}} = 2k_b(\theta_{N_2,2}R_2^2 - \theta_{1,3}R_2 R_3)$$

$$\frac{\partial V_b}{\partial \theta_{1,3}} = 2k_b(-\theta_{N_2,2}R_2 R_3 + \theta_{1,3}R_2^2) \tag{3.21}$$

where $R_j \equiv$ pulley radius for the jth rotor, $V_b \equiv$ TRV potential energy in belt and $N_2 =$ Number of stations on rotor-2 = Station number rotor-2's last station.

At this point, all components needed to write the equations of motion for the TRV system in Figure 3.3 are ready for implementation.

3.4.2.3 Complete Equations of Motion

For the individual rotors, the *distributed mass* approach is used here simply because it is better than the lumped mass approach, as discussed earlier. Thus, Equation 3.7 is applied for construction of the three single-rotor mass matrices, $[M_1]$, $[M_2]$, and $[M_3]$. Equation 3.9 is used to construct the three single-rotor stiffness matrices $[K_1]$, $[K_2]$, and $[K_3]$, adding any *to-ground* flexible connections to the *free–free* TRV stiffness matrices from

Equation 3.8. At this point, constructing the total system mass and stiffness matrices only entails catenating the single-rotor matrices and implementing the already developed modifications to the matrices dictated by the rigid and flexible connections. Employing modifications extracted from Equations 3.15 and 3.17, $[M_1]$ is augmented as follows. Superscript "rc" refers to *rigid connection*.

$$[M_1^*] = [M_1] + [M_1^{rc}], \text{ where } [M_1^{rc}] = \begin{bmatrix} 0 & 0 \\ \hline 0 & n_{21}^2 I_{1,2}^{(d)} + \frac{1}{3} n_{21}^2 I_{1,2}^{(s)} \end{bmatrix}_{N_1 \times N_1} \quad (3.22)$$

All elements in $[M_1^{rc}]$ are zero except element (N_1, N_1).

Eliminating its first row and first column, $[M_2]$ is reduced to $[M_2^*]$. The complete system mass matrix can be assembled at this point, catenating $[M_1^*]$, $[M_2^*]$, and $[M_3^*]$, and adding the cross-coupling terms contained in Equation 3.17, as follows:

$$[M] = \begin{bmatrix} [M_1^*]_{M_{cc}} & & \\ & M_{cc}[M_2^*] & \\ & & [M_3] \end{bmatrix}_{N \times N} \quad (3.23)$$

Subscript "cc" refers to *cross-coupling*.

$$M_{N_1, N_1+1} = M_{N_1+1, N_1} = \tfrac{1}{6} n_{21} I_{1,2}^{(s)} \equiv M_{cc}, \quad N = N_1 + (N_2 - 1) + N_3$$

The complete system stiffness matrix $[K]$ is similarly constructed. Employing modifications extracted from Equation 3.19, $[K_1]$ is augmented as follows:

$$[K_1^*] = [K_1] + [K_1^{rc}], \text{ where } [K_1^{rc}] = \begin{bmatrix} 0 & 0 \\ \hline 0 & n_{21}^2 K_{1,2} \end{bmatrix}_{N_1 \times N_1} \quad (3.24)$$

All elements in $[K_1^{rc}]$ are zero except element (N_1, N_1).

Eliminating its first row and first column, $[K_2]$ is reduced to $[K_2^\#]$, which is augmented to form $[K_2^*]$ as follows. Superscript "fc" refers to *flexible connection*.

$$[K_2^*] = [K_2^\#] + [K_2^{fc}], \text{ where } [K_2^{fc}] = \begin{bmatrix} 0 & 0 \\ \hline 0 & 2k_b R_2^2 \end{bmatrix}_{N_2^* \times N_2^*} \quad (3.25)$$

All elements in $[K_2^{fc}]$ are zero except element (N_2^*, N_2^*).

$N_2^* = N_2 - 1$

$[K_3]$ is augmented to form $[K_3^*]$ as follows.

$$[K_3^*] = [K_3] + [K_3^{fc}], \text{ where } [K_3^{fc}] = \begin{bmatrix} 2k_bR_3^2 & 0 \\ \hline 0 & 0 \end{bmatrix}_{N_3 \times N_3} \quad (3.26)$$

All elements in $[K_3^{fc}]$ are zero except element (1,1).

The complete system stiffness matrix can be assembled at this point, catenating $[K_1^*]$, $[K_2^*]$, and $[K_3^*]$, and adding the cross-coupling terms contained in Equations 3.19 and 3.21, as follows:

$$[K] = \begin{bmatrix} \begin{bmatrix} K_1^* \end{bmatrix} & & & & \\ & K_{cc}^{1,2} & & & \\ & & K_{cc}^{1,2} & & \\ & & & \begin{bmatrix} K_2^* \end{bmatrix} & & \\ & & & & K_{cc}^{2,3} & \\ & & & & & K_{cc}^{2,3} \\ & & & & & & \begin{bmatrix} K_3^* \end{bmatrix} \end{bmatrix}_{N \times N} \quad (3.27)$$

$K_{cc}^{1,2} \equiv -n_{21}K_{1,2}$; extracted from Equation 3.19

$K_{cc}^{2,3} \equiv -2k_bR_2R_3$; extracted from Equation 3.21

The complete TRV equations of motion for the Figure 3.3 system are thus expressible in the same matrix format as Equation 3.10, that is, $[M]\{\ddot{\theta}\} + [K]\{\theta\} = \{m(t)\}$. The $[M]$ and $[K]$ matrices here are tridiagonal, which is consistent with the designation of *unbranched*. The formulations developed here are readily applicable to any *unbranched* TRV system of coupled rotors.

3.4.3 Branched Systems with Rigid and Flexible Connections

The system shown in Figure 3.4 bears a close similarity to the system in Figure 3.3, except that its gear set and pulley set are located inboard of their respective rotor ends, each of these connections thus making it a *branched* system. Its $[M]$ and $[K]$ matrices are therefore not tridiagonal, as now shown.

Torsional Rotor Vibration Analysis Models

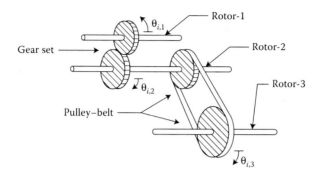

FIGURE 3.4 Branched three-rotor system with a gear set and a pulley–belt set.

Constructing mass and stiffness matrices for the Figure 3.4 system follows the same procedures of the previous subsection for *unbranched* systems. For the individual rotors, the *distributed mass* approach is again used, applying Equation 3.7 for construction of the single-rotor mass matrices, $[M_1]$, $[M_2]$, and $[M_3]$. Also, Equation 3.9 is again used to construct the single-rotor stiffness matrices $[K_1]$, $[K_2]$, and $[K_3]$, adding any *to-ground* flexible connections to the *free–free* TRV stiffness matrices from Equation 3.8. Using the standard substructuring approach previously applied here to *unbranched* systems, constructing the $[M]$ and $[K]$ matrices for the Figure 3.4 system is only slightly more involved than for the Figure 3.3 system.

3.4.3.1 Rigid Connections

The two gears joining rotor-1 to rotor-2 are assumed here to be a perfectly rigid torsional connection between the two rotors. Accordingly, the equation of motion for the rotor-2 gear station (N_{G2}) is absorbed into the equation of motion for the rotor-1 gear station (N_{G1}), with the eliminated rotor-2 gear DOF ($\theta_{N_{G2},2}$) expressed by the constant speed ratio (n_{21}) times the rotor-1 gear coordinate ($\theta_{N_{G1},1}$), as follows ($n_{21} \equiv \omega_2/\omega_1$).

$$\theta_{N_{G2},2} = n_{21}\theta_{N_{G1},1} \tag{3.28}$$

Similar to Equation 3.14, the TRV kinetic energy of the two rigidly coupled gears is thus expressible as follows:

$$T_{12}^{\text{gears}} = \tfrac{1}{2} I_{N_{G1},1}^{(d)} \dot{\theta}_{N_{G1},1}^2 + \tfrac{1}{2} I_{N_{G2},2}^{(d)} \dot{\theta}_{N_{G2},2}^2 = \tfrac{1}{2}\left(I_{N_{G1},1}^{(d)} + n_{21}^2 I_{N_{G2},2}^{(d)} \right) \dot{\theta}_{N_{G1},1}^2 \tag{3.29}$$

The combined TRV nonstructural inertia of the two gears is thus lumped in the motion equation for station N_{G1} of rotor-1 as follows:

$$\frac{d}{dt}\left(\frac{\partial T_{12}^{gears}}{\partial \dot\theta_{N_{G1},1}}\right) = \left(I_{N_{G1},1}^{(d)} + n_{21}^2 I_{N_{G2},2}^{(d)}\right)\ddot\theta_{N_{G1},1} \qquad (3.30)$$

Using the *distributed mass* approach, the TRV kinetic energy of the rotor-2 shaft element just to the left of rotor-2's station N_{G2} and of the element just to the right of station N_{G2} are derived to be the following, similar to Equation 3.16:

$$T_{N_{G2}-1,2}^{(s)} = \frac{I_{N_{G2}-1,2}^{(s)}}{6}\left(n_{21}^2\dot\theta_{N_{G1},1}^2 + n_{21}\dot\theta_{N_{G1},1}\dot\theta_{N_{G2}-1,2} + \dot\theta_{N_{G2}-1,2}^2\right)$$

$$T_{N_{G2},2}^{(s)} = \frac{I_{N_{G2},2}^{(s)}}{6}\left(n_{21}^2\dot\theta_{N_{G1},1}^2 + n_{21}\dot\theta_{N_{G1},1}\dot\theta_{N_{G2}+1,2} + \dot\theta_{N_{G2}+1,2}^2\right) \qquad (3.31)$$

The following *distributed mass* matrix contributions of these two rotor-2 shaft elements are thus obtained, similar to Equation 3.17:

$$\frac{d}{dt}\left(\frac{\partial T_{N_{G2}-1,2}^{(s)}}{\partial \dot\theta_{N_{G1},1}}\right) = \tfrac{1}{3}n_{21}^2 I_{N_{G2}-1,2}^{(s)}\ddot\theta_{N_{G1},1} + \tfrac{1}{6}n_{21} I_{N_{G2}-1,2}^{(s)}\ddot\theta_{N_{G2}-1,2}$$

$$\frac{d}{dt}\left(\frac{\partial T_{N_{G2}-1,2}^{(s)}}{\partial \dot\theta_{N_{G2}-1,2}}\right) = \tfrac{1}{6}n_{21} I_{N_{G2}-1,2}^{(s)}\ddot\theta_{N_{G1},1} + \tfrac{1}{3} I_{N_{G2}-1,2}^{(s)}\ddot\theta_{N_{G2}-1,2}$$

$$\frac{d}{dt}\left(\frac{\partial T_{N_{G2},2}^{(s)}}{\partial \dot\theta_{N_{G1},1}}\right) = \tfrac{1}{3}n_{21}^2 I_{N_{G2},2}^{(s)}\ddot\theta_{N_{G1},1} + \tfrac{1}{6}n_{21} I_{N_{G2},2}^{(s)}\ddot\theta_{N_{G2}+1,2}$$

$$\frac{d}{dt}\left(\frac{\partial T_{N_{G2},2}^{(s)}}{\partial \dot\theta_{N_{G2},2}}\right) = \tfrac{1}{6}n_{21} I_{N_{G2},2}^{(s)}\ddot\theta_{N_{G1},1} + \tfrac{1}{3} I_{N_{G2},2}^{(s)}\ddot\theta_{N_{G2}+1,2} \qquad (3.32)$$

Equations 3.30 and 3.32 contain all the terms needed to merge rotor-1 and rotor-2 mass matrices.

The following formulation details for merging rotor-1 and rotor-2 stiffness matrices are developed using the potential energy term of the Lagrange formulation, the same procedure as used to develop Equations 3.18 and 3.19.

$$V_{N_{G2}-1,2} = \tfrac{1}{2}K_{N_{G2}-1,2}(\theta_{N_{G2},2} - \theta_{N_{G2}-1,2})^2$$

$$V_{N_{G2},2} = \tfrac{1}{2}K_{N_{G2},2}(\theta_{N_{G2}+1,2} - \theta_{N_{G2},2})^2 \qquad (3.33)$$

Substituting from Equation 3.28 for $\theta_{NG2,2}$ into Equation 3.33 thus leads to the following terms for merging rotor-1 and rotor-2 stiffness matrices:

$$\frac{\partial V_{NG2-1,2}}{\partial \theta_{NG1,1}} = K_{NG2-1,2}\left(n_{21}^2 \theta_{NG1,1} - n_{21}\theta_{NG2-1,2}\right)$$

$$\frac{\partial V_{NG2-1,2}}{\partial \theta_{NG2-1,2}} = K_{NG2-1,2}\left(-n_{21}\theta_{NG1,1} + \theta_{NG2-1,2}\right) \quad (3.34)$$

$$\frac{\partial V_{NG2,2}}{\partial \theta_{NG1,1}} = K_{NG2,2}\left(n_{21}^2 \theta_{NG1,1} - n_{21}\theta_{NG2+1,2}\right)$$

$$\frac{\partial V_{NG2,2}}{\partial \theta_{NG2,2}} = K_{NG2,2}\left(-n_{21}\theta_{NG1,1} + \theta_{NG2+1,2}\right)$$

3.4.3.2 Flexible Connections

As a torsionally flexible connection between rotor-2's station N_{P2} and rotor-3's station N_{P3}, the pulley–belt set in Figure 3.4 needs no corresponding modifications to the mass matrix of either of the two rotors. Following the identical procedure used to develop Equations 3.20 and 3.21, the formulation details for merging the rotor-2 and rotor-3 stiffness matrices are as follows:

$$V_b = \tfrac{1}{2}(2k_b)\left(\theta_{NP2,2}R_2 - \theta_{NP3,3}R_3\right)^2$$
$$= k_b\left(\theta_{NP2,2}^2 R_2^2 - 2\theta_{NP2,2}\theta_{NP3,3}R_2 R_3 + \theta_{NP3,3}^2 R_3^2\right) \quad (3.35)$$

$$\frac{\partial V_b}{\partial \theta_{NP2,2}} = 2k_b\left(\theta_{NP2,2}R_2^2 - \theta_{NP3,3}R_2 R_3\right)$$

$$\frac{\partial V_b}{\partial \theta_{NP3,3}} = 2k_b\left(-\theta_{NP2,2}R_2 R_3 + \theta_{NP3,3}R_3^2\right) \quad (3.36)$$

As explained in Equation 3.20, this formulation is applicable to *flexible gear sets*.

At this point, all components needed to write the equations of motion for the TRV system in Figure 3.4 are ready for implementation.

3.4.3.3 Complete Equations of Motion

Assembling $[M]$ and $[K]$ for the complete system in Figure 3.4 follows the same procedures used in the previous subsection for *unbranched* systems. The main difference with *branched* systems is that the matrices of the respective coupled rotors are not just catenated with tridiagonal splicing. With *branched* systems, the catenated matrices are instead spliced at their respective intermediate connection coordinates; hence the *bandwidths* of [M] and [K] are not limited to being tridiagonal. Directions for the attendant matrix bookkeeping are clearly indicated by Equations 3.30, 3.32, and 3.34 for *rigid connections* between rotors, and Equation 3.36 for *flexible connections* between rotors. In the interest of space, the complete [M] and [K] matrices for Figure 3.4 system are not written here.

3.5 Semidefinite Systems

Some of the problem exercises at the end of this chapter entail what are called *semidefinite systems*, for example, Problems 1 and 2. Synonymous with this designation is a system having a singular stiffness matrix, or a system with a natural frequency equal to zero. When a natural frequency is zero, the corresponding mode shape is a rigid body one. The other modes of such a system are commonly referred to as *free–free modes*, which are relative to the rigid-body mode(s). Modern finite-element codes automatically handle systems with zero natural frequencies. However, for semidefinite systems when setting up motion equations external to some automatic code, one may first reduce the number of motion equations by the degree of singularity of the stiffness matrix. That is, the rigid-body modes are constraint equations that reduce the number of generalized coordinates needed to express the motion equations relative to the rigid-body modes. The resulting reduced stiffness matrix is then nonsingular, that is, $|K| \neq 0$. Any modern text devoted entirely to vibration theory will provide a detailed treatment on semidefinite systems and their handling.

3.6 Examples

3.6.1 High-Capacity Fan for Large Altitude Wind Tunnel

The large two-stage fan illustrated in Figure 3.5 has an overall length of 64 m (210 ft) and requires several thousand horsepower from its two electric drive motors to operate at capacity. This machine fits into the previously designated TRV category of *coaxial same-speed coupled rotors*. The drive-motor portion of this machine is constructed with an extended shaft length adjacent to each motor so that the two motor stators can easily be moved horizontally to expose motor internals for inspection and service. This need to provide unobstructed space adjacent to each motor, as shown in Figure 3.5, adds considerably to the motor shaft overall length. This

Torsional Rotor Vibration Analysis Models

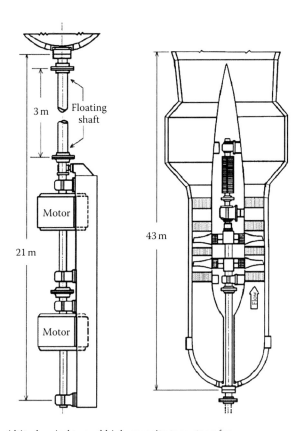

FIGURE 3.5 Altitude wind tunnel high-capacity two-stage fan.

complete turbo-machine is substantially longer, but of substantially less power and torque (i.e., smaller diameter shafting) than typical large steam turbo-generator sets. These factors combine to produce a quite torsionally flexible rotor, giving rise to the highest potential for serious TRV resonance problems that must be addressed at the design stage for the machine to operate successfully.

Since this fan powers an *altitude wind tunnel*, the internal air pressure of the wind tunnel, and therefore of the fan, is controlled to pressures below outside ambient pressure. A close inspection of the fan portion of this machine reveals that the drive shaft passes into the fan/wind tunnel envelope within a coaxial nonrotating cylindrical section which is internally vented to the outside ambient pressure. The fan shaft and its bearings are thus at outside ambient pressure. Controlled-leakage dynamic seals are therefore located on the fan-blade hubs, thus requiring axially accurate positioning of the seal parts on the rotor with respect to close-proximity nonrotating seal components. As a consequence, the fan and the motor shafts each have their own double-acting oil-film axial thrust bearing and the two shafts are connected by a 3 m "floating shaft" with couplings that allow enough free axial relative displacement between fan and motor shafts to accommodate differential thermal expansion or other such relative movements. The central of three oil-film radial bearings on the fan shaft also houses the double-acting oil-film thrust bearing. Being located axially close to the two fan-blade rows, this thrust bearing provides the required axially accurate positioning of the fan rotor with respect to the controlled-leakage dynamic seals at the fan-blade hubs. The motor shaft's double-acting oil-film thrust bearing is housed with the motor-shaft oil-film radial bearing closest to the fan shaft.

The torsionally soft floating-shaft connection between the motor and fan rotors essentially also provides a TRV *low-pass filter* that isolates the two rotors insofar as most important TRV modes are concerned. This does not in any way lessen the need for extensive TRV design analyses for this machine, but it does isolate any significant motor-produced torque pulsations from "infesting" the fan rotor and any of its lightly damped fan-blade natural-frequency modes. Both the motor and the fan rotors each must be analyzed in their own right to determine the possibilities for TRV forced resonances from torque pulsations and the corresponding potential need for TRV damping to be designed into one or more of the five shaft couplings.

3.6.2 Four-Square Gear Tester

This is a well-known type of test machine in the gear industry for testing high-torque capacity gears. The basic principle is quite simple. Two gear sets of the same speed ratio and pitch diameters are mounted on two parallel shafts as illustrated in Figure 3.6. One of the shafts or one of the gear-to-shaft mountings is made so that its torsional characteristic is relatively quite flexible. The gears are meshed with a prescribed pretwist in the torsionally flexible component, thereby "locking in" a prescribed test torque that the two gear sets apply against each other. The torque and power required of the drive motor is then only what is needed to accommodate the relatively small nonrecoverable power losses in the gears, shaft bearings, couplings, seals, windage, and so on. Clearly, this type of test machine eliminates the large expense of a drive motor with torque and power ratings of the gears. Only one of the gear sets needs to be tested gears, whereas the other gear set may be viewed as the energy regenerative set, designed to the machine's maximum capacity.

The configuration in Figure 3.6 is clearly a TRV *branched* system. The primary need for TRV design analyses of this type of machine stems from the low torsional stiffness inherent in its basic operating principle. The drive motor is most likely of a controlled variable speed type and may inherently produce torque pulsations. Also, as mentioned in the introductory

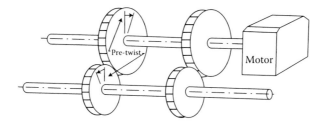

FIGURE 3.6 Conceptual illustration of a four-square gear tester.

comments of this chapter, if the speed controller employs speed feedback, then the differential equations of state for the motor controller should be coupled to the TRV equations of motion of the rotor system to analyze the potential for instability type self-excited TRV. Furthermore, geometric inaccuracies are inherent in all gear sets and are a potential source for resonance excitation of TRV modes. It is likely that a special coupling with prescribed damping characteristics is a prudent component for such a machine as a preventative measure against any of these potentially serious TRV problems from occurring.

Because of the potential use of a torsionally flexible gear-to-shaft mounting for a four-square gear tester, proper modeling formulation for TRV-flexible gear sets is reiterated here. The modeling formulations given in Equations 3.21 and 3.36 were derived using the *flexible* pulley–belt connection of Figures 3.3 and 3.4 *unbranched* and *branched* configurations, respectively. These equations also provide the proper formulation for gear sets deemed *flexible* instead of *rigid*, as stated in the previous section. To model a TRV-flexible gear set, the factor $2k_b$ in Equations 3.21 and 3.36 is replaced with k_g, the equivalent tangential translation stiffness of the gear set. This is illustrated in Figure 3.7.

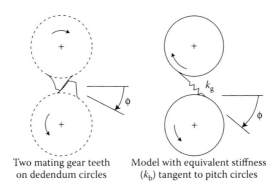

Two mating gear teeth on dedendum circles

Model with equivalent stiffness (k_b) tangent to pitch circles

FIGURE 3.7 TRV-flexible gear-set model ($\phi \equiv$ gear pressure angle).

3.6.3 Large Steam Turbo-Generator Sets

Modern single-drive-line large steam turbo-generators are in the power range of approximately 300–800 megawatts (MW). The rotor shown in Figure 2.1 is for a unit in the lower half of this power range since it has only one low-pressure (LP) turbine, whereas the largest single-drive-line units typically have two or even three LP turbines. Single-boiler *cross-compound* configurations as large as 1300 MW are in service, but are actually two 650 MW side-by-side drive lines, having a high-pressure (HP) turbine + 2 LP turbines + generator/exciter on one rotor and an intermediate-pressure (IP) turbine + 2 LP turbines + generator/exciter on the second rotor, with interconnecting steam lines (for further description, see Section 11.2).

The primary TRV problems concerning large turbo-generator units are the high alternating stresses caused by transient TRV that occurs as a result of the electrical connection transients associated with power-line transmission interruption and restoration. Weather storms, severe lightning, and malfunctions of protective systems are the prominent causes of these harmful interactions between the electrical and the mechanical systems due to the switching procedures used to restore the network transmission lines. Faced with the possibility of *system collapse* (i.e., major regional blackout), the power generation industry desires increased switching speeds. This runs counter to the turbo-generator manufacturers' efforts to minimize the problem of *cumulative fatigue damage* accrued at critical rotor locations in each such transient disturbance. There are several categories of electrical network disturbances including the following that are most prominent: (a) transmission line switching, (b) high-speed reclosing of circuit breakers after fault clearing on transmission lines leaving power stations, (c) single-phase operation that produces alternating torques at twice the synchronous frequency, (d) out-of-phase synchronization, (e) generator terminal faults, and (f) full load trips.

A comprehensive set of TRV analyses are reported by Maghraoui (1985, see Bibliography at end of this chapter), whose 93 DOF TRV model for an actual 800 MW single-drive-line turbo-generator is shown here in Figure 3.8 (illustration not to scale). The turbine section of the turbo-generator unit modeled is similar to that illustrated in Figure 3.9, except for the rigid couplings. The primary focus of Maghraoui's work is to model and compute transient TRV for an HSR event on the 800 MW unit modeled in Figure 3.8. In that work, the model is used to accurately extract the eight lowest-frequency undamped modes (i.e., natural frequencies and corresponding mode shapes). Using these modes, the TRV transient response through a typical HSR event is computed by superimposing the contributions of all the included modes, using the methods given in Section 1.3. Maghraoui gives a comprehensive bibliography on the overall topic of electrical network disturbances, from which the following formula

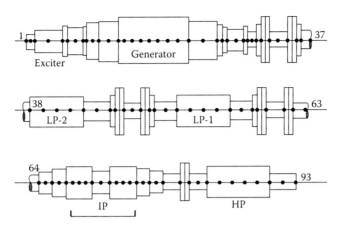

FIGURE 3.8 Layout for 93-DOF model of 800 MW 3600 rpm turbo-generator.

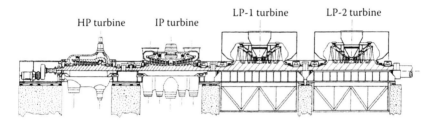

FIGURE 3.9 Turbine section of an 800 MW 3600 rpm turbo-generator (generator not shown).

is obtained for electrically imposed generator torque fluctuations caused by various transient electrical disturbances, such as those previously listed:

$$T_E = A_o + A_1 e^{-\alpha_1 t} \cos(\omega_o t + \delta_1) + A_2 e^{-\alpha_2 t} \cos(2\omega_o t + \delta_2)$$
$$+ A_3 e^{-\alpha_3 t} \cos(\omega_n t + \delta_3) \quad (3.37)$$

The system's synchronous frequency is given by (ω_o). Appropriate input values of disturbance electromechanical frequency (ω_n), the phase angles δ_j, the damping exponents (α_j), and amplitude coefficients (A_j) are tabulated by Maghraoui for successful and unsuccessful HSR from mild to severe conditions.

3.7 Summary

Table 3.1 summarizes interesting and important contrasts between TRV and LRV. TRV is often not an important consideration in many rotating

TABLE 3.1

Contrasting Characteristics of TRV and LRV in Single Rotors

Lateral Rotor Vibration	Torsional Rotor Vibration
Always an important consideration	Often not an important consideration
Resonant modes are usually sufficiently damped by bearings, seals, and so on	Modes are very lightly damped, so resonance avoidance is a "must"
More difficult to accurately model and computationally simulate because of the uncertainties in rotor-dynamic properties of bearings and seals	Relatively easy to accurately model and simulate because of decoupling from LRV modes
Easy to measure and monitor, thus does not become dangerously excessive with no warning, making LRV monitoring of rotating machinery now common	Can become excessive with no obvious outward symptoms or readily monitored motion. First sign of trouble can be seen when the shaft fails from material fatigue

machinery types, especially in machines with single uncoupled rotors. However, in contrast to LRV modes, TRV modes are nearly always very lightly damped, unless special design measures such as a flexible coupling with TRV damping capacity are taken. Therefore, if torque fluctuations with a substantial frequency content of one or more TRV modes are present, shaft failure from material fatigue can readily occur after only a relatively short time period of machine operation. Because TRV modes are usually uncoupled from LRV modes, TRV modes can be continuously or intermittently undergoing large amplitude forced resonance without the machine exhibiting any readily monitored or outward signs of distress or "shaking." The first sign of distress may be when a material-fatigue-initiated shaft failure occurs. When single rotors are coupled together, the possibility for excitation of coupled-system torsional natural frequency modes is greater. In most coupled drive trains, it is the characteristics of the couplings, gear trains, and electric motors or generators that instigate TRV problems.

Although rotating machinery TRV problems are less amenable to monitoring and early detection than LRV problems, TRV characteristics can generally be more accurately modeled for predictive analyses than LRV. This is because TRV is usually uncoupled from the characteristics (i.e., bearings, seals, and other rotor-casing interactions) that make LRV model-based predictions more uncertain and challenging to perform. Furthermore, since TRV modes of primary importance (i.e., those in the lower frequency range of the system) are almost always quite lightly damped, accurate prediction of TRV natural frequencies and corresponding mode shapes is further enhanced. That is, the actual system's TRV characteristics are essentially embodied in its model's mass and stiffness matrices, which are accurately extractable from the detailed rotor geometry through the modern finite-element modeling procedures developed and explained in this chapter.

The primary focus of this book and its subsequent chapters is on LRV. So to make this chapter on TRV more "stand alone" than its LRV counterpart (Chapter 2), Section 3.6 on TRV *Examples* has been included in this chapter.

Bibliography

Maghraoui, M., *Torsional Vibrations of a Large Turbine-Generator Set*, MS Thesis, Case Western Reserve University, May 1985, pp. 104.

Vance, J. M., *Rotordynamics of Turbomachinery*, Wiley, New York, 1988, pp. 388.

PROBLEM EXERCISES

1. A single drive-line pump and coupled drive motor are modeled as a torsionally twistable shaft of relatively small mass and torsional stiffness K connecting two disks of polar moments of inertia, J_M and J_P, for the motor rotor and pump impeller, respectively. Derive the equations of motion, natural frequencies, and mode shapes. Assume that torsional stiffness connections to ground are negligible.

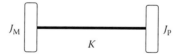

2. An improved model for the system of Problem 1 accounts for the polar moment of inertia of the coupling J_C that joins the pump and motor shafts. Again, assuming that torsional stiffness connections to ground are negligible, derive the equations of motion, natural frequencies, and mode shapes.

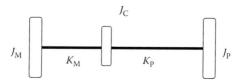

3. A modified version of the Problem 1 model has added a significantly stiff torsional connection to the ground to represent a

stiff motor speed control. Derive the equations of motion, natural frequency, and mode shapes.

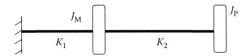

4. A modified version of the Problem 2 model has added a significantly stiff torsional connection to the ground to represent a stiff motor speed control. Derive the equations of motion, natural frequencies, and mode shapes.
5. The system shown is a model for a two-shaft-geared transmission. Geared connections are typically modeled as torsionally rigid (see Figure 3.7) and thus rigidly couple the shafts together. Neglect shaft inertias. Derive the equations of motion and the natural frequencies. The tooth numbers of the gears on shaft 1 and shaft 2 are N_1 and N_2, respectively.

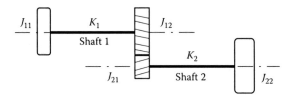

6. *Project on torsional rotor vibration software.* Develop a MATLAB® computer code for general multirotor branched linear systems with rigid and flexible connections as covered in Section 3.4.3 of this chapter. A complete version of the code should include the determination of (a) undamped natural frequencies and mode shapes; (b) steady-state single-frequency harmonic excitation torques with a modal damping option; and (c) time-transient resonance build-up simulation.

Part II

Rotor Dynamic Analyses

4

RDA Code for Lateral Rotor Vibration Analyses

4.1 Introduction

The RDA Fortran computer code is a general purpose tool for linear rotor vibration analyses. It is developed on the FE formulations derived in Chapter 2, Section 2.3. First written for use on early generation PCs, it was initially limited to fairly simple rotor–bearing configuration models with 10 or less mass stations (40 DOFs or less) because of the memory limitations of early PCs. RDA was initially written to simulate rotor–bearing systems as part of research efforts on *active control of rotor vibration* in the author's group at Case Western Reserve University (CWRU). Validation tests and other background information for RDA are provided by Maghraoui (1989) in his PhD dissertation (see Bibliography at the end of this chapter). RDA has been distributed and used by the author in *machinery dynamics* courses and student research projects at CWRU for over 20 years and in professional short courses in the United States and Europe. The current enlarged version supplied with this book, RDA99, has now been exercised by countless users since being provided free with the 2001 first edition of this book. RDA99 has been successfully used by the author in modeling several large power plant machinery, in vibration troubleshooting missions (see Part 4 of this book). It has also been successfully used by the author in troubleshooting and redesigning a high-speed vertical spin-pit test rig specially configured for research on aircraft jet engine blade-on-casing tip-rub-induced blade vibrations and transmitted blade-casing interaction dynamic forces.

The compiled code included here has been dimensioned to accommodate up to 99 rotor mass stations (396 DOF rotor), making it suitable for virtually any single-drive-line rotor–bearing system, including large steam turbo-generator rotors as subsequently demonstrated in Part 4 of this book. The author and his troubleshooting associates still use this newer RDA99 as the primary rotor vibration analysis tool both for troubleshooting work in plants as well as research.

As demonstrated in this chapter, RDA99 is a *user-interactive* code and thus does not utilize the *batch-mode* input approach typical of older computer codes written in the era of older mainframe computers. RDA99 has interactive input and output selection menus, each with several options. Not all these options are demonstrated here. Only the ones that are the most expedient for design or troubleshooting applications are demonstrated here.

There are many quite useful PC codes that were initially developed to run in the DOS environment prior to the introduction of Windows. The RDA executable code (RDA99.exe), supplied with this book, is but one example. The DOS operating system, developed for first-generation PCs and the forerunner of Windows, has therefore naturally been retained as an application within Windows. Earlier versions of Windows are actually an application within DOS. **RDA99.exe** will execute successfully on any PC as a DOS application within Windows.

Within the DOS operation mode, RDA99 is accessed simply by entering the appropriate drive and folder. Execution is then initiated simply by entering **RDA99**. The monitor then displays the following main menu.

```
^^^^^^^^^^^^^^^^^^^^^^^^^^^^^^^^^^^^^^^^^^^
     ROTOR DYNAMICS ANALYSIS
vvvvvvvvvvvvvvvvvvvvvvvvvvvvvvvvvvvvvvvvvv
              MAIN MENU
 1. Solve the Undamped Eigenvalue Problem Only
 2. Solve for Damped Eigenvalues Only
 3. Solve Both Damped and Undamped Eigenvalue Problems
 4. Perform a Stability Analysis of the System
 5. Obtain the Steady-State Unbalance Response
 6. Active Control Simulation
 7. Data Curve Fitting By Cubic Spline
 8. Exit
Choose Option <1-8> ...
```

All the MAIN MENU options are covered in Maghraoui (1989). When accessed by entering its number, each displays the DATA MENU from which the INPUT OPTIONS menu is accessed. Vibration specialists may wish to use options 1, 2, and 3 of the MAIN MENU to construct maps of eigenvalues as functions of rotor spin speed, and these are demonstrated in Maghraoui (1989). Options 6 and 7 may be ignored. MAIN MENU options 4 and 5 are the most important and useful ones. Therefore, the detailed instructions covered in this chapter are focused exclusively on options 4 and 5.

4.2 Unbalance Steady-State Response Computations

Referring to Equation 2.69, a synchronous (i.e., at rotor-speed frequency) corotational rotating radial force may be appended to any rotor model mass station to simulate the effect of a mass unbalance at that mass station. Thus, with an analysis algorithm such as contained in the RDA code, virtually any realistic rotor unbalance distribution can be accordingly postulated and the resulting steady-state vibration at all rotor mass stations computed. The standard algorithm for this is the solution of the corresponding set of simultaneous complex algebraic equations, Equation 1.52. From the MAIN MENU, **enter option 5**. The monitor then displays the following menu.

DATA MENU
1. Input/Read Data
2. Print Data on the Screen
3. Save Data in a File
4. Edit Data
5. Run Main Menu Option
6. Return to Main Menu

Choose Option <1–6> ...

To initiate input for a case, enter option 1 from the above DATA MENU. The monitor then displays the following input options.

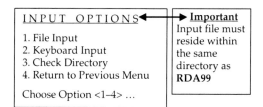

For a completely new model, the INPUT OPTIONS menu would appear to indicate that option 2 is the only route since no previously saved input file would yet exist in the RDA directory for the model to be run. In fact, for a completely new model there are actually two options: 1 and 2. For a simple model with a relatively small number of mass stations and bearings, the Keyboard Input route (option 2) is a satisfactory option that any new RDA user should try out just to be familiar with it. As shown subsequently, the user is prompted at each step of the Keyboard Input option. The drawback with this option is that while it does recover and appropriately prompt interactively for many types of inadvertent keyboard errors, it does not recover from all types of keyboard errors. In such an unrecoverable error occurrence, one may make corrections by accessing option 4 (Edit Data) at the completion of the Keyboard Input cycle. However, for large DOF models, one has the option to first create the new input file using a full-screen editor (e.g., Notepad) outside the RDA environment, employing

RDA's own output format allocations for saving an input file. This will avoid unrecoverable input errors that necessitate starting the input creation all over again. However, in employing the full-screen editor option, the user must externally calculate and then enter the rotor element weights into the input file, since that is programmed into the RDA Keyboard option 2.

Once an input file already exists for the model in the **RDA99** directory, the File Input (option 1) is naturally used. Entering number 1 from the INPUT OPTIONS menu, the monitor prompts for the input file name. Upon entering the input file name, the monitor returns to the DATA MENU, from which option 5 is entered if the named input file is ready to run. A comprehensive analysis almost always entails computing several different operating cases, for example, different unbalance conditions, different speed range and speed increments, different bearing properties, and so on. Thus, if as usual, the named input file is to be first modified before executing the run, the user again has two options. First, option 4 on the DATA MENU may be accessed. Or as just recommended for creating large new input files, input file modifications are more conveniently implemented using a full-screen editor outside the RDA environment. If the input file modifications are fairly short, Edit Data (option 4) is a reasonable route which when accessed displays the EDITOR OPTIONS menu of 16 different user options, each specific to the type of modification to be made. When accessed, each of these options on the EDITOR OPTIONS menu prompts the user for the information necessary to implement the desired input file changes. In the interest of space, these will not be individually covered here. In fact, the author practically never uses the EDITOR OPTIONS menu, preferring the previously indicated full-screen editor route.

One may reasonably ask why RDA has not been updated to streamline its use, like a GUI Windows version. Since RDA is not a commercially marketed code, there is no group of dedicated programmers to do this. When commercially marketed codes are updated, a battery of test cases must be run to ensure that no new bugs have found their way into the code. The version of RDA supplied with this book has been in use for many years, but has not been modified in any way in that interim. This eliminates the possibility of any program bugs acquired over its many years of use, internationally. The degree of confidence inherent in that approach surely surpasses in importance the niceties of streamlining, bells and whistles.

In matters of rotor unbalance analysis and rotor balancing procedures, there are a number of parameters that need precise clarification, most notably a clear explanation of *phase angle* and *direction of rotation*. However, before covering such clarifications, some simple examples are worked through first, to acclimate the new RDA user on how to get started in running RDA. Input and output files for all shown examples are included with the here-supplied RDA software, so users can readily check their own input/output work.

4.2.1 3-Mass Rotor Model + 2 Bearings and 1 Disk

The *simple nontrivial 8-DOF* model illustrated in Figure 2.4 consists of two identical rotor elements, giving three mass stations, with one concentrated disk mass at the middle mass station. Furthermore, it has a radial bearing at each end, both modeled with the standard 8-coefficient linear bearing model introduced by Equation 2.2 and further explained in Section 2.3.9. An example with numerical inputs for this model is employed here to provide the new user an expedient first exercise using RDA to compute unbalance response. RDA includes transverse moment of inertia not only for specified disks, but also automatically for every shaft element, as detailed in Section 2.3 of Chapter 2. Therefore, the RDA equivalent model to that in Figure 2.4 has 12 DOFs, since each of the three mass stations has 4 DOFs. Employing the user menu and input prompts explained thus far in this chapter, the following model data are used to construct an input file from the Keyboard Input option.

Input Title: 3-Mass, 1 Disk, 2 Bearing Sample No. 1	
Number of stations	3
Number of disks	1
Number of bearings	2
Number of pedestals	0
Number of extra weights	0
Modulus of elasticity and weight density	30,000,000 psi, 0.285 lb/in.^{3}

Shaft Element Data	OD (in.)	ID (in.)	Length (in.)	Inertia (lb s^2/in.)	Weight (lb)
Element no. 1	0.5	0.0	10	0.0	0.0
Element no. 2	0.5	0.0	10	0.0	0.0

Typically, shaft element inertia and weight are input as "zero," like in this example, and RDA then calculates them from dimensions and input weight density. Input of "nonzero" values overrides the RDA calculated ones.

Disc Data	Station No.	OD (in.)	ID (in.)	Length (in.)	Weight	I_P	I_T
Disc no. 1	2	5.0	0.5	1.0	0.0	0.0	0.0

Similarly, disc weight (lb), polar moment of inertia I_P, and transverse moment of inertia I_T (lb in.^{2}) are typically input as "zero" and RDA calculates them from disc dimensions and the default weight density input.

For components that are not *disc like* (e.g., impellers, couplings, etc.), appropriate "nonzero" weight, I_P and I_T overriding values may be input instead, accompanied by zero inputs for the disc OD, ID, and length.

Bearing Data	Bearing No. (Prompt)	Station No.	Weight
	1	1	0.0
	2	3	0.0

Typically, bearing weight is also input as "zero." For a rolling-element bearing, the added weight of the inner raceway on the shaft may be input here as a nonzero bearing weight. However, the same identical effect may alternatively be incorporated into the model using the extra weights option.

	Bearing Properties Speed Dependent Y/N? N			
	Bearing Stiffness and Damping Coefficients			
Bearing no. 1	K_{xx} 2000 lb/in.	K_{xy} 0.0	C_{xx} 5.0 lb s/in.	C_{xy} 0.0
	K_{yx} 0.0	K_{yy} 2000 lb/in.	C_{yx} 0.0	C_{yy} 5.0 lb s/in.
Bearing no. 2	K_{xx} 2000 lb/in.	K_{xy} 0.0	C_{xx} 5.0 lb s/in.	C_{xy} 0.0
	K_{yx} 0.0	K_{yy} 2000 lb/in.	C_{yx} 0.0	C_{yy} 5.0 lb s/in.

Unbalance Data	Station No.	Amplitude (lb in.)	Phase Angle (°)
	1	0.0	0.0
	2	0.005	0.0
	3	0.0	0.0

At this point, RDA returns the user to the DATA MENU, where option 3 is entered to save the input file just created using the Keyboard Input option, and the user is again returned to the DATA MENU. At that point, option 5 "Run Main Menu Option" may be entered to execute the run. However, one may first wish to check the input file by entering option 2 "Print Data on the Screen," and if input errors are detected, enter option 4 "Edit Data" to access the EDITOR OPTIONS. Corrections to the input file may also be done outside of RDA using a full-screen editor on the saved input file.

Upon entering option 5 "Run Main Menu Option," a PLOT OPTIONS output menu is displayed with several self-explanatory options, and the new RDA user may wish to explore all of them. Option 11, which produces a complete labeled input/output file but no plots, is chosen here. After an output option is entered, the user is prompted for the following:

Input Speed Range Data for Unbalance Response	
Enter input starting speed (rpm)	100
Enter input ending speed (rpm)	2100
Enter input speed increment (rpm)	200

Shaft Mass Model Options
1. Lumped Mass
2. Distributed Mass
3. Consistent Mass

The *Consistent Mass* option is usually the preferred choice, and is chosen here, that is, **enter option 3**. For any given rotor, prudent users may compare model resolution accuracy and convergence of these three options by varying the number of shaft elements.

The last prompt is to specify the "output file name." Entering the file name (e.g., **sample01.out**), RDA executes the run to completion. If output option 11 is specified, the complete labeled output information may be viewed by opening the output file in a full-screen editor like Notepad. The abbreviated output shown below does not include the input review and does not show response for station 3 since it is the same as station 1 due to symmetry. Unbalance at station 2 has $\theta_2 = 0$, and is thus a *reference signal*.

	Response of Rotor Station No. 1			
	X-Direction		Y-Direction	
Speed (rpm)	Amplitude (mils)	Phase Angle (°)	Amplitude (mils)	Phase Angle (°)
100.0	0.000	−1.5	0.000	−91.5
300.0	0.003	−4.5	0.003	−94.5
500.0	0.010	−7.6	0.010	−97.6
700.0	0.021	−10.7	0.021	−100.7
900.0	0.040	−13.9	0.040	−103.9
1100.0	0.073	−17.6	0.073	−107.6
1300.0	0.143	−22.2	0.143	−112.2
1500.0	0.360	−31.0	0.360	−121.0
1700.0	**1.897**	**−129.6**	**1.897**	**140.4**
1900.0	0.440	167.5	0.440	77.5
2100.0	0.264	160.0	0.264	70.0

Maximum amplitudes of station 1 occurred at
   1700.0 rpm for the X-direction with 1.9 mils and a phase angle of −129.6°.
   1700.0 rpm for the Y-direction with 1.9 mils and a phase angle of 140.4°.

<table>
<tr><th colspan="5">Response of Rotor Station No. 2</th></tr>
<tr><th></th><th colspan="2">X-Direction</th><th colspan="2">Y-Direction</th></tr>
<tr><th>Speed (rpm)</th><th>Amplitude (mils)</th><th>Phase Angle (°)</th><th>Amplitude (mils)</th><th>Phase Angle (°)</th></tr>
<tr><td>100.0</td><td>0.003</td><td>−.2</td><td>0.003</td><td>−90.2</td></tr>
<tr><td>300.0</td><td>0.027</td><td>−.6</td><td>0.027</td><td>−90.6</td></tr>
<tr><td>500.0</td><td>0.080</td><td>−1.0</td><td>0.080</td><td>−91.0</td></tr>
<tr><td>700.0</td><td>0.173</td><td>−1.5</td><td>0.173</td><td>−91.5</td></tr>
<tr><td>900.0</td><td>0.331</td><td>−2.2</td><td>0.331</td><td>−92.2</td></tr>
<tr><td>1100.0</td><td>0.615</td><td>−3.3</td><td>0.615</td><td>−93.3</td></tr>
<tr><td>1300.0</td><td>1.212</td><td>−5.5</td><td>1.212</td><td>−95.5</td></tr>
<tr><td>1500.0</td><td>3.080</td><td>−11.9</td><td>3.080</td><td>−101.9</td></tr>
<tr><td>**1700.0**</td><td>**16.388**</td><td>**−108.1**</td><td>**16.388**</td><td>**161.9**</td></tr>
<tr><td>1900.0</td><td>3.843</td><td>−168.7</td><td>3.843</td><td>101.3</td></tr>
<tr><td>2100.0</td><td>2.327</td><td>−174.0</td><td>2.327</td><td>96.0</td></tr>
</table>

Maximum amplitudes of station 2 occurred at
  1700.0 rpm for the X-direction with 16 mils and a phase angle of −108.1°.
  1700.0 rpm for the Y-direction with 16 mils and a phase angle of 161.9°.

RDA output tabulates *single-peak* vibration amplitudes in thousandths of an inch (*mils*) for both x and y directions. The abbreviated output here clearly shows a first (lowest) *critical speed* near 1700 rpm where the synchronous unbalance vibration amplitude passes through a maximum value as a function of rotor speed. Comparing results at the bearings (station 1 results same as 3) with results at the disc (station 2) shows that the rotor undergoes a significant relative amount of bending vibration at the critical speed. With RDA it is quite easy to "zoom in" on the critical speed by using a finer speed increment (resolution) in order to accurately capture it and its maximum value. For this sample, simply repeat the run (using the saved input file) with a start speed and end speed inputs near 1700 rpm and a significantly reduced speed increment, as demonstrated with the following inputs.

Input Speed Range Data for Unbalance Response	
Enter input starting speed (rpm)	1600
Enter input ending speed (rpm)	1800
Enter input speed increment (rpm)	20

Output for this revised speed range and increment is tabulated for station 2 as follows. It shows that the critical-speed peak is between 1680 and 1700 rpm. One could "zoom in" further.

Response of Rotor Station No. 2

	X-Direction		Y-Direction	
Speed (rpm)	Amplitude (mils)	Phase Angle (°)	Amplitude (mils)	Phase Angle (°)
1600.0	6.785	−25.0	6.785	−115.0
1620.0	8.474	−31.3	8.474	−121.3
1640.0	10.880	−41.1	10.880	−131.1
1660.0	14.062	−57.0	14.062	−147.0
1680.0	**16.795**	**−81.0**	**16.795**	**−171.0**
1700.0	16.388	−108.1	16.388	161.9
1720.0	13.580	−129.0	13.580	141.0
1740.0	10.856	−142.2	10.856	127.8
1760.0	8.848	−150.5	8.848	119.5
1780.0	7.421	−155.9	7.421	114.1
1800.0	6.385	−159.8	6.385	110.2

Maximum amplitudes of station 2 occurred at
  1680.0 rpm for the X-direction with 17 mils and a phase angle of −81.0°.
  1680.0 rpm for the Y-direction with 17 mils and a phase angle of −171.0°.

In this simple example, the bearing inputs are all radial isotropic (see Section 2.4 of Chapter 2) and thus the rotor vibration orbits are all circular. This is indicated by the x and y vibration amplitudes being equal and 90° out of phase (x *leading* y, therefore *forward whirl*). With one or more anisotropic bearings, the rotor orbits are ellipses.

4.2.2 Phase Angle Explanation and Direction of Rotation

Before demonstrating additional sample cases, the phase-angle convention employed in RDA is given a careful explanation at this point because of the confusion and errors that frequently occur in general where rotor vibration phase angles are involved. Confusion concerning rotor vibration phase angles stems from a number of sources. The first source of confusion, common to harmonic signals in general, is the sign convention, (i.e., is the phase angle defined positive when the signal *leads* or *lags* the *reference signal*?). The second source of confusion stems from the visual similarity between the *complex plane* illustration of harmonic signals as *rotating vectors* and the actual rotation of fixed points or force vectors on the rotor, for example, *high spot, heavy spot* (or unbalance mass).

On real machines, the most troublesome consequence of *phase-angle confusion* occurs when balance correction weights are placed at incorrect angular locations on a rotor. Similar mistakes often result from the fact that the rotor must spin clockwise (cw) when viewed from one end and counterclockwise (ccw) when viewed from the other end. Consequently, it is far less confusing to have the analysis model consistent with the actual rotor's rotational direction and this is accomplished by starting the shaft element inputs from the *proper end* of the rotor. As shown in Section 2.3 of Chapter 2, RDA is formulated in a standard *xyz right-hand* coordinate system where *x* and *y* define the radial directions and positive *z* defines the axis and direction of positive rotor spin velocity. Thus, if one views the rotor from the end where the rotation is ccw, the positive z-axis should point toward them and the rotor model shaft elements' input should start from the other end of the rotor. The *proper end* of the rotor to start RDA shaft element inputs is accordingly demonstrated in Figure 4.1 for a three-element (four-mass-station) example.

The RDA phase angle sign convention is defined by the following specifications for unbalance force and vibration displacement components:

$$F_x = m_u r_u \omega^2 \cos(\omega t + \theta), \quad m_u = \text{unbalance mass}, \quad x = X \cos(\omega t + \phi_x)$$
$$F_y = m_u r_u \omega^2 \sin(\omega t + \theta), \quad r_u = \text{unbalance radius}, \quad y = Y \cos(\omega t + \phi_y)$$

(4.1)

These specifications define a phase angle (θ, ϕ_x, and ϕ_y) as positive when its respective harmonic signal *leads* the reference signal. A commonly used convenient way to visualize this full complement of synchronous harmonic signals is the *complex plane* representation, which illustrates each harmonic signal as a *rotating vector*. Figure 4.2 shows this for the RDA unbalance force and vibration displacement components.

The three complex vectors (X, Y, and F) shown in Figure 4.2 are conceived rotating at the angular velocity ω in the ccw direction, thus maintaining their relative angular positions to each other. However, this is not to be confused with points or vectors fixed on the rotor that also naturally rotate ccw at ω. It is only that the mathematics of *complex numbers* has long been recognized and used as a convenient means of representing a group of related harmonic signals all having the same frequency, such as the various

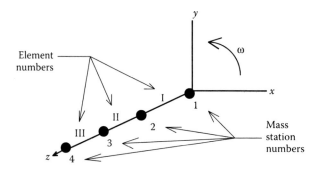

FIGURE 4.1 Proper shaft element and station input ordering.

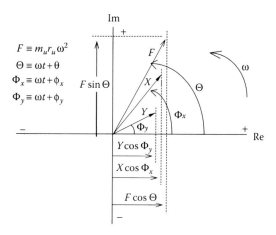

FIGURE 4.2 Complex plane representation of synchronous harmonic signals.

components of voltage- and current-related signals in alternating-current electricity. Since the unbalance force is purely a synchronous rotating vector, it is easy to view it as a complex entity since its x-component projects onto the real axis, while its y-component projects onto the imaginary axis. The same could be said of the rotor orbits for the *simple three-mass rotor model* given in the previous subsection, because the bearing stiffness and damping inputs are radially isotropic and thus yield circular orbits. But this is not typical.

To avoid confusion when applying the complex plane approach to rotor vibration signals, it is essential to understand the relationship between the standard complex plane illustration and the position coordinates for the orbital trajectory of rotor vibration. First and foremost, the real (Re) and imaginary (Im) axes of the standard complex plane shown in Figure 4.2 are not the x and y axes in the plane of radial orbital rotor vibration trajectory. There are a few rotor vibration academics who have joined the complex plane and the rotor x–y trajectory into a single illustration and signal management method by using the real axis for the x-signal and the imaginary axis for the y-signal. This can be accomplished by having the component $Y \cos \Phi_y$ projected onto the imaginary axis by defining Φ_y relative to the imaginary axis instead of the real axis. The author does not embrace this approach since all the rotor orbital trajectory motion coordinates then entail complex arithmetic. The author does embrace the usefulness of the complex plane, as typified by Figure 4.2, to illustrate the steady-state rotor vibration harmonic signals specified by Equation 4.1.

Figure 4.3 is an addendum to Figure 4.2, illustrating the x-displacement, x-velocity, and x-acceleration in the complex plane. The same can be done for the y-direction signals. As in the previous complex plane illustration,

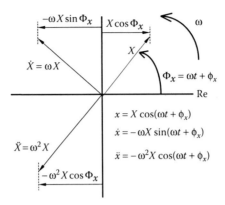

FIGURE 4.3 Complex plane view of x-displacement, velocity, and acceleration.

all the shown *vectors* in Figure 4.3 rotate ccw at the angular velocity ω, thus maintaining their relative angular positions to each other.

4.2.3 3-Mass Rotor Model + 2 Bearings/Pedestals and 1 Disk

The previous 3-mass model is augmented here with the addition of bearing pedestals, as formulated in Section 2.3 of Chapter 2. The inputs here differ from those in the previous example only by the addition of a pedestal at each bearing. For creating this input file from the Keyboard Input option, only the following inputs are added to the previous sample's input as prompted by RDA. The following numerical inputs are used in this example.

	Number of Pedestals: 2		
Pedestal Data	Pedestal No. (Prompt)	Station No.	Weight (lb)
	1	1	5.0
	2	3	5.0

	Pedestal Stiffness and Damping Coefficients			
Pedestal No. (Prompt)	K_{xx} (lb/in.)	K_{yy} (lb/in.)	C_{xx} (lb s/in.)	C_{yy} (lb s/in.)
1	2000	2000	0.5	0.5
2	2000	2000	0.5	0.5

This example has 16 DOFs, four more than the previous example, because each of the two pedestals has two DOF, that is, x and y. An abbreviated output summary follows. Since the model is symmetric about station 2, rotor and pedestal responses at station 3, being the same as at station 1, are not shown here.

	Response of Rotor Station No. 1			
	X-Direction		Y-Direction	
Speed (rpm)	Amplitude (mils)	Phase Angle (°)	Amplitude (mils)	Phase Angle (°)
100.0	0.001	−0.8	0.001	−90.8
300.0	0.007	−2.5	0.007	−92.5
500.0	0.020	−4.2	0.020	−94.2
700.0	0.044	−5.9	0.044	−95.9
900.0	0.087	−7.9	0.087	−97.9
1100.0	0.174	−10.4	0.174	−100.4
1300.0	0.398	−14.7	0.398	−104.7
1500.0	**1.803**	**−36.5**	**1.803**	**−126.5**
1700.0	1.196	−173.9	1.196	96.1

Maximum amplitudes of station 1 occurred at
  1500.0 rpm for the X-direction with 1.8 mils and a phase angle of −36.5°.
  1500.0 rpm for the Y-direction with 1.8 mils and a phase angle of −126.5°.

	Response of Rotor Station No. 2			
	X-Direction		Y-Direction	
Speed (rpm)	Amplitude (mils)	Phase Angle (°)	Amplitude (mils)	Phase Angle (°)
100.0	0.003	−0.2	0.003	−90.2
300.0	0.031	−0.6	0.031	−90.6
500.0	0.091	−1.0	0.091	−91.0
700.0	0.201	−1.5	0.201	−91.5
900.0	0.396	−2.3	0.396	−92.3
1100.0	0.780	−3.7	0.780	−93.7
1300.0	1.762	−7.1	1.762	−97.1
1500.0	**7.843**	**−28.0**	**7.843**	**−118.0**
1700.0	5.083	−164.6	5.083	105.4

Maximum amplitudes of station 2 occurred at
  1500.0 rpm for the X-direction with 7.8 mils and a phase angle of −28.0°.
  1500.0 rpm for the Y-direction with 7.8 mils and a phase angle of −118.0°.

	Response of Pedestal No. 1			
	Located at Station No. 1			
	X-Direction		Y-Direction	
Speed (rpm)	Amplitude (mils)	Phase Angle (°)	Amplitude (mils)	Phase Angle (°)
100.0	0.000	−0.2	0.000	−90.2
300.0	0.003	−0.5	0.003	−90.5

continued

Continued				
500.0	0.010	−0.9	0.010	−90.9
700.0	0.023	−1.4	0.023	−91.4
900.0	0.046	−2.2	0.046	−92.2
1100.0	0.093	−3.7	0.093	−93.7
1300.0	0.219	−7.2	0.219	−97.2
1500.0	**1.025**	**−28.3**	**1.025**	**−118.3**
1700.0	0.704	−165.2	0.704	104.8

Maximum amplitudes of pedestal 1 occurred at
  1500.0 rpm for the X-direction with 1.0 mils and a phase angle of −28.3°.
  1500.0 rpm for the Y-direction with 1.0 mils and a phase angle of −118.3°.

A number of observations can immediately be made from this abbreviated output summary. First, the addition of pedestals has dropped the *first critical speed* from about 1680 rpm (previous example) to about 1500 rpm, as all the response signals here peak at approximately 1500 rpm. Second, the orbital trajectories of rotor stations as well as the pedestal masses are all circular and corotational. This is shown by the x and y amplitudes for a given rotor station or pedestal mass being equal, with the x-signal leading the y-signal by 90°. This is the result of all bearing and pedestal stiffness and damping coefficients being radial isotropic, otherwise the trajectories would be ellipses. Third, the total response of rotor station 1 is almost twice its pedestal's total response. Relative rotor-to-bearing/pedestal motions are now continuously monitored on nearly all large power plant and process plant rotating machinery using noncontacting inductance-type proximity probes mounted in the bearings and targeting the rotor (journals). Part 3 of this book, *Monitoring and Diagnostics*, describes this in detail. Since a bearing is held in its pedestal, *bearing motion* and *pedestal motion* are synonymous here within the context of an RDA model. The corresponding additional computation of rotor (journal) orbital trajectory relative to the bearing can be derived directly with the aid of the previously introduced complex plane, wherein the standard rules for vector addition and subtraction apply. This is illustrated in Figure 4.4 and specified by Equations 4.2:

$$\begin{aligned} x_R &= X_R \cos(\omega t + \phi_{RX}), & y_R &= Y_R \cos(\omega t + \phi_{RY}) \\ x_B &= X_B \cos(\omega t + \phi_{BX}), & y_B &= Y_B \cos(\omega t + \phi_{BY}) \\ x_{rel} &= x_R - x_B & y_{rel} &= y_R - y_B \\ &\equiv X_{rel} \cos(\omega t + \phi_{Xrel}), & &\equiv Y_{rel} \cos(\omega t + \phi_{Yrel}) \end{aligned} \quad (4.2)$$

All the vectors in Figure 4.4 maintain their relative angular position to each other and rotate ccw at ω. By considering the view shown to be at time $t = 0$, it is clear from standard vector arithmetic that the single-peak amplitudes and phase angles for the relative rotor-to-bearing orbital

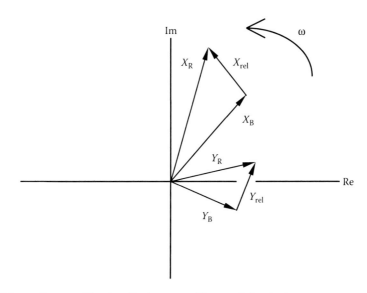

FIGURE 4.4 Rotor and bearing displacements (R, rotor; B, bearing).

trajectory harmonic signals are given as follows:

$$X_{rel} = \sqrt{(X_R \cos \phi_{RX} - X_B \cos \phi_{BX})^2 + (X_R \sin \phi_{RX} - X_B \sin \phi_{BX})^2}$$
$$Y_{rel} = \sqrt{(Y_R \cos \phi_{RY} - Y_B \cos \phi_{BY})^2 + (Y_R \sin \phi_{RY} - Y_B \sin \phi_{BY})^2}$$
$$\phi_{Xrel} = \tan^{-1}\left(\frac{X_R \sin \phi_{RX} - X_B \sin \phi_{BX}}{X_R \cos \phi_{RX} - X_B \cos \phi_{BX}}\right) \quad (4.3)$$
$$\phi_{Yrel} = \tan^{-1}\left(\frac{Y_R \sin \phi_{RY} - Y_B \sin \phi_{BY}}{Y_R \cos \phi_{RY} - Y_B \cos \phi_{BY}}\right)$$

Equations 4.3 are general, and thus applicable to the RDA outputs for any case. Substituting outputs from the simple isotropic bearing/pedestal example problem here, one may confirm that the relative rotor-to-bearing orbits are circles since the individual rotor and pedestal orbits are circles. For general anisotropic systems, the total-motion and relative-motion orbits are ellipses.

4.2.4 Anisotropic Model: 3-Mass Rotor + 2 Bearings/Pedestals and 1 Disk

The previous model is modified here to provide an example with bearing and pedestal dynamic properties that are not isotropic and thus more realistic. Starting with the input file from the previous example, the bearing and pedestal inputs are modified according to the following input specifications.

Bearing Stiffness and Damping Coefficients				
Bearing No. 1	K_{xx} 1500 lb/in.	K_{xy} 750.0	C_{xx} 20.0 lb s/in.	C_{xy} 5.0
	K_{yx} 50.0	K_{yy} 5000 lb/in.	C_{yx} 5.0	C_{yy} 30.0 lb s/in.
Bearing No. 2	K_{xx} 1500 lb/in.	K_{xy} 750.0	C_{xx} 20.0 lb s/in.	C_{xy} 5.0
	K_{yx} 50.0	K_{yy} 5000 lb/in.	C_{yx} 5.0	C_{yy} 30.0 lb s/in.

Note that the bearing stiffness coefficient matrices while symmetric are postulated as anisotropic (nonisotropic) and thus provide a more realistic example for fluid-film journal bearings, as dissected in Section 2.4 of Chapter 2 and more fully developed in Chapter 5.

Pedestal Stiffness and Damping Coefficients				
	K_{xx} (lb/in.)	K_{yy} (lb/in.)	C_{xx} (lb s/in.)	C_{yy} (lb s/in.)
Pedestal no. 1	2000	3000	10	10
Pedestal no. 2	2000	3000	10	10

In an RDA model, the connection between horizontal and vertical is essentially through the bearing and pedestal stiffness and damping inputs. It is typical for horizontal-rotor machines that pedestal vertical stiffness is approximately 50% or more larger than the pedestal horizontal stiffness and the inputs in this example emulate that (x horizontal, y vertical).

The steady-state unbalance response for this third sample case is driven by the same single unbalance at station 2 (the disc) of the previous two examples with its phase angle input as "zero." Therefore, as with the previous two examples, the unbalance at station 2 is a *reference signal* to which all phase angles in the output are referenced. This example is also symmetric about the rotor mid-plane (station 2) in all details, and thus its abbreviated output, tabulated as follows, does not include here bearing and pedestal responses for station 3.

Response of Rotor Station No. 1				
	X-Direction		Y-Direction	
Speed (rpm)	Amplitude (mils)	Phase Angle (°)	Amplitude (mils)	Phase Angle (°)
1500.0	0.677	−60.4	0.494	−131.7
1550.0	0.915	−72.6	0.692	−142.2
1600.0	1.249	−92.6	1.007	−160.1

continued

Continued

1650.0	1.509	−124.3	1.369	169.3
1700.0	1.292	−157.6	1.383	130.7
1750.0	0.945	−173.8	1.036	101.8
1800.0	0.772	178.2	0.748	87.4
1850.0	0.661	171.5	0.582	80.2
1900.0	0.577	166.0	0.482	75.7

Maximum amplitudes of station 1 occurred at
   1650.0 rpm for the X-direction with 1.5 mils and a phase angle of $-124.3°$.
   1700.0 rpm for the Y-direction with 1.4 mils and a phase angle of $130.7°$.

	Response of Rotor Station No. 2			
	X-Direction		Y-Direction	
Speed (rpm)	Amplitude (mils)	Phase Angle (°)	Amplitude (mils)	Phase Angle (°)
1500.0	3.699	−33.1	3.412	−109.3
1550.0	5.077	−44.7	4.783	−117.9
1600.0	7.054	−64.7	7.003	−133.7
1650.0	**8.640**	**−97.4**	**9.715**	**−162.1**
1700.0	7.206	−133.3	10.197	160.6
1750.0	4.791	−148.6	7.867	131.2
1800.0	3.802	−151.5	5.688	116.1
1850.0	3.315	−154.8	4.383	109.0
1900.0	2.956	−158.2	3.596	105.1

Maximum amplitudes of station 2 occurred at
   1650.0 rpm for the X-direction with 8.6 mils and a phase angle of $-97.4°$.
   1700.0 rpm for the Y-direction with 10 mils and a phase angle of $160.6°$.

	Response of Rotor Station No. 1			
	Located at Station No. 1			
	X-Direction		Y-Direction	
Speed (rpm)	Amplitude (mils)	Phase Angle (°)	Amplitude (mils)	Phase Angle (°)
1500.0	0.407	−56.3	0.336	−124.2
1550.0	0.556	−68.9	0.474	−134.8
1600.0	0.766	−89.4	0.694	−153.0
1650.0	**0.932**	**−121.8**	**0.949**	**176.0**
1700.0	0.795	−156.3	0.959	136.9
1750.0	0.572	−172.6	0.714	107.6
1800.0	0.466	179.9	0.514	93.3
1850.0	0.402	173.4	0.400	86.3
1900.0	0.354	167.9	0.333	81.9

Maximum amplitudes of pedestal 1 occurred at
   1650.0 rpm for the X-direction with 0.93 mils and a phase angle of $-121.8°$.
   1700.0 rpm for the Y-direction with 0.96 mils and a phase angle of $136.9°$.

The response outputs here all show a critical speed near 1650 rpm. A more precise critical speed value and corresponding response-peak values may of course be obtained by speed "zooming in" around 1650 rpm. The main feature that distinguishes this example from the previous two is that the bearings/pedestals are anisotropic and thus the orbits are ellipses, not circles.

4.2.5 Elliptical Orbits

As typified by the last example, when one or more bearings and/or pedestals have anisotropic stiffness and/or damping coefficient matrices, the steady-state unbalance response orbits are ellipses. Size, shape, and orientation of the elliptical response orbits change from station to station (refer to Figure 2.1). Furthermore, depending on the difference $(\phi_x - \phi_y)$, an orbit's trajectory direction can be corotational (forward whirl) or counter-rotational (backward whirl). Whirl direction at a given rotor station (absolute or relative to bearing) may be ascertained directly from the corresponding RDA response output using the following:

$$\text{Forward-Whirl Orbit} \longrightarrow 0 < (\phi_x - \phi_y) < 180°$$
$$\text{Backward-Whirl Orbit} \longrightarrow -180° < (\phi_x - \phi_y) \quad (4.4)$$
$$\text{Straight-Line Orbit} \longrightarrow (\phi_x - \phi_y) = 0, 180°$$

In long slender rotors, such as for large steam turbo-generator units, the whirl direction along the rotor can change direction as a function of axial position. That is, some portions of the rotor steady-state response can be in forward whirl, while the other portions are in backward whirl. Troubleshooting cases in Part 4 of this book deal with several large turbo-generators. In addition, the steady-state response orbit at a given rotor station changes with speed, as typified by the example illustrated in Figure 4.5, which shows the progressive change in orbit size, shape, and orientation as a critical speed is traversed.

The geometric properties of an orbital ellipse can be computed directly from the x and y harmonic displacement signals. With the aid of the complex-plane representation of harmonic signals previously introduced, the x and y displacement signals are first transformed in the following standard way:

$$\begin{aligned} x &= X\cos(\omega t + \theta_x) = X_1 \sin \omega t + X_2 \cos \omega t \\ y &= Y\cos(\omega t + \theta_y) = Y_1 \sin \omega t + Y_2 \cos \omega t \\ X_1 &\equiv X \sin \theta_x \quad X_2 \equiv X \cos \theta_x \\ Y_1 &\equiv Y \sin \theta_y \quad Y_2 \equiv Y \cos \theta_y \end{aligned} \quad (4.5)$$

RDA Code for Lateral Rotor Vibration Analyses

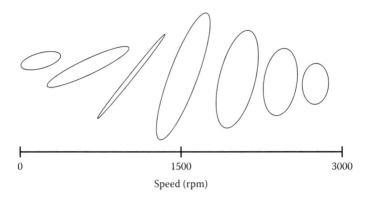

FIGURE 4.5 Orbit at a rotor station versus speed; critical speed at 1500 rpm.

Here it is advantageous to handle the orbital position vector as a complex entity, as follows ($i \equiv \sqrt{-1}$):

$$r(t) = x(t) + iy(t) \tag{4.6}$$

The complex exponential forms for the sine and cosine functions are as follows:

$$\sin \omega t = -\tfrac{i}{2}\left(e^{i\omega t} - e^{-i\omega t}\right), \quad \cos \omega t = \tfrac{1}{2}\left(e^{i\omega t} + e^{-i\omega t}\right) \tag{4.7}$$

First substitute the components of Equations 4.7 into Equations 4.5 and then substitute the results into Equation 4.6 to yield the following:

$$\begin{aligned} r(t) = &\tfrac{1}{2}\left[(X \cos \theta_x + Y \sin \theta_y) + i(-X \sin \theta_x + Y \cos \theta_y)\right] e^{i\omega t} \\ &+ \tfrac{1}{2}\left[(X \cos \theta_x - Y \sin \theta_y) + i(X \sin \theta_x + Y \cos \theta_y)\right] e^{-i\omega t} \end{aligned} \tag{4.8}$$

$r(t)$ is thus expressed in terms of two rotating vectors, as follows:

$$r(t) = R_1 e^{i(\omega t + \beta_1)} + R_2 e^{-i(\omega t - \beta_2)} \tag{4.9}$$

$$R_1 \equiv \tfrac{1}{2}\sqrt{(X \cos \theta_x + Y \sin \theta_y)^2 + (-X \sin \theta_x + Y \cos \theta_y)^2}$$

$$R_2 \equiv \tfrac{1}{2}\sqrt{(X \cos \theta_x - Y \sin \theta_y)^2 + (X \sin \theta_x + Y \cos \theta_y)^2}$$

$$\beta_1 = \arctan\left(\frac{-X \sin \theta_x + Y \cos \theta_y}{X \cos \theta_x + Y \sin \theta_y}\right)$$

$$\beta_2 = \arctan\left(\frac{X \sin \theta_x + Y \cos \theta_y}{X \cos \theta_x - Y \sin \theta_y}\right)$$

Equation 4.9 shows that the elliptical orbit decomposes into two synchronously rotating vectors, one corotational of radius R_1 and the other counterrotational of radius R_2, both with an angular speed magnitude of ω. At $t = 0$, these two vectors are positioned relative to the x-axis by their respective angles β_1 and β_2. It is then clear, as Figure 4.6 illustrates at $t = 0$, that the angle Ψ from the x-axis to the major ellipse axis is the average of these two angles, as follows:

$$\Psi = \frac{\beta_1 + \beta_2}{2} \qquad (4.10)$$

When $R_1 > R_2$ their vector sum produces *forward whirl*, and conversely when $R_1 < R_2$ their vector sum produces *backward whirl*. The orbit is a straight line when $R_1 = R_2$. Furthermore, the semimajor axis (b) and semiminor axis (a) of the orbit ellipse are given by the following expressions:

$$b = |R_1| + |R_2|, \quad a = ||R_1| - |R_2|| \qquad (4.11)$$

All the results developed here for the orbit ellipse properties in terms of the x and y harmonic displacement signals are applicable for steady-state *unbalance response* signals as well as for *instability threshold* modal orbits.

As mentioned in explaining gyroscopic effects, Section 2.4.2 of Chapter 2, even the best known twentieth century vibrations engineer, MIT Professor J. P. Den Hartog (1901–1989) "*Mechanical Vibrations,*" McGraw-Hill, 1940, had not yet fully appreciated the possibility of *backward whirl* actually occurring as he described the gyroscopic effect. In that connection, Professor Den Hartog expressed that while *backward whirl* appeared to be a valid motion solution of a rotor with a gyroscopic component, he did not see how it could be excited (by residual rotor unbalance). At that time (1940, birth year of this book's author) Professor Den Hartog apparently did not think of elliptical orbits, but considered only circular orbits.

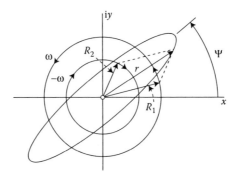

FIGURE 4.6 Elliptical orbit: the sum of two counter whirling circular orbits.

In that case the forward rotating unbalance force could not put energy into a circular *backward whirl*. However, elliptical orbits decompose into a corotational circular orbit and a counterrotational circular orbit. So a corotational unbalance force does put energy into the corotational circular orbit portion of a backward elliptical whirl orbit.

Once a steady-state response is computed, the task of visually presenting the results depends on how much detail the user requires. A multi-DOF version of Figure 1.5, with plots of amplitudes and phase angles at selected rotor stations as functions of speed, is often all that may be needed. However, to appreciate the potentially complex contortions the complete rotor undergoes in one cycle of motion requires that the orbital trajectories are pictured as a function of axial position at selected rotor speeds. Special animation software can be employed to construct an *isometric-view* "movie" of rotor orbital trajectories along the rotor. Such animations show the greatly slowed-down and enlarged whirling rotor centerline position by line connecting the instantaneous rotor radial (x, y) coordinates on the elliptical orbits axially positioned along the rotor. Animations for a flexible rotor on anisotropic bearings/pedestals clearly show that the rotor "squirms" as part of a complete cycle of motion. This is because of the response phase angle changes along the rotor, which give rise to the size, shape, and orientation of the elliptical response orbits changing from station to station at a given speed.

Thoughtfully prepared still-picture presentations can provide much of the visual communication of animations. The most extensive and informative compilation of axially distributed rotor orbits on still-picture presentations is given in the book by Lalanne and Ferraris (1998) listed in this chapter's Bibliography. Figures 4.7 through 4.9 provide a few such examples to emphasize how the fundamental rotor-orbit characteristics can be delineated by whether or not the model has *damping* and by whether or not the model has one or more *anisotropic* bearings and/or pedestals.

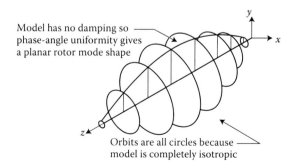

FIGURE 4.7 Isotropic model with no damping and very near resonance gives circular orbits that are all in phase.

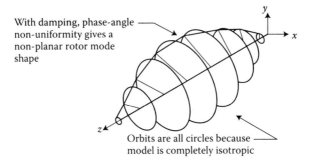

FIGURE 4.8 Isotropic model with damping gives circular orbits.

The example shown in Figure 4.7 typifies the nature of near resonance orbits at selected rotor stations for a case where all bearings and pedestals are isotropic and the model contains no damping. Naturally the orbit amplitudes (typically in the range of a few thousandths of an inch) are illustrated here as greatly enlarged. As with any harmonically excited linear vibration model, RDA response computed exactly at a natural frequency without any damping will exhibit numerical difficulties because theoretical amplitudes approach infinity with zero damping. The case illustrated in Figure 4.7 is not exactly at a critical speed. As shown, since the model is completely isotropic, the orbits are all circular and thus the additional feature of "no damping" makes the rotor mode shape planar.

The addition of isotropic bearing damping to the case in Figure 4.7 maintains the orbits as circular but causes the rotor response shape to be nonplanar, as shown in Figure 4.8. In both these isotropic cases, the respective rotor response shapes are fixed and simply precess synchronously at ω.

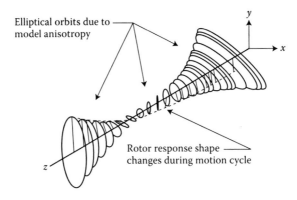

FIGURE 4.9 Response orbits of an anisotropic model with damping.

For general unbalance response linear models such as with RDA, the bearing/pedestal properties are usually anisotropic and there is nearly always bearing damping included in the model. In such general models, the synchronous orbital response trajectories are ellipses that progressively change in size, shape, and orientation both as functions of axial rotor position and speed. The case shown in Figure 4.9 illustrates such a general unbalance response output at a given speed. In such a general case, the previously mentioned animation adds significantly to the visualization to show the rotor "squirming." That is, the rotor response shape is not fixed as it is in the previous examples of Figures 4.7 and 4.8.

Unbalance response computation is one of the two most important and necessary types of rotor vibration analyses, providing a number of valuable pieces of information about the analyzed system. Unbalance response analyses show the speeds (i.e., critical speeds) where unbalance produces forced resonance responses and also how sensitive the critical-speed vibration peaks are to residual rotor unbalance magnitude and axial location. Unbalance response analyses also show if postulated damping (e.g., at the bearings) is adequate for a reasonable tolerance at resonance to residual rotor unbalance. Lastly, unbalance response analyses can be used to supplement actual balancing *influence coefficients* from trial weights (see Section 12.11 of Chapter 12).

4.2.6 Campbell Diagrams

Some vibratory systems are generically characterized by *resonance frequencies* that are *strong functions of some parameter* that can significantly change during normal operation. The most prominent and important example is the resonance frequencies of the vibration modes of blades in axial flow turbo-machinery, such as in power generation turbo-machinery and aircraft gas turbine jet engines. Typically, turbo-blade resonance frequencies increase quite strongly with rotor speed because *blade stress-stiffening* strongly increases with rotor speed.

Some rotor–bearing systems are also inherently characterized by a speed dependence of resonance frequencies. Of somewhat less criticality than axial flow turbo-machinery blade design analyses, successful design and operation for rotor dynamical performance can benefit by mapping the influence of rotor speed on analysis predicted resonance frequency magnitudes.

Presentations of resonance frequencies as functions of rotor speed are called *Campbell diagrams.* For rotor–bearing systems, application of the Campbell diagram arises when speed-dependent bearing stiffness and/or gyroscopic influences make rotor–bearing system lateral rotor vibration resonance frequencies significantly speed dependent. For fairly simple 2 bearing rotors with strong gyroscopic influences, a Campbell diagram is also instructional in clearly demonstrating spectral bifurcations of resonance frequencies along forward-whirl and backward-whirl mode branches (see Section 2.4.2 of Chapter 2). In this respect, Campbell diagrams are an excellent insight tool. Lalanne and Ferraris (1998) provide several examples demonstrating insight gleaned from Campbell diagrams.

Figure 4.10 shows an illustration that typifies large multibearing machines supported on oil-film bearings where rotor speed influence upon bearing stiffness and thus resonance frequencies is pronounced. Predicted critical speeds are indicated by the intersection ($\omega = \Omega_j$'s) of the once-per-revolution synchronous line and the resonance frequencies as functions of rotor speed. Campbell diagram assessments are useful for configuration screening at the early design phase of rotor–bearing systems when a number of design configurations are competing. Campbell diagrams can also be helpful in understanding what detailed model-based unbalance response predictions like those from RDA yield. However, when troubleshooting already existing operating machines exhibiting excessive vibration, calibrating a detailed RDA-type prediction model with reliable vibration measurements off the troubled machine is the more likely route to mission success.

Figure 4.10 superimposes a Campbell diagram with the corresponding unbalance excited rotor vibration amplitudes, thus correlating the two. The author much prefers focusing upon unbalance response over the operating speed range because it not only shows where the critical speeds are located, but also predicts vibration amplitudes at the critical speeds. Campbell diagrams only provide the critical speed locations.

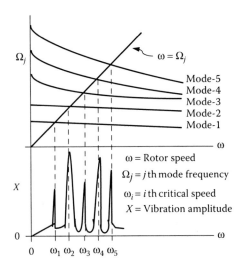

FIGURE 4.10 Campbell diagram combined with unbalance vibration.

In long flexible multibearing drivelines, such as large steam turbine-generator sets, some of the critical speeds indicated by a corresponding Campbell diagram do not show up as significant vibration peaks. This is because of a mode being heavily damped and/or not sensitive to the predominant unbalance axial locations, that is, nearness of unbalance to a mode's nodal point (see Figure 4.13). Typically a Campbell diagram may indicate say eight or nine critical speeds for a large steam turbine-generator set between 0 and 3600 rpm, but the unit only exhibits four or five of these critical speeds. In this regard, RDA-type unbalance response predictions emulate the actual machine, with the important resonance speeds evidenced by pronounced vibration peaks.

A Campbell diagram superimposed on predicted vibration amplitudes (Figure 4.10) provides engineering insight at the design phase of a machine by identifying modes that are overly sensitive to residual unbalance magnitude and axial location. Acceptable modification of features like bearing spans, bearing configurations and preloads, balance drum dynamic characteristics, support structure rigidity, bearing alignment tolerance, and process fluid vibration excitation sources can be evaluated by presenting extensive computational parametric studies in the manner of the example of Figure 4.10.

4.3 Instability Self-Excited-Vibration Threshold Computations

Avoiding self-excited rotor vibration is an absolute necessity because in most occurrences the resulting vibration levels are dangerously high, potentially causing severe machine damage within a relatively short interval of time. Even with the best of design practices and most effective methods of avoidance, self-excited rotor vibration causes are so subtle and pervasive that incidents continue to occur. Thus, a major task for the vibrations engineer is *diagnosis and correction*. Crandall (1983) provides physical descriptions for sources of dynamic destabilizing forces that are known to energize self-excited lateral rotor vibrations. Crandall describes how each of the various destabilizing mechanisms have one thing in common: a dynamic force component that is perpendicular to the instantaneous rotor dynamic radial displacement vector and thus at least partially colinear with the orbital trajectory, that is, colinear with the instantaneous trajectory velocity. Since *force × velocity = power*, such a dynamic force is nonconservative and thus potentially destabilizing. As described in Section 2.4 of Chapter 2 for linear LRV models, such destabilizing forces are embodied model-wise within the skew-symmetric portion of the stiffness matrix that operates upon a radial displacement vector to produce a force vector perpendicular to the radial displacement, consistent with Crandall's physical descriptions.

RDA utilizes the standard formulation for the extraction of eigenvalues covered in Section 1.3 and prescribed by Equation 1.59. In assessing the potential for self-excited LRV, computations are performed to locate boundaries for operating parameters (e.g., speed, power output) where a mode's complex conjugate set of eigenvalues transitions from *positive damping* to *negative damping*. In Table 1.1 (Section 1.3), this corresponds to case 1, which is the transition boundary between case 2 and case 3. In self-excited LRV, such a boundary is usually referred to as an *instability threshold*. Self-excited vibration resulting from a negatively damped 1-DOF system is formulated in Section 1.1. The initial transient vibration build-up of a self-excited unstable LRV mode occurs just like its 1-DOF counterpart illustrated in Figure 1.3. A typical journal orbital vibration transient build-up for a rotor speed above the instability threshold speed is shown in Figure 4.11. Although the nonlinear barrier presented by the bearing clearance limits the vibration amplitude, in most cases a machine would not tolerate such a high vibration level for a long period of time without sustaining significant damage. The simple examples that follow are used to demonstrate RDA's use for predicting *instability threshold speeds*.

4.3.1 Symmetric 3-Mass Rotor + 2 Anisotropic Bearings (Same) and Disk

The *simple nontrivial 8-DOF* model illustrated in Figure 2.4 is again used, here as a basis for new-user RDA demonstrations on computations to predict *instability threshold speeds*. Chapter 5 is devoted to formulations, computations, and experiments to determine bearing and seal dynamic properties. In this example, bearing dynamic properties will be used that are typical for fluid-film journal bearings, and are scaled to be consistent with the relatively small dimensions of the rotor in this example. The same 3-mass rotor model used in the three previous examples, for unbalance response, is also used here. The bearing stiffness coefficient matrices are *anisotropic* and *nonsymmetric* and are the same for both bearings to preserve symmetry about the rotor mid-plane. For this first instability threshold example, pedestals are not included.

In the previous three examples, for unbalance response, the bearing properties were contrived to be independent of speed just to keep the input shorter. In actual applications involving journal bearings, the dynamic properties of the bearings are usually quite speed dependent and should thus be input as such even for unbalance response computations. In the examples here for instability threshold speed prediction, speed-dependent bearing properties are not optional since they are required to demonstrate the computations. RDA uses bearing dynamic property inputs at a user-selected number of appropriate speeds (maximum of 10) to interpolate

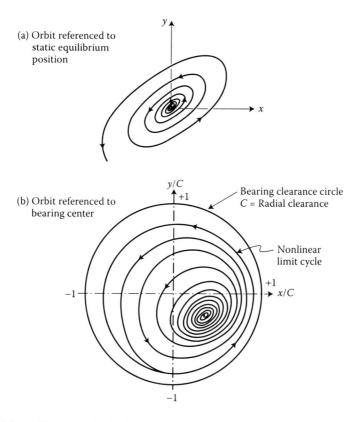

FIGURE 4.11 Transient orbital vibration build-up in an unstable condition: (a) initial linear transient build-up and (b) growth to nonlinear limit cycle.

for intermediate speeds using a cubic-spline curve fit, both for unbalance response as well as instability threshold speed computations.

From the RDA MAIN MENU, option 4 initiates an instability threshold speed computation, and the DATA MENU shown earlier in this chapter appears. Using bearing property inputs at five or more speeds is not unusual and the considerable amount of corresponding input certainly suggests that the user use a full-screen editor outside the RDA environment, at least for the speed-dependent bearing properties. That following full-screen input process is applicable for both *unbalance response* and *instability thresholds*. Inputs are in *free format*.

Input Title: 50 spaces for any alpha-numeric string of characters (1 line)
No. of: Stations, Disks, Bearings, Pedestals, Extra Weights (*integer*)
 (1 line)
Units Code: "**1**" for inches & pounds

Shaft Elements: OD, ID, Length, Inertia, Weight (1 line for each element)
Disks: Station No. (*integer*), OD, ID, Length, Weight, I_P, I_T (1 line for each disc)
Bearings: Station No. (*integer*), Weight (1 line for each bearing)
Pedestals: Station No. (*integer*), Weight (1 line for each pedestal)
Pedestals: $K_{xx}, K_{yy}, C_{xx}, C_{yy}$ (1 line for each pedestal)
Added Rotor Weights: Station No. (*integer*), Weight (1 line for each weight)
Shaft Material: Modulus of elasticity, Poisson's ratio (1 line)
No. of Speeds for Bearing Dynamic Properties: (*integer*) (1 line)
Bearing Dynamic Properties:

> RPM (1 line)
> $K_{xx}, K_{xy}, C_{xx}, C_{xy}, K_{yx}, K_{yy}, C_{yx}, C_{yy}$
> (1 line for each bearing)
⎤ Sequence for each RPM

Unbalances: Station No., Amplitude, Phase Angle (1 line for each station).

This last input group of lines (unbalances) is ignored by RDA when executing *threshold speed* runs, but may be retained in the input file. It can therefore also be excluded when executing *threshold-speed* runs. The input file for this sample can be viewed in file **sample04.inp**, but is not printed here in the interest of space.

Entering option 1 in the DATA MENU produces the INPUT OPTIONS menu, from which option 1 (file input) prompts the user for the input file name, which is **sample04.inp** for this example. **Input file must reside in RDA99 directory**. Upon entering the input file, the user is returned to the DATA MENU, where the user can select any of the six options, including option 5 that executes the previously designated Main Menu option 4 for stability analyses. Three stability analysis options are displayed as follows:

STABILITY ANALYSIS
****
The Options Are:

1. Do not iterate to find threshold speed.
   Energy check will not be performed.
   Store the eigenvalues for plotting.

2. Find the threshold speed of instability.
   Perform energy check at threshold speed.
   Store the eigenvalues for plotting.

3. Find the threshold speed of instability.
   Perform energy check at threshold speed.
   Do not store the eigenvalues for plotting.

The new user should explore all three of these options. Option 1 provides the complex eigenvalues for a speed range and increment prompted from the user. Plotting the real eigenvalue parts as functions of speed is one way of determining the instability threshold speed, that is, by finding the lowest speed at which one of the eigenvalue real parts changes from negative (positively damped) to positive (negatively damped). At this negative-to-positive crossover speed, the two eigenvalues for the threshold (zero-damped) mode are imaginary conjugates and thus provide the natural frequency of the unstable mode. Plotting the first few lowest frequency modes' eigenvalues versus speed can provide information (see Campbell diagram in Section 4.2.6) to corroborate which modes are shown to be sensitive to rotor unbalance. However, option 3 is more expedient since it automatically "halves in" on the positive-to-negative crossover *threshold speed* to within the user supplied *speed convergence tolerance*. In this demonstration example, option 3 is selected. Although not often experienced by the author, option 3 may skip over an instability threshold speed due to the fact that the bearing coefficients are provided at distinct speeds between which curve fitting of bearing coefficients is used. Small errors stemming from this curve fitting can significantly corrupt the instability threshold search algorithm. Rather than reflecting algorithm shortcomings, this potential difficulty is a result of the extreme sensitivity of the balance of positive and negative energy right at an instability threshold. If this difficulty occurs, use option 1 and plot the real part versus speed for the lower frequency modes and thereby graphically capture the zero crossover speed.

The unstable mode theoretically has exactly zero net damping at the instability threshold, so its eigenvector at the threshold speed (and only at the threshold speed) is not complex. Thus, a real mode shape can be extracted from the threshold-speed eigenvector. The "energy check" referenced in the STABILITY ANALYSIS menu uses the eigenvector components for the mode at the determined stability threshold speed to construct that mode's normalized x and y harmonic signals at the bearings to perform an energy-per-cycle computation at each bearing, as provided by Equation 2.81. This computation provides a potential side check for solution convergence of the threshold speed, because exactly at a threshold of instability the sum of all energy-per-cycle "in" should exactly cancel all energy-per-cycle "out." However, the second example in this section demonstrates that in some cases inherent computational tolerances in eigenvector extraction can make the energy-per-cycle residual convergent to a relatively small but nonzero limit. STABILITY ANALYSIS option 3 prompts for the following (inputs shown):

Input lower speed (rpm)	0 (RDA starts at the lowest bearing data speed)
Input upper speed (rpm)	4000
Desired accuracy (rpm)	1

The user is next prompted with an option to change the speed tolerance. With present PCs being so much faster than the early PCs for which RDA was originally coded, the user should answer the prompt with "N" for "No." The user is next prompted to select from the following three choices.

The bearing coefficients will be fitted by a cubic spline.

Three types of end conditions could be used:

1. Linear
2. Parabolic
3. Cubic

Option 1 (Linear) is used in this demonstration example.

The user is next prompted to select from the following three choices pertaining to shaft mass model formulation, just as in unbalance response cases.

Shaft mass model options:

1. Lumped mass
2. Distributed mass
3. Consistent mass

The *consistent mass* option is usually the preferred choice, and is chosen here. For any given rotor, curious users may compare model resolution accuracy or convergence of these three options by varying the number of shaft elements.

The last user prompt is to give a name to the output file that will be generated (here **sample04.out** is provided). The complete output file for this example is provided with the CD-ROM that comes with this book. An abbreviated portion of that output file is given here as follows:

Stability Analysis Results	
Threshold speed	2775.6 rpm ± 1.00 rpm
Whirl frequency	1692.5 cpm
Whirl ratio	0.6098

		Energy Per Cycle at the Onset of Instability		
Bearing No.	Rotor Location	Damping Part, C_{ij}^s	Stiffness Part, K_{ij}^{ss}	Net Energy
1	1	−349.9	350.0	0.148
2	3	−348.5	348.6	0.143
Energy/cycle of the bearings total				0.291

As can be observed from the quite small bearing energy-per-cycle residual, the user provided 1-rpm convergence criteria for the instability threshold speed provides an eigenvector indicative of a zero-damped mode. The energy-per-cycle output tabulations reflect that the model (including bearing coefficient inputs) is symmetric about the rotor midplane. In the next example, where the bearings are somewhat different, it is seen that the total energy per cycle residual does not approach "smallness" to the same degree as this example, even though the threshold speed iteration has essentially converged to the solution. One can conclude that the energy-per-cycle criteria for convergence are much more stringent than the speed tolerance. The "whirl ratio" (whirl frequency/threshold speed) is always less than "one" for this type of instability, that is, the associated self-excited vibration is always *subsynchronous*.

The normalized threshold (zero-damped) mode used for the energy-per-cycle computations is essentially planar, which can be deduced from the following RDA output for this example. The mode for this example is indicative of the typical nearly circular orbit shapes at instability thresholds.

	Normalized Self-Excited Vibration Mode			
	Coordinate	Amplitude	Phase (rad)	Phase Angle (°)
x_1	1	0.1016590	0.9276607E−03	0.0
y_1	2	0.8729088E−01	−1.822578	−104.4
θ_{x1}	3	0.1160974	1.317930	75.5
θ_{y2}	4	0.1352273	−0.9837417E−04	0.0
x_2	5	1.000000	0.0000000	0.0
y_2	6	0.8586232	−1.823513	−104.5
θ_{x2}	7	0.2455765E−03	1.304664	74.8
θ_{y2}	8	0.2896766E−03	0.9317187E−02	0.0
x_3	9	0.1014572	0.8898759E−03	0.0
y_3	10	0.8711492E−01	−1.822611	−104.4
θ_{x3}	11	0.1163277	−1.823615	−104.5
θ_{y3}	12	0.1355006	3.141485	180.0

Namely, the orbits are "fat ellipses" or "almost circular," and there is an insight to be gleaned from this. Referring to Equation 2.79 for the energy-per-cycle input from the skew-symmetric part of the bearing stiffness matrix, the integrated expression is the *orbit area*. Thus, the destabilizing energy is proportional to the normalized orbit area, which is a maximum for a purely circular orbit. A major European builder of large steam turbogenerator units used this idea "in reverse" by making the journal bearings much stiffer in the vertical direction than in the horizontal direction, to create "very flat" modal orbit ellipses (i.e., small normalized orbit areas), with the objective of increasing the *instability threshold power* for steam-whirl-induced self-excited vibration. This design feature unfortunately made these machines difficult to balance well, and was thus subsequently "reversed" in the power plants as per customers' request (Adams and Makay, 1981).

4.3.2 Symmetric 3-Mass Rotor + 2 Anisotropic Bearings (Different) and Disk

The model for this example differs from the previous model only in the bearing coefficients for bearing 2, which are somewhat different from those of bearing 1. This example demonstrates instability threshold output for the more typical machine configuration where perfect symmetry is not preserved. The input file (**sample5.inp**) and output file (**sample5.out**) are on the CD-ROM that comes with this book. Below is an abbreviated output summary with a speed tolerance of ±1 rpm.

Stability Analysis Results	
Threshold speed	2017.4 rpm ± 1.00 rpm
Whirl frequency	1455.7 cpm
Whirl ratio	0.7216

	Energy Per Cycle at the Onset of Instability			
Bearing No.	Rotor Location	Damping Part, C_{ij}^s	Stiffness Part, K_{ij}^{ss}	Net Energy
1	1	−36,041	30,865	−5176
2	3	−157	167	10
Energy/cycle of the bearings total				−5166

A first impression of the energy-per-cycle residual here might induce one to question the quality of solution convergence. However, the following abbreviated output summary from a rerun of this example with the significantly smaller speed tolerance of ±0.1 rpm does not support such a first impression.

Stability Analysis Results	
Threshold speed	2016.5 rpm ± 0.10 rpm
Whirl frequency	1455.2 cpm
Whirl ratio	0.7216

Energy Per Cycle at the Onset of Instability				
Bearing No.	Rotor Location	Damping Part, C_{ij}^s	Stiffness Part, K_{ij}^{ss}	Net Energy
1	1	−36,035	30,852	−5183
2	3	−158	167	9
Energy/cycle of the bearings total				−5174

For practical purposes, the threshold speed answer here is the same as computed in the initial run that used a ±1 rpm speed tolerance. In contrast to the previous example, which is symmetric about the mid-plane, the threshold mode in this example has its largest modal motion at station 1 (bearing 1) and its smallest motion at station 2 (disc). In the previous example, the disc's threshold modal orbit is about 10 times as large as at the bearings. The difference in energy-per-cycle residual convergence characteristics between these two examples, one symmetric and one not, invites further research. Clearly, the energy-per-cycle criteria for threshold speed convergence are more stringent than speed tolerance, but fortunately threshold speed is the answer sought.

In all RDA examples thus far presented, the bearing damping coefficient arrays used are symmetric. For the examples in this section, bearing stiffness and damping coefficients originate from standard computations for fluid-film hydrodynamic journal bearings, that is, using "small" radial position and velocity perturbations on the solution of the Reynolds lubrication equation (RLE). As shown in Section 2.4 of Chapter 2, a skew-symmetric portion of a bearing "damping" matrix is not really damping since it embodies a conservative force field, and thus should be present only if needed to capture bearing (or seal) fluid inertia effects. Chapter 5 more thoroughly develops this and other aspects of bearing dynamic properties, but it is relevant to mention here that the classical RLE encompasses only the viscous effects of the lubricant fluid with no account of the fluid inertia effects. Thus, journal bearing dynamic properties obtained from Reynolds-equation-based perturbations must have symmetric damping coefficient arrays.

This first group of RDA examples should provide one with a background to begin analyses of other cases. Like any computer code, RDA is just a tool and thus can be used properly or improperly. As will be exposed more fully in Part 4, proper use of an LRV code like RDA demands that the user apply good engineering judgment and care in devising models that adequately portray the important modes and responses of the system.

4.4 Additional Sample Problems

The sample problems of the previous two sections were devised primarily to give one a primer on use of the RDA code. The new RDA user is encouraged to analyze variations of that initial batch of samples. That is, to perform basic parametric studies on the model to study the influence of input variations (bearing, pedestal, shaft, etc.) on the results, such as *critical speeds* and attendant *amplitude peaks*, and *instability threshold speeds*. The sample problems provided in this section are an extension of the RDA *primer* begun with the previous sample problems.

4.4.1 Symmetric 3-Mass Rotor + 2 Anisotropic Bearings and 2 Pedestals

The inputs for this example are the same as the previous example except that pedestals are added at each of the two bearings. Input file **sample06.inp** for this sample contains the following requisite input modifications to input file **sample05.inp**.

	Number of Pedestals: 2		
Pedestal Data:	Pedestal No.	Station No.	Weight (lb)
	1	1	25.0
	2	3	25.0

	Pedestal Stiffness and Damping Coefficients			
	K_{xx} (lb/in.)	K_{yy} (lb/in.)	C_{xx} (lb s/in.)	C_{yy} (lb s/in.)
Pedestal no. 1:	15,000	25,000	1	1
Pedestal no. 2:	15,000	25,000	1	1

The following is an abbreviated results summary for this example.

Stability Analysis Results	
Threshold speed	2004.1 rpm ± 0.10 rpm
Whirl frequency	1530.0 cpm
Whirl ratio	0.7634

The new RDA user is encouraged at this point to explore moderate input variations, specifically for the pedestal parameters. For example, a small reduction in both pedestal x-stiffness inputs may provide the surprise of eliminating a threshold of instability from the speed range below the maximum speed of bearing input stiffness and damping coefficients (i.e., no threshold speed below 8015 rpm). The new RDA user should attempt to explain such dramatic changes in the results.

4.4.2 Nine-Stage Centrifugal Pump Model, 17-Mass Stations, 2 Bearings

The input file name for this model is **pump17.inp**. Both *unbalance response* and *instability threshold speed* cases are included here for the main purpose of comparison with the next example which is a five-mass-station model of the same pump. The rotor model for this example is based on a pump rotor quite similar to that shown in Figure 4.12. It has two oil-film journal bearings and nine impeller stages to produce a very high pump pressure. The model here is purely for RDA demonstration purposes. It does not account for the quite significant effects of any of the inter-stage close-clearance sealing gaps and end seals, all of which have their own *bearing-like* rotor dynamic coefficients that can be entered into RDA just like the coefficient inputs for journal bearings. In Part 4, such effects were included in the models used for troubleshooting case studies presented.

4.4.2.1 Unbalance Response

This pump could be driven either by a constant speed driver (e.g., induction motor) or a variable speed driver (e.g., frequency-inverter drive motor, induction motor through fluid coupling, or an auxiliary steam turbine).The main advantages of variable-speed drive for such a pump include operation at a "best efficiency point" (BEP) over a wide flow range, and avoidance of intense flow-induced vibration at flows significantly below BEP flow. In any case, it is prudent practice to analyze the unbalance response over a speed range that is significantly higher than the anticipated maximum operation speed. This is to insure the detection of any unbalance-sensitive critical speeds of the model that might be located just above the maximum operating speed. Given the possible inaccuracies of any model, such critical speeds that are computed to be only marginally above the maximum operating speed could in fact intrude into the upper range of the operating speed on the actual machine. Such pumps are generally driven through a so-called flexible coupling, which provides a tolerance of angular as well as parallel misalignment between the driver and pump. As a consequence, LRV characteristics of the pump are essentially decoupled from the driver.

FIGURE 4.12 Rotor of a nine-stage double-case centrifugal pump.

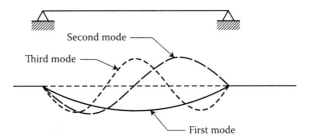

FIGURE 4.13 First three planar mode shapes for a simply supported uniform beam.

This example involves a relatively long flexible rotor with nine impeller stages inboard of two journal bearings that are located near their respective ends of the rotor. Thus, one should anticipate the possibility of more than one bending-type critical speed existing within the operating speed range. To insure the potential for exciting multiple bending critical speeds with unbalances, the axial location and phasing of the unbalance inputs should be properly configured, as is demonstrated in this example. For an indication of where to place unbalances in the model, one should be guided by the mode shapes for a uniform beam with appropriate approximate boundary conditions. In this example the so-called *simple support* case of a uniform beam, illustrated in Figure 4.13, is appropriate.

In Section 1.3 of Chapter 1 it was shown that the influence of a force on a particular mode is proportional to the *participation factor* of the mode at the point of the force's application, that is, proportional to the relative displacement magnitude at the point of application. Accordingly, an unbalance placed near the axial midpoint of this example rotor can be expected to provide near-maximum effect on the first and third modes, whereas unbalances placed near the 1/4 and 3/4 axial locations at 180° out of phase can be expected to provide near-maximum effect on the second mode. Accordingly, input file **pump17.inp** is configured with three such unbalances. Also, to make the problem a bit more interesting, the two 180°-out-of-phase unbalances are placed 90° out of phase with the axial midpoint unbalance. The full output file is **pump17ub.out**, from which the following abbreviated output summary is extracted.

	Response of Rotor Station No. 5 (Near 1/4 Axial Position)			
	X-Direction		Y-Direction	
Speed (rpm)	Amplitude (mils)	Phase Angle (°)	Amplitude (mils)	Phase Angle (°)
1200.0	0.333	−1.9	0.327	−93.9
1400.0	0.678	−1.5	0.676	−93.0

continued

Continued

1600.0	2.138	−1.1	2.106	−93.0
1800.0	**4.593**	**179.8**	**4.759**	**91.8**
2000.0	1.410	−179.8	1.416	90.5
2200.0	0.943	−179.2	0.946	90.9
6200.0	0.786	−131.7	0.786	138.3
6600.0	1.162	−120.1	1.164	149.9
7000.0	2.636	−107.3	2.634	162.7
7200.0	7.237	−101.0	7.234	169.0
7400.0	**10.493**	**85.2**	**10.256**	**−4.7**
7600.0	3.159	91.3	3.162	1.3
8000.0	1.452	102.1	1.453	12.1

	Response of Rotor Station No. 9 (Near 1/2 Axial Position)			
	X-Direction		Y-Direction	
Speed (rpm)	Amplitude (mils)	Phase Angle (°)	Amplitude (mils)	Phase Angle (°)
1200.0	0.442	−0.7	0.434	−92.5
1400.0	0.896	−0.7	0.894	−92.1
1600.0	2.812	−0.8	2.770	−92.6
1800.0	**6.004**	**179.6**	**6.223**	**91.6**
2000.0	1.830	179.4	1.838	89.8
2200.0	1.214	179.4	1.219	89.5
2400.0	0.963	179.3	0.964	89.4
6200.0	0.428	175.5	0.428	85.5
6400.0	0.419	174.2	0.419	84.2
6600.0	0.411	172.4	0.410	82.3
6800.0	0.403	168.8	0.403	78.8
7000.0	0.402	161.0	0.402	71.0
7200.0	0.499	133.1	0.499	43.0
7400.0	**0.696**	**−130.0**	**0.684**	**139.4**
7600.0	0.421	−157.9	0.421	112.1
7800.0	0.384	−165.6	0.384	104.4
8000.0	0.366	−169.1	0.366	100.9

	Response of Rotor Station No. 13 (Near 3/4 Axial Position)			
	X-Direction		Y-Direction	
Speed (rpm)	Amplitude (mils)	Phase Angle (°)	Amplitude (mils)	Phase Angle (°)
1200.0	0.312	0.4	0.306	−91.9
1400.0	0.634	0.1	0.631	−91.6
1600.0	1.997	−0.4	1.964	−92.4
1800.0	**4.282**	**179.4**	**4.435**	**91.4**
2000.0	1.312	178.6	1.317	88.9

continued

Continued

2200.0	0.876	177.9	0.879	88.0
2400.0	0.699	177.1	0.700	87.2
6200.0	0.687	121.1	0.686	31.2
6400.0	0.823	114.1	0.823	24.1
6600.0	1.047	107.2	1.049	17.1
6800.0	1.480	99.9	1.480	9.9
7000.0	2.492	93.0	2.490	3.0
7200.0	7.020	86.3	7.017	−3.7
7400.0	**10.438**	**−99.8**	**10.204**	**170.2**
7600.0	3.215	−105.5	3.218	164.5
7800.0	2.018	−110.6	2.018	159.4
8000.0	1.531	−115.2	1.532	154.8

The results summarized here clearly show two critical speeds, the first near 1800 rpm and the second near 7400 rpm. One may of course "zoom in" on these two speeds to more accurately acquire the model's critical speeds and associated amplitude peaks. As the full unabridged results output on file **pump17ub.out** show, the motion at the two journal bearings is vanishingly small over the complete computed speed range, indicating that the rotor locations at the bearings are virtual *nodal points* for both critical speeds. And that is consistent with the mode shapes at the first and second critical speeds, albeit nonplanar, closely resembling the corresponding mode shapes shown in Figure 4.13 for the simply supported uniform beam. For example, the vibration level at station 9 (near the rotor mid-plane) shows virtually no sensitivity to the second critical speed, indicating that station 9 is practically a *nodal point* for the second critical speed. Furthermore, the relative amplitudes for the second critical speed at stations 5, 9, and 13 are in qualitative agreement with the second mode shape for the simply supported uniform beam. This example clearly shows that vibration measurements near the bearings may not correlate well at all with *rotor vibration at the mid-span zone*.

4.4.2.2 Instability Threshold Speed

The same input file from the previous example is used here to perform a computation for determining if a threshold speed is predicted in the speed range below 8000 rpm for the nine-stage centrifugal pump. The full unabridged results are on output file **pump17ts.out.** An abbreviated output summary is given as follows:

Stability Analysis Results	
Threshold speed	2648.1 rpm ± 0.10 rpm
Whirl frequency	1417.0 cpm
Whirl ratio	0.5351

This result shows a couple of features that are typical for this type of instability (commonly called *oil whip*). First, the whirl ratio at the *oil-whip threshold speed* is close to 1/2. Second, the mode that is self-excited is quite similar to the first-critical-speed mode excited by unbalance in the previous example, but with two notable differences: (*i*) The motion at the bearings is approximately 5% of the maximum (at rotor mid-plane) instead of being vanishingly small. This is because the journal bearings are providing the self-exciting destabilizing mechanism. (*ii*) The unstable mode's natural frequency (1417 cpm) is noticeably lower than the first critical speed (~1800 rpm). This is because the bearing's hydrodynamic oil films become thicker, and thus less stiff, as rotational speed is increased. Consequently, the first mode's natural frequency at the threshold speed (2648 rpm) is 1417 cpm, not 1800 cpm.

4.4.3 Nine-Stage Centrifugal Pump Model, 5-Mass Stations, 2 Bearings

This 5-mass-station rotor model has been configured to provide a best effort at approximating the previous 17-mass-station rotor model. The input file is **pump5.inp**. Both the *unbalance response* case and *instability threshold speed* case have been rerun with the 5-mass rotor model. Bearing inputs are the same as the **pump17.inp** file. A brief summary for unbalance response output from file **pump5ub.out** is presented as follows. The first critical speed is reasonably close to that with **pump17** but the second critical speed differs considerably from that of **pump17**, as should be expected.

$$\text{1st critical speed} \cong 1800 \text{ rpm with } x_{\frac{1}{2}}\text{-axial-position amplitude}$$
$$\cong 6 \text{ mils}$$
$$\text{2nd critical speed} \cong 6400 \text{ rpm with } x_{\frac{3}{4}}\text{-axial-position amplitude}$$
$$\cong 37 \text{ mils}$$

The *instability threshold speed* case computed with **pump17.inp** is repeated here using **pump5.inp**. The following output summary is extracted from the full output file **pump5ts.out**.

Threshold speed	2449.5 rpm ± 10 rpm
Whirl frequency	1512.3 rpm
Whirl ratio	0.6174

As these results show, the threshold speed computed here is approximately 200 rpm lower than that computed from **pump17.inp** and the whirl frequency is approximately 100 cpm higher. The reason the threshold results from **pump5.inp** are this close to those from **pump17.inp** is because the mode at instability threshold is just a somewhat "softer version" of the mode at the first critical speed, as previously explained. As explained in Chapter 1, the higher the mode number needed, the more the DOF (i.e., the more finite elements) necessary to accurately portray the actual continuous media body with a discrete model. The comparisons between the **pump17** and **pump5** results are completely consistent with that axiom.

4.5 Summary

The primary focus of this chapter is to provide a primer on using the RDA code for LRV analyses. Several carefully configured examples are presented for that purpose. In addition to the "how to" instructions, attention is given to important issues needed to make comprehensive use of what "comes out" of RDA. This includes showing that *unbalance response* results are the best approach to determine the so-called *critical speeds* at which sensitivity to residual rotor unbalance can produce significant resonant vibration peaks. Also, the confusing topic of rotor vibration *phase angles* is clearly and comprehensively covered. The explanation of *elliptical orbits* and their changing size, shape, and orientation as functions of rotor axial position and speed is provided as an in-depth treatment. The important topic *instability self-excited rotor vibration* is both analyzed and explained. In Part 4 (*Troubleshooting*), a constructive interplay between the analysis types covered in this chapter and the *Monitoring and Diagnostics* methods covered in Part 3 provide serious vibration analysts and troubleshooters a broad picture of the methods used to solve rotating machinery vibration problems.

When using a code like RDA for design prediction analyses of rotor dynamic behavior, it is important to keep in mind the inherent uncertainties in the inputs, most notably bearing stiffness and damping coefficients. Even under controlled laboratory testing, these uncertainties remain significant as Adams and Falah (2004) show with their laboratory 3-bearing flexible-rotor test rig. They show comparisons between test results and RDA predictions, both for rotor unbalance response through critical speed and oil whip thresholds speeds, for a wide range of bearing loads.

Bibliography

Adams, M. L. and Makay, E., *How to Apply Pivoted-Pad Journal Bearings*, Power Magazine, McGraw-Hill, New York, October 1981.

Adams, M. L. and Falah, A. H., *Experiments and Modeling of a Three- Bearing Flexible Rotor for Unbalance Response and Instability Thresholds*, Eighth IMechE

International Conference on Vibrations in Rotating Machinery Swansea, Wales, UK, 2004.

Crandall, S. H., *The Physical Nature of Rotor Instability Mechanisms*, ASME Applied Mechanics Division Symposium on Rotor Dynamical Instability, AMD Vol. 55, 1983, pp. 1–18.

Lalanne, M. and Ferraris, G., *Rotordynamics Prediction in Engineering*, Wiley, England, 1998, pp. 254.

Maghraoui, M., *Control of Rotating Machinery Vibration Using an Active Multi-Frequency Spectral Optimization Strategy*, PhD Thesis, Case Western Reserve University, January 1989.

PROBLEM EXERCISES

1. Using RDA, model the simple rotor configuration of Problem 1 in the problem exercises of Chapter 2 with a 2-element (3-mass) model. Compute its response to a central unbalance (at the disk) up to and somewhat beyond the lowest critical speed. This will necessitate adding bearing stiffness and damping at the two ends of the shaft, to replace the two simple supports of the prior problem. Start with a low bearing isotropic stiffness (like 1000 lb/in.) and low isotropic bearing damping (like 0.005 lbs/in.). Progressively increase the bearing stiffness until the output does not appreciably change. Then adjust the unbalance force magnitude to achieve single peak critical speed vibration amplitude of about 0.010 in. at the disk. Remember, the system is linear. Now model and simulate this configuration using the 8-DOF nontrivial model (Equation 2.18) and compare its predictions with RDA results for critical speed and peak vibration magnitude. Compare the RDA critical speed with the natural frequency determined in Problems 1 and 2 of the problem exercises in Chapter 2.

2. Using RDA, devise a 2-element model of the rotor shown in Problem 3 in the problem exercises of Chapter 2. As indicated in Problem 1 here, provide isotopic bearing and stiffness coefficients, sufficiently stiff to emulate the rigid simple end supports and lightly damped so that critical speed vibration does not seek infinite amplitude. Compare the critical speed(s) with the natural frequency(s) of the configuration in Chapter 2.

3. With the RDA model developed in Problem 2 here as a starting model, progressively increase the number of shaft elements to determine how many elements are required to converge to best achievable model resolution. Perform this parametric study for all RDA's three mass matrix options (lumped, distributed, and consistent mass models). Do this parametric study for the two configurations (a) shaft with disk and (b) shaft alone without disk. Plot results for both (a) and (b), showing predicted critical speed values as a function of number of elements.

4. With the RDA model developed in Problem 1 here as a starting model, progressively increase bearing stiffness to establish

the maximum bearing stiffness value before RDA computations exhibit numerical difficulties. Explain the fundamental reason for these numerical difficulties. Try to establish an approximate guideline for maximum bearing-to-shaft stiffness ratio that will not cause such numerical problems.
5. In a manner similar to Problem 4 and using the same RDA model, progressively decrease bearing stiffness to establish the minimum bearing stiffness value before RDA computations exhibit numerical difficulties. Similar to Problem 4, try to establish an approximate guideline for minimum bearing-to-shaft stiffness ratio that will not cause such numerical problems. Clearly, Problems 4 and 5 are important to demonstrate how to use RDA for rotor systems where bearings are tantamount to rigid simple supports and the other extreme case where the rotor is tantamount to being a free–free flexible body of revolution.
6. Conduct a computational research investigation to seek an explanation for the apparent discrepancy of instability threshold speed energy balance in the **sample5** example problem (see Section 4.3.2).
7. Conduct a research investigation to seek an explanation for the extreme sensitivity of predicted threshold speed to bearing pedestal stiffness in the **sample6** example problem (see Section 4.4.1).

5
Bearing and Seal Rotor Dynamics

5.1 Introduction

RDA, the modern FE-based PC code supplied with this book, is presented from a fundamentals perspective in Chapter 2 and a user's perspective in Chapter 4. There are a number of commercially available codes with similar capabilities. Engineering analysis codes in general and rotor dynamics codes in particular nearly always have one tacit fundamental trait in common. That trait is as follows:

> Those aspects of the problem class that are reasonably well defined and modeled by first principles are what is "inside" the computer code. Whereas, those aspects which are not as well defined and modeled by first principles show up as some of the "inputs" to the computer code.

With this approach, the typical computer code developer and marketer has long been quick to tout their code as capable of handling "any" conceivable problem within the code's intended range of usage, *as long as one has all the "correct" inputs.*

For LRV analyses, those important inputs that present the biggest challenge are the dynamic properties (stiffness, damping, and inertia coefficients) for the components that *dynamically connect* the rotor to the stator (stator ≡ everything that does not rotate). These components include first and foremost the *radial bearings*. In many rotating machinery types (e.g., turbo-machinery) other liquid- and gas-filled internal close-clearance annular gaps, such as seals, are also of considerable LRV importance. Furthermore, the confined liquid or gas that surrounds a rotor component (e.g., centrifugal pump impeller and balancing drum) may also significantly contribute to the basic vibration characteristics of a rotating machine, both in an interactive way much like bearings and seals, and as explicit time-dependent unsteady-flow forces (e.g., hydraulic instability in centrifugal pumps, rotating stall in turbo-compressors). Motor and generator electromagnetic forces also contribute. Most modern LRV research has been devoted to all these rotor–stator effects. One could justifiably devote an entire book just to this single aspect of LRV. This chapter focuses on bearing and seal LRV dynamic properties. Small clearances critical to these properties are of significant uncertainty because of manufacturing tolerances.

Thus, LRV characteristics are really *stochastic* rather than *deterministic*. That is, if significant inputs are random-variable distributions, then so are the outputs.

5.2 Liquid-Lubricated Fluid-Film Journal Bearings

5.2.1 Reynolds Lubrication Equation

The strong urge to rigorously derive the classic Reynolds lubrication equation (RLE) is here resisted in the interest of space and because the RLE is so aptly derived in several references (e.g., Szeri, 1998). To facilitate the serious reader's understanding of available derivations of the RLE, the following perspective is provided. Figure 5.1 provides an elementary illustration of a journal bearing.

The general starting point for modeling fluid mechanics problems is encompassed in the three coupled *fluid-momentum PDEs* [Navier–Stokes (N–S) equations] plus the single *conservation-of-mass PDE* (continuity equation). The three scalar N–S equations (which are nonlinear) are obtained by applying Newton's Second Law $\Sigma \vec{F} = d(m\vec{v})/dt$ to an inertial differential control volume (CV) of a continuum flow field. Attempting to solve these equations for 2D and 3D problems has historically been the challenge to occupy the careers of fluid mechanics theoreticians, because these equations are nonlinear and coupled. The ingenious contributions of the precomputer age fluid mechanics "giants" (like Osborne Reynolds) sprang from the application of their considerable physical insight into specific problems, leading them to make justifiable simplifying assumptions, thereby producing important solvable formulations. This was tantamount

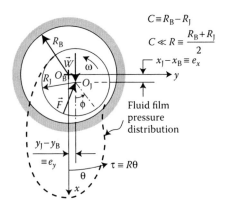

FIGURE 5.1 Generic journal bearing configuration and nomenclature.

to identifying and excising those terms in the N–S equations of secondary importance for a specific problem. In this regard, Reynolds' (1886) original paper on development of the RLE is nothing short of a masterpiece.

In a "nutshell" the RLE applies to an incompressible laminar (no turbulence) strictly viscous (no fluid inertia) thin fluid film between two closely spaced surfaces in relative motion. Because of the neglect of fluid inertia, all the nonlinearities (convective inertia terms) are deleted from the N–S equations. Because of the close spacing of the two surfaces, a number of further simplifying assumptions are invoked. These include neglect of local surface curvature and neglect of gradients of fluid shear stress components in the local plane of the thin fluid film, because they are much smaller than the gradients across the thin fluid film. The simplifying assumptions additionally include neglect of the fluid velocity and the change in local pressure normal to the local plane of the film. When all these simplifying assumptions are implemented, the N–S equation for the direction normal to the film is eliminated. The other two N–S equations (for the two in-plane directions) are decoupled from each other and are left with only shear stress and pressure terms for their respective directions. Integrating these two differential equations and applying the surface velocity boundary conditions yield solutions for the two in-plane velocity distributions in the film in terms of the local in-plane pressure gradient terms and relative velocity components between the surfaces.

These velocity solutions substituted into the conservation-of-mass condition yield the Reynolds equation. Originally, only sliding velocity between the two surfaces was considered. Much later, as Equation 5.1 for the RLE reflects, the so-called *squeeze-film* term was added to handle local relative velocity of the surfaces perpendicular to their local plane. Within the context of rotor dynamics, it is the *sliding velocity* term that gives rise to the bearing *stiffness* coefficients and the *squeeze-film velocity* term that gives rise to the bearing *damping* coefficients:

Sliding velocity term   Squeeze-film term

$$\frac{\partial}{\partial \tau}\left(\frac{h^3}{\mu}\left(\frac{\partial p}{\partial \tau}\right)\right) + \frac{\partial}{\partial z}\left(\frac{h^3}{\mu}\left(\frac{\partial p}{\partial z}\right)\right) = 6\omega R \frac{dh}{d\tau} + 12 \frac{dh}{dt}$$

$p = p(\tau, z), \quad h = h(\tau, z), \quad 0 \leq \tau \leq 2\pi R, \quad -L/2 \leq z \leq L/2, \quad \mu = \text{viscosity}$

(5.1)

Here, $p(\tau, z)$ is the film pressure distribution, $h(\tau, z)$ is the film thickness distribution, and L is the hydrodynamic-active axial length of the journal bearing. $p(\tau, z)$ is the unknown parameter and all other parameters are specified.

It was Reynolds' objective to explain then recently published experimental results for rail-locomotive journal bearings, which showed a capacity to generate film pressures in order to keep the rotating journal from contact rubbing of the bearing. Reynolds' derivation showed how the sliding action of the rotating journal surface, shearing oil into the converging thin gap between an eccentric journal and bearing, produced a *hydrodynamic pressure distribution* that could support static radial loads across the oil film without the journal and bearing making metal-to-metal contact. This is one of the most significant discoveries in the history of engineering science. Reynolds' derivation clearly showed that this hydrodynamic load capacity was in direct proportion to the sliding velocity (rotational speed) and the lubricant viscosity. Virtually every first-level undergraduate text in Machine Design has a chapter devoted to hydrodynamic journal bearing design based on the Raimondi and Boyd (1958) computer-generated nondimensional solutions to the RLE for static load capacity (see A. A. Raimondi in the Acknowledgment section of this book). The focus here is primarily on how solutions of the RLE are used to determine journal bearing *stiffness and damping* coefficients.

Before the existence of digital computers, Equation 5.1 was solved by neglecting either the axial pressure flow term ("long bearing" solution) or the circumferential pressure flow term ("short bearing" solution). With either approximation, the RLE is reduced to an ODE (i.e., one independent spatial coordinate) and thus solvable without computerized numerical methods. These two approximate solutions provide an upper bound and lower bound, respectively, for the "exact" 2D solutions to Equation 5.1. Whether using one of these approximate solution approaches or a full 2D numerical solution algorithm, pressure boundary conditions must be specified with Equation 5.1 in order to have a well-posed mathematical problem. The generic circumferential view of a journal bearing hydrodynamic pressure distribution in Figure 5.1 is for the typical ($p = 0$) boundary condition in which cavitation or film rupture in the diverging portion of the fluid film gap is handled by imposing the additional boundary condition $\vec{\nabla}p = 0$ at the interface between the full-film region (in which the RLE is used) and the ruptured region (in which the pressure distribution is set equal to vapor pressure ≈ 0). The $\vec{\nabla}p = 0$ condition imposes the physical requirement that lubricant mass flow is conserved across the interface boundary separating the full-film and ruptured-film regions.

To show how journal bearing *stiffness and damping* coefficients are obtained from the RLE, it is necessary to first show how solution of the RLE is used to generate static load capacity design curves like those originally published by Raimondi and Boyd (1958). The sequence of computational steps for obtaining solutions to Equation 5.1 is exactly the reverse of the sequence of steps when pre-existing RLE solutions are subsequently used

5.2.1.1 For a Single RLE Solution Point

1. Specify $e \equiv \sqrt{(e_x^2 + e_y^2)}$, $\phi = \arctan(e_y/e_x)$, $e_x = x_J - x_B$, $e_y = y_J - y_B$

   With journal-to-bearing axial alignment, $h = C - e_x \cos(\tau/R) - e_y \sin(\tau/R)$ giving $(dh/d\tau) = (e_x/R)\sin(\tau/R) - (e_y/R)\cos(\tau/R)$, $\dot{h} = -\dot{e}_x \cos(\tau/R) - \dot{e}_y \sin(\tau/R)$

2. Solve the RLE for the pressure distribution $p = p(\tau, z)$
3. Integrate $p(\tau, z)$ over the journal cylindrical surface to get x and y forces upon the journal:

$$F_x = -\int_{-L/2}^{L/2}\int_0^{2\pi R} p(\tau, z) \cos(\tau/R)\, d\tau\, dz$$

$$F_y = -\int_{-L/2}^{L/2}\int_0^{2\pi R} p(\tau, z) \sin(\tau/R)\, d\tau\, dz \qquad (5.2)$$

   In a numerical finite-difference solution for $p(\tau, z)$, the pressure is determined only at the grid points of a 2D rectangular mesh. The above integrations are then done numerically, such as by using Simpson's rule.

4. Calculate resultant radial load and its angle:

$$W = \sqrt{F_x^2 + F_y^2}, \quad \theta_W = \arctan(F_y/F_x) \qquad (5.3)$$

By performing the above steps, 1 through 4, over a suitable range of values for $0 \le e/C < 1$ and ϕ, enough solution points are generated to construct design curves similar to those of Raimondi and Boyd. As stated earlier, the sequence of computations in design analyses is the reverse of the above sequence. That is, one starts by specifying the bearing load, W, and its angle θ_W, and uses design curves preassembled from many RLE solutions to determine the corresponding journal eccentricity, e, and attitude angle, ϕ.

5.2.2 Journal Bearing Stiffness and Damping Formulations

Solutions to the RLE are a nonlinear function of the journal-to-bearing radial displacement or eccentricity, even though the RLE itself is a linear differential equation. Thus, F_x and F_y given by Equations 5.2 are nonlinear (but continuous) functions of journal-to-bearing motion. Therefore, they may each be expanded in a Taylor series about the static equilibrium position. For sufficiently "small" motions, the corresponding changes in the journal fluid-film force components about equilibrium can thus be linearized for displacement and velocity perturbations, as indicated by Equations 2.60.

Since solutions for the fluid-film radial force components F_x and F_y are usually obtained through numerical integration on $p(\tau, z)$ as it is obtained from numerical solution of the RLE, the *partial derivatives* of F_x and F_y that are the bearing *stiffness and damping* coefficients must also be numerically computed. This is shown by the following equations:

$$-k_{xx} \equiv \frac{\partial F_x}{\partial x} \simeq \frac{\Delta F_x}{\Delta x} = \frac{F_x(x+\Delta x, y, 0, 0) - F_x(x, y, 0, 0)}{\Delta x}$$

$$-k_{yx} \equiv \frac{\partial F_y}{\partial x} \simeq \frac{\Delta F_y}{\Delta x} = \frac{F_y(x+\Delta x, y, 0, 0) - F_y(x, y, 0, 0)}{\Delta x}$$

$$-k_{xy} \equiv \frac{\partial F_x}{\partial y} \simeq \frac{\Delta F_x}{\Delta y} = \frac{F_x(x, y+\Delta y, 0, 0) - F_x(x, y, 0, 0)}{\Delta y}$$

$$-k_{yy} \equiv \frac{\partial F_y}{\partial y} \simeq \frac{\Delta F_y}{\Delta y} = \frac{F_y(x, y+\Delta y, 0, 0) - F_y(x, y, 0, 0)}{\Delta y}$$

$$-c_{xx} \equiv \frac{\partial F_x}{\partial \dot{x}} \simeq \frac{\Delta F_x}{\Delta \dot{x}} = \frac{F_x(x, y, \Delta \dot{x}, 0) - F_x(x, y, 0, 0)}{\Delta \dot{x}}$$

$$-c_{yx} \equiv \frac{\partial F_y}{\partial \dot{x}} \simeq \frac{\Delta F_y}{\Delta \dot{x}} = \frac{F_y(x, y, \Delta \dot{x}, 0) - F_y(x, y, 0, 0)}{\Delta \dot{x}}$$

$$-c_{xy} \equiv \frac{\partial F_x}{\partial \dot{y}} \simeq \frac{\Delta F_x}{\Delta \dot{y}} = \frac{F_x(x, y, 0, \Delta \dot{y}) - F_x(x, y, 0, 0)}{\Delta \dot{y}}$$

$$-c_{yy} \equiv \frac{\partial F_y}{\partial \dot{y}} \simeq \frac{\Delta F_y}{\Delta \dot{y}} = \frac{F_y(x, y, 0, \Delta \dot{y}) - F_y(x, y, 0, 0)}{\Delta \dot{y}}$$

(5.4)

Here, $x \equiv e_x$, $y \equiv e_y$, $\dot{x} \equiv \dot{e}_x$, $\dot{y} \equiv \dot{e}_y$.

The definitions contained in Equations 5.4 for the eight *stiffness and damping* coefficients are compactly expressed using subscript notation, as follows:

$$k_{ij} \equiv -\frac{\partial F_i}{\partial x_j} \quad \text{and} \quad c_{ij} \equiv -\frac{\partial F_i}{\partial \dot{x}_j} \qquad (5.5)$$

It is evident from Equations 5.4 that the journal radial force components F_x and F_y are expressible as continuous functions of journal-to-bearing radial displacement and velocity components, as follows:

$$F_x = F_x(x, y, \dot{x}, \dot{y})$$
$$F_y = F_y(x, y, \dot{x}, \dot{y})$$
(5.6)

It is also evident from Equations 5.4 that for each selected static equilibrium operating condition $(x, y, 0, 0)$, five solutions of the RLE are required to compute the eight *stiffness and damping* coefficients. These five slightly different solutions are tabulated as follows:

$(x, y, 0, 0)$, Equilibrium condition,
$(x + \Delta x, y, 0, 0)$, x—displacement perturbation about equilibrium,
$(x, y + \Delta y, 0, 0)$, y—displacement perturbation about equilibrium,
$(x, y, \Delta \dot{x}, 0)$, x—velocity perturbation about equilibrium,
$(x, y, 0, \Delta \dot{y})$, y—velocity perturbation about equilibrium.

5.2.2.1 Perturbation Sizes

Because of the highly nonlinear nature of Equations 5.6, special care must be taken when computing the numerically evaluated partial derivatives in Equations 5.4. That is, each of the displacement and velocity perturbations $(\Delta x, \Delta y, \Delta \dot{x}, \Delta \dot{y})$ must be an appropriate value (neither too large nor too small) for the particular equilibrium condition to ensure reliable results. This point is apparently not adequately appreciated by some who have developed computer codes to perform the calculations implicit in Equations 5.4. The author's approach to handle this problem, while possibly not original, is explained as follows.

For the sake of explanation, it is assumed that the individual computations are accurate to eight significant digits. The basic approach is to program for automatic adjustment of each perturbation size based on the number of digits-of-agreement between unperturbed and perturbed force components. This approach is quite versatile and can be calibrated for any specific application involving the computation of derivatives by numerical differences. The author has found the following guidelines to work well. If the difference between the unperturbed and perturbed force components originates between the third and the fifth digit, the perturbation size is accepted. If the difference invades the first three digits, then the particular perturbation is reduced, dividing it by 10. If the difference originates beyond the first five digits, then the particular perturbation is increased, multiplying it by 10. In this manner, the displacement and velocity force derivatives are tangent (not secant) gradients, and are accurate to at least three significant digits. This is enumerated by the following:

$$F_i(x, y, 0, 0) = 0.a_1 a_2 a_3 a_4 a_5 a_6 a_7 a_8 \times 10^n, \quad \text{unperturbed force component}$$

$$F_i(x, y, 0, 0) + \Delta F_i = 0.b_1 b_2 b_3 b_4 b_5 b_6 b_7 b_8 \times 10^n, \quad \text{perturbed force component}$$

$a_6 a_7 a_8 \neq b_6 b_7 b_8,$ To ensure at least 3-digit accuracy

$a_1 a_2 a_3 = b_1 b_2 b_3,$ To ensure tangent gradients

5.2.2.2 Coordinate Transformation Properties

With few exceptions, journal bearing *stiffness and damping* coefficient arrays are *anisotropic*. This means that the individual array elements change when the orientation of the x–y coordinate system is changed. It is therefore quite useful to be aware of the coordinate transformation properties of radial bearing and seal rotor dynamic coefficients. For example, if available *stiffness and damping* coefficient data are referenced to a coordinate system orientation not convenient for a given LRV model, the available coefficient arrays can be easily transformed to the desired coordinate system orientation. A similar example is when one wishes to rotate a bearing orientation in a pre-existing LRV model, say for the purpose of analyzing the potential improvement in vibration behavior such a change might accomplish.

With journal-to-bearing eccentricity, the predominant anisotropic character of journal bearing dynamic arrays is in contrast to the isotropic assumption usually invoked for radial seals that are more reasonably conceptualized with a rotationally symmetric flow field than are bearings.

The rotor dynamic coefficient arrays defined in Equations 5.4 and 5.5 are in fact quite properly categorized as single-point *second rank tensors*, being mathematically just like the components for *stress at a point* and the components for the *mass moment-of-inertia* of a rigid body with respect to a point. The defining property of a tensor entity is its orthogonal transformation properties, that is how its individual scalar components transform when the coordinate system orientation is rotated from that in which the tensor's components are initially specified. The most common application of tensor transformation is in stress analysis when the coordinate system rotation is into the *principal coordinate system*, in which all the shear stresses disappear and the normal stresses are the *principal stresses*. Radial bearing and seal rotor dynamic coefficients involving two spatial coordinates

Bearing and Seal Rotor Dynamics

x and y are thus comparable to bi-axial stress. Therefore, the same *direction cosines* relating two orthogonal coordinate systems used to transform bi-axial stress components also apply to the rotor dynamic coefficients of radial bearings and seals. For the unprimed and primed coordinate systems shown in Figure 5.2, the following transformations apply:

$$k'_{ij} = Q_{ir}Q_{js}k_{rs} \quad \text{and} \quad c'_{ij} = Q_{ri}Q_{js}c_{rs} \quad \text{(in tensor notation) or}$$
$$[k'] = [Q][k][Q]^T \quad \text{and} \quad [c'] = [Q][c][Q]^T \quad \text{(in matrix notation)} \tag{5.7}$$

where

$$[Q] = \begin{bmatrix} \cos\gamma & \sin\gamma \\ -\sin\gamma & \cos\gamma \end{bmatrix}$$

Unlike the component arrays for the stress and mass moment-of-inertia tensors, bearing and seal rotor dynamic coefficient arrays are not necessarily symmetric. Recalling the decomposition of bearing and seal arrays into symmetric and skew-symmetric parts (see Section 2.4 of Chapter 2), the skew-symmetric part is an *isotropic* tensor, that is, it is invariant to coordinate system angular orientation.

As described in Section 5.3, these stiffness and damping coefficients are not directly measurable, but are extracted from measured time-varying force and displacement signals. That is, each coefficient is the derivative of one measured signal with respect to another measured signal. Thus, the challenge of achieving coefficient accuracy from such tests is obvious. It is suggested here that the tensor transformation property of the stiffness and damping arrays could be utilized to combat various sources of error in extracting these coefficient arrays from laboratory measured force and displacement signals. Specifically, it is proposed that the measured displacement and force signals be simultaneously measured in a multitude of orthogonal coordinate systems as illustrated in Figure 5.2. The stiffness

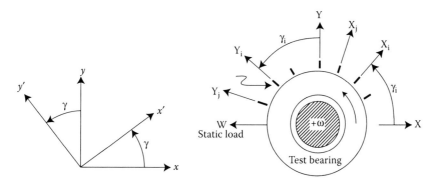

FIGURE 5.2 Tensor transformation of bearing coefficients, Equations 5.7.

and damping coefficient arrays are first extracted in each coordinate system. An optimization algorithm is then employed to merge the coefficient arrays from the different coordinate systems imposing their tensor transformation property. In theory, the measurement errors do not possess the tensor transform property and thus should be so minimized. This manner of experimental error minimization could be referred to as *tensor filtering*.

5.2.2.3 Symmetry of Damping Array

It is evident from Equation 5.1 that the RLE retains certain pressure and viscous fluid effects in the thin lubricant film while all fluid inertia effects are absent, as described in the perspective on simplifying assumptions leading to Equation 5.1. As rigorously developed in Section 2.4 of Chapter 2 and briefly mentioned at the end of Section 4.3 of Chapter 4, the skew-symmetric portion of an unsymmetric array added to the damping array embodies a *conservative force field* and thus must reflect inertia effects, similar to gyroscopic moment effects. Thus, solution perturbations from the RLE must yield symmetric damping arrays ($c_{ij} = c_{ji}$).

5.2.3 Tilting-Pad Journal Bearing Mechanics

Also referred to as *pivoted-pad journal bearings* (PPJB), this style of hydrodynamic fluid-film radial bearing is now used in a multitude of rotating machinery types, for example, turbines, compressors, pumps, motors, and generators. When properly designed and appropriately employed, this style of bearing yields distinct advantages in rotor vibration control over the purely cylindrical journal bearing. But a lack of fundamental understanding of this type of bearing has led to misapplications where rotor dynamical performance has turned out worse rather than better. If proper design precautions are not taken, this bearing type can be quite sensitive to the static load direction.

The basic understanding of this type of bearing is facilitated with its comparison to the purely cylindrical journal bearing. Figure 5.3 illustrates a fundamental difference between the two. Figure 5.3a illustrates a single partial-arc cylindrical bearing as it responds with stability to an incremental change in static load. In contrast, Figure 5.3b illustrates how a single tilting-pad arc cannot support a static load with stability, since the freely pivoting pad must have a film pressure distribution with the pressure center directed through the pivot point, about which the sum of the moments must be zero. As Figure 5.3b further illustrates, static stability dictates that at least two tilting pads must share the static radial bearing load.

One ideal design condition is to have nominal bearing load directed between two pads. The least desirable is to have the bearing load supported mostly by one pad. Note that as long as the tilting-pad bearing has at least

Bearing and Seal Rotor Dynamics

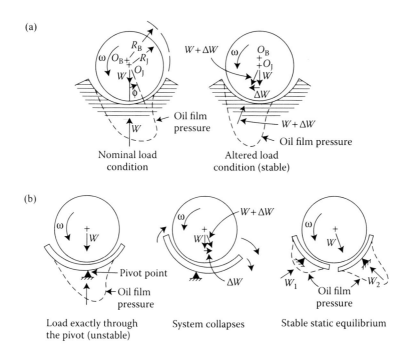

FIGURE 5.3 Comparison between cylindrical and tilting-pad journal bearings: (a) cylindrical bearing arc and (b) tilting-pad bearing.

three equally spaced pads, a load passing directly through one of the pivot points does not cause the bearing to collapse because the three or more pads at least capture the journal. But even with three or more pads, a load that is supported mostly by a single pad can produce poor rotor dynamical characteristics. By restraining a single pad pivot and journal (test setup or simulation), the load capacity of a single pad is obtained as a function of journal pivot radial eccentricity as typified by Figure 5.4, where the slope is pad pivot *radial film stiffness*.

It is seen that the pad radial film stiffness acts as a nonlinear spring in compression. Consider the four-pad bearing, illustrated in Figure 5.5 for two loading conditions: (a) load between two pads and (b) load on a

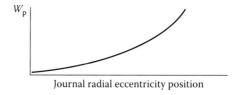

FIGURE 5.4 Tilting pad load W_P versus journal pivot radial displacement.

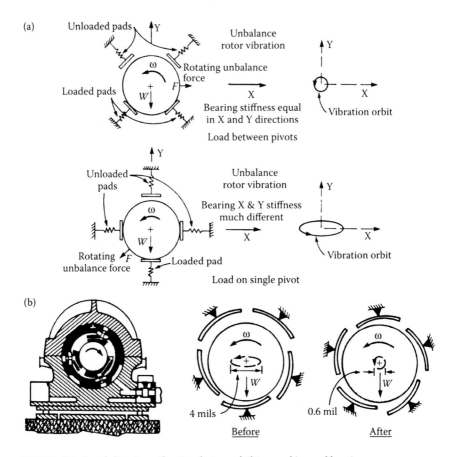

FIGURE 5.5 Load-direction vibration factors of tilting-pad journal bearings.

single pad. Utilizing the generic information provided by Figure 5.4, it is clear from the rotor vibration orbits as to why it is better to have the load directed between two pads rather than directed on a single pad.

Figure 5.5 also shows illustrations based on two actual power plant machine case studies. The first shows a strange looking three-pad bearing of a European design steam turbine generator. The second shows a "before" and "after" for a boiler feed water pump (BFP). In both machines, vibration problems and balancing difficulties necessitated rotating the inner bearing shell to more evenly distribute the bearing static load on two pads.

The sensitivity to static load direction illustrated in Figure 5.5 can be greatly reduced by adding preload to the bearing. Illustrated in Figure 5.6, preload is achieved by assembling the bearing with the pivot radial clearance smaller than the concentric radial clearance. Similar to ball bearings, preloading also has the added benefit of achieving higher bearing film

Bearing and Seal Rotor Dynamics

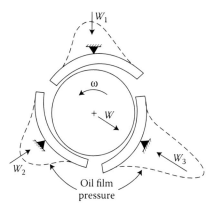

FIGURE 5.6 Illustration of preloaded tilting-pad journal bearing.

stiffness, even when the static load is zero, like machines with vertical centerlines (see Section 12.2 of Chapter 12). A moderate preload is $C'/C = 0.7$, where C is the concentric bearing radial clearance (C = radius of pad surface minus radius of journal) and C' is the radial clearance between pad surface pivot circle and journal. Excessive preload ($C'/C = 0.3$) risks excessive bearing temperatures and intolerance to thermal differential growth.

When performing rotor vibration simulation analyses (Chapter 4) in cases where one or more tilting-pad journal bearings are in play, one must be aware of the frequency dependence of the eight stiffness and damping coefficients, as here expressed for the nondimensional journal bearing stiffness and damping coefficients (see tilting-pad bearing tables in electronic folder BEARCOEF):

$$\bar{k}_{ij} = \bar{k}_{ij}(S, \bar{\Omega}) \quad \bar{c}_{ij} = \bar{c}_{ij}(S, \bar{\Omega}) \quad S = \text{Sommerfeld No.} \quad \bar{\Omega} = \Omega/\omega$$

ω = rotational speed Ω = vibration frequency.

The reason for this frequency dependence is that additional DOFs are present for each tilting pad, since at least 1-DOF for each pad's pitching motion must be included. However, the 2×2 journal bearing stiffness and damping arrays only account for the bearing-local two-DOFs (x, y) of rotor orbital motion. As is the standard means to correctly absorb "unseen" DOFs in structural mechanical impedance, the coefficients of the retained DOFs must then be frequency dependent (see Section 5.3.1).

Versatility of the tilting-pad journal bearing concept is extended by the novel configuration developed by Adams and Laurich (2005). This bearing was developed, built, and rigorously tested by the author in his university laboratory. Its unique inside-out configuration is devised for next-generation centerless grinder high-speed spindles (Figure 5.7).

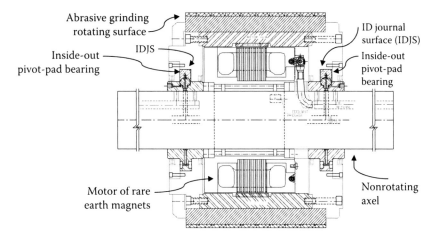

FIGURE 5.7 (See color insert following page 262.) Next-generation centerless grinder spindle.

Recent grinding research has demonstrated the feasibility of finish grinding ceramics with considerable improvements in throughput, costs, and quality. To implement this advancement in ceramic grinding necessitates the use of high-speed high-power grinding spindles (7000 rpm, 50 hp). Also needed are radial spindle bearings with real-time adjustable stiffness. Extensive development devoted to achieve a bearing design to accomplish the requirements resulted in a predicted optimum design. Based on extensive advanced analyses, a three-pad inside-out tilting-pad bearing promised to fulfill the demanding requirements. Subsequent laboratory testing validated the predicted bearing performance, as reported by Adams and Laurich.

This novel inside-out three-pad bearing is pictured in Figure 5.8. All three pads are supported on spherical seats, allowing pad self-alignment in addition to the generic tilting characteristic. Two of the three pads support the spindle grinding force, rotor weight, and radial preload imparted by the hydraulically actuated third pad. It is well known among bearing specialists that preload can be set by a single pad since three pivot points define a circle. This is often utilized by conventional three-pad bearings such as illustrated in Figure 5.6.

5.2.4 Journal Bearing Stiffness and Damping Data and Resources

Early LRV investigators modeled flexible rotors as circular flexible beams carrying concentrated masses and supported on rigid points at the bearings. The importance of gyroscopic effects was identified quite early, by Stodola (1927), as was the self-excited vibration induced with

FIGURE 5.8 (**See color insert following page 262.**) Photo of three-pad inside-out PPJB with three copper pads to facilitate heat removal; three steel thrust sectors.

hydrodynamic oil-film journal bearings, by Newkirk and Taylor (1925). However, it was not until 1956 that Hagg and Sankey identified the need to model journal bearings as radial springs and dampers. Sternlicht (1959) and others generalized the Hagg and Sankey idea to formulate the linear model given in Equations 2.60.

Raimondi and Boyd (1958) were among the first to use the digital computer to obtain "exact" 2D numerical solutions of the RLE. They and others soon thereafter applied the same computerized numerical RLE solution algorithms to begin providing journal bearing *stiffness and damping* coefficients by applying the numerical partial differentiation approach shown in Equations 5.4.

The earliest major compendium of such journal bearing rotor dynamic property coefficients was published by Lund et al. (1965), and it is still a significant resource for rotor vibration specialists. It contains nondimensional *stiffness and damping* coefficients plotted against bearing nondimensional speed (the Sommerfeld number) for several types of journal bearing configurations, including 360° cylindrical, axially grooved, partial-arc, lobed, and tilting pad, for both laminar and turbulent films. The most significant recent compendium of journal bearing rotor dynamic properties is provided by Someya et al. (1988). It is based on data contributed by technologists from several Japan-based institutes and companies participating in a joint project through the Japanese Society of Mechanical Engineers (JSME). It contains not only computationally generated data but also corresponding data from extensive laboratory testing. Although many industry and university organizations now have computer codes that can generate bearing dynamic properties for virtually any bearing configuration and operating condition, urgent on-the-spot rotor vibration analyses in troubleshooting circumstances are more likely to necessitate the use of existing available bearing dynamic coefficient data. The published data, such as by Lund et al. (1965) and Someya et al. (1988), are thus invaluable to the successful troubleshooter and designers.

For use with LRV computer codes such as RDA, tabulated bearing dynamic properties are more convenient than plotted curves since the bearing input for such codes is tabulated data at specific rotational speeds, as demonstrated with some of the RDA sample problems in Chapter 4. Furthermore, tabulations are more accurate than reading from plotted curves, especially semilog and log–log plots spanning several powers of 10. This accuracy issue is particularly important concerning *instability threshold* predictions. In fact, if a next generation of LRV code is to be written, it should be directly integrated with a companion journal bearing dynamic coefficient code. In this manner, at every speed (or speed iteration) where eigenvalues (stability analyses) or unbalance responses are computed, the bearing coefficients are generated exactly for that condition, instead of using curve-fit interpolations between data points at a limited number of input speeds. The CD-ROM that accompanies this book contains a directory, in MS-WORD (portions can be electronically copied) and pdf, named **BearCoef** in which there are several files, each providing a tabulation of bearing stiffness and damping coefficients for a particular bearing type and geometry. Space here is thereby not wasted on several hard-to-read graphs of *stiffness and damping* coefficients.

5.2.4.1 Tables of Dimensionless Stiffness and Damping Coefficients

The bearing data files in the directory **BearCoef** use the standard non-dimensionalization most frequently employed for journal bearing rotor dynamic coefficients, as defined by the following dimensionless parameters for stiffness ($\bar{k}_{ij}$) and damping ($\bar{c}_{ij}$) as functions of a dimensionless speed, S:

$$\bar{k}_{ij} \equiv \frac{k_{ij}C}{W}, \quad \bar{c}_{ij} \equiv \frac{c_{ij}\omega C}{W}, \quad S \equiv \frac{\mu n}{P}\left(\frac{R}{C}\right)^2 \tag{5.8}$$

where C is the radial clearance, W is the static load, S is the Sommerfeld number, μ is the lubricant viscosity, $P = W/DL$, the unit load, R is the nominal radius, $D = 2R$, L is the length, and n (revs/s) $= \omega/2\pi$.

5.2.5 Journal Bearing Computer Codes

There are now several commercially available PC codes to analyze all aspects of journal bearings, including stiffness and damping coefficients. Most of these codes are older main-frame computer codes that have been adapted to run on a PC, while a few are more recently developed specifically for PC usage, as was the RDA code. The author uses primarily two journal bearing codes, one that is an in-house code and a quite similar code that is commercially available. Since these two codes have very similar features, the commercially available one is described here.

The COJOUR journal bearing code was originally developed for main frame computers by Mechanical Technology Incorporated (MTI) under sponsorship of the Electric Power Research Institute (EPRI). It is documented in a published EPRI report authored by Pinkus and Wilcock (1985). The COJOUR code is now commercially available in a PC version, and it has two attractive features that set it apart from other competing codes.

The first of these attractive features is a user option to specify the bearing *static load* magnitude and direction. COJOUR then iterates to determine the corresponding static-equilibrium *radial eccentricity* magnitude and direction of the journal relative to the bearing, as illustrated in Figure 5.1. Most other journal bearing codes only function in the opposite sequence outlined by the four-step approach leading to Equations 5.3, but COJOUR functions either way at the user's option. As implicit in Equations 5.4, for any specified bearing operating point, determining the *static equilibrium* position is clearly a prerequisite to generating the *stiffness and damping* coefficients about that operating point. COJOUR is thus quite convenient for this purpose.

The second attractive feature of COJOUR is that the user may choose either the *uniform-viscosity* solution inherent in all journal bearing codes or a *variable-viscosity* solution based on a noniterative *adiabatic* formulation that assumes that all the viscous drag losses progressively accumulate as film heating in the direction of sliding with no heat transfer (via bearing or journal) to or from the lubricant film. This is a first-order approximation of a much more computationally intensive formulation (not in any commercially available software) that couples the RLE to the energy and heat transfer equations. The *variable-viscosity* option in COJOUR is particularly relevant to large turbo-generator bearings. For most case studies in PART-4, the COJOUR code was used to obtain the journal bearing dynamic coefficients.

5.2.6 Fundamental Caveat of LRV Analyses

The RDA example problems given in Chapter 4 provide a suitable basis for one to explore the considerable sensitivity of important output answers to variations in bearing dynamic coefficient inputs, such as those arising from manufacturing tolerances. As implied in Section 5.1, rotor vibration computer code vendors are not necessarily attuned to the considerable uncertainties that exist regarding radial bearing rotor dynamic coefficients. Uncertainties arise from a number of practical factors that are critical to bearing dynamic characteristics, but controllable only to within statistical measures. The most prominent example is journal bearing clearance, which is a small difference between two relatively large numbers. Referring to Equations 5.8, the dimensionless bearing speed (S) varies with C^{-2}, where $C = R_B - R_J$ is the radial clearance. The following is a realistic example of bearing and journal manufacturing dimensions.

5.2.6.1 Example

Bearing bore diameter, $5.010^{\pm 0.001}$ in.
Journal diameter, $5.000^{\pm 0.001}$ in.
Radial clearance = $\begin{pmatrix} 0.004'' \text{ min.} \\ 0.006'' \text{ max.} \end{pmatrix}$

$$\frac{S_{max}}{S_{min}} = \left(\frac{0.006}{0.004}\right)^2 = 2.25$$

This is more than a 2-to-1 range of dimensionless speed, which can be related to parameter ranges such as 2-to-1 in rpm or lubricant viscosity or static load. This provides a sizable variation in journal bearing dynamic coefficients, to say the least. This is just one of many factors that prove the old power plant adage that *no two machines are exactly alike*. Other prominent factors that add uncertainty to journal bearing characteristics include the following:

- Large variations in *oil viscosity* from oil temperature variations.
- Journal-to-bearing angular *misalignment* (see Figure 2.14).
- Uncertainties and operating variations in bearing *static load*.
- Bearing surface *distortions* from loads, temperature gradients, wear, and so on.
- Basic simplifying assumptions leading to the RLE.

These revelations are not intended to show LRV analyses to be worthless, because they most assuredly are of considerable value. But, the savvy analyst and troubleshooter must keep these and any other sources of uncertainty uppermost in their mind when applying LRV computations. As described in the next section, laboratory experimental efforts are at least as challenging.

5.3 Experiments to Measure Dynamic Coefficients

Bearing and seal rotor dynamic characteristics are of overwhelming importance to the success of modern high-performance rotating machinery, especially turbo-machinery. A review of the technical literature on this subject shows that a keen recognition of this fact dates back to the 1950s, for example, Hagg and Sankey (1956) and Sternlicht (1959). Several serious experimental efforts were subsequently undertaken. The most impressive of these was the work of Morton (1971) and his coworkers at General Electric Co. (GEC) in Stafford, England. They devised a test apparatus to measure *stiffness and damping* coefficients on full size journal bearings of large turbo-generators. The other major world manufacturers of large turbo-generators developed similar test machines at their respective research facilities and/or collaborating universities, albeit on smaller scaled-down journal bearings.

About the same time, a general recognition emerged that many types of annular seals and other fluid-annulus gaps also inherently possess rotor dynamic characteristics of considerable importance. Black (1969, 1971, 1974), recognized as one of the first to intensively research these fluid dynamical effects, provided a major initial contribution to this aspect of rotor dynamics technology. Over the last 30 years, precipitated by the high-energy-density turbo-machines developed for NASA's space flight programs, the major portion of significant experimental and computational development work on rotor dynamic properties of *seals* has been conducted at Texas A&M University's Turbomachinery Laboratory under the direction of Professor Dara Childs. A comprehensive treatise of this work is contained in his book, Childs (1993).

A major ($10 million) EPRI-sponsored multiyear research project on improving reliability of *BFPs* was started in the early 1980s at the Pump Division of the Sulzer Company in Winterthur, Switzerland. One of the major tasks in this research project was to build a quite elaborate experimental test apparatus to measure the rotor dynamic coefficients of a complete impeller–diffuser stage of a high-head centrifugal pump, as reported by Bolleter et al. (1987). The final report covering all the tasks of this EPRI project, Guelich et al. (1993), is a major technical book in itself.

Under laboratory conditions, sources of uncertainty in bearing rotor dynamic characteristics (enumerated at the end of the previous section) can be minimized but not eliminated. Measurement uncertainties arise. Bearing and seal rotor dynamic array coefficients are not directly measurable quantities because they are based on a mathematical decomposition of a single *interactive radial force vector* into several parts, as clearly delineated by Equations 2.60, 2.70, and 5.4. The parameters that can be directly measured are the *x* and *y* *orbital displacement* signals and the corresponding *x* and *y* interactive *radial force* signals. One can thus begin to appreciate the inherent challenge in bearing and seal rotor dynamic coefficient "measurements" just from Equations 2.70, which show the following obvious fact:

> Bearing and seal rotor dynamic coefficients are each a derivative of one measured signal (a force component) with respect to one of the other measured signals (a displacement component) or its first or second derivative in time (velocity or acceleration).

As technologists of many fields know, extracting derivatives of one set of time-varying measured signals with respect to a second set of time-varying measured signals and their derivatives in time is a challenge, to say the least. The significance of measurement accuracy and signal corruption (noise-to-signal ratio) issues is substantial. The yet-to-be-tried *tensor filtering* method, proposed by the author in Section 5.2.2 (Figure 5.2), entices the author as a worthy research mission: that is, filtering out effects and signal corruption that do not conform to the tensor transformation property of radial bearing and seal linearized rotor dynamic coefficient arrays.

As described in Section 2.4.4, the number of decomposition parts of the *interactive radial force vector* depends on what physical assumptions are invoked for the fluid flow within the bearing or seal. For a mathematical model consistent with the RLE (no fluid inertia effects are retained), the rotor dynamic coefficients (eight) can capture only displacement gradient (stiffness) and velocity gradient (damping) parts of the radial force vector, and furthermore the array of damping coefficients are symmetric.

For high Reynolds number film bearings and most annular radial clearance sealing gaps, fluid inertia effects should be included because of their importance. The mathematical model must then also capture the acceleration gradient (virtual mass or inertia) parts, giving rise to four more coefficients, for a total of 12 rotor dynamic coefficients. As fully explained in Section 2.4.4, the highest-order rotor dynamic coefficient array should be symmetric so as to avoid physical inconsistencies in the model. Thus, when stiffness, damping, and inertia coefficients are all employed, the inertia coefficient array should be constrained to symmetry, whereas the stiffness and damping arrays do not have to be symmetric.

In a practical sense, bearing and seal rotor dynamic coefficients are really *curve-fit coefficients* that exist as entities primarily in the minds of rotor vibration analysts. When the radial pressure field within a bearing or seal is perturbed in response to relative orbital vibration of its close-proximity rotating component, the pressure-field interactive radial force vector is of course correspondingly perturbed. To model and analyze LRV with reasonable accuracy, such interactive radial force perturbations must be modeled in a mathematical format that is compatible with $\vec{F} = m\vec{a}$ based equations of motion. This means that the model must accommodate no higher than second derivatives of displacements with time, that is, accelerations. Furthermore, the inclusion of such rotor–stator interactive forces must fit a linear mathematical model in order to facilitate most vibration analysis protocols. Thus, the "die is cast" for the mathematical model of bearing and seal rotor–stator radial interactive forces. The required model is given by Equation 5.9, which is a restatement of Equation 2.70:

$$\begin{Bmatrix} f_x \\ f_y \end{Bmatrix} = -\begin{bmatrix} k_{xx} & k_{xy} \\ k_{yx} & k_{xx} \end{bmatrix} \begin{Bmatrix} x \\ y \end{Bmatrix} - \begin{bmatrix} c_{xx} & c_{xy} \\ c_{yx} & c_{yy} \end{bmatrix} \begin{Bmatrix} \dot{x} \\ \dot{y} \end{Bmatrix} - \begin{bmatrix} m_{xx} & m_{xy} \\ m_{yx} & m_{yy} \end{bmatrix} \begin{Bmatrix} \ddot{x} \\ \ddot{y} \end{Bmatrix} \quad (5.9)$$

$$k_{ij} \equiv -\frac{\partial F_i}{\partial x_j}, \quad c_{ij} \equiv -\frac{\partial F_i}{\partial \dot{x}_j}, \quad \text{and} \quad m_{ij} \equiv -\frac{\partial F_i}{\partial \ddot{x}_j}$$

are defined at static equilibrium.

There are a number of experimental procedures that have been employed to extract some or all of the coefficients in Equation 5.9. The degree of success or potential success varies, depending on which procedure is used in a given application. Mechanical impedance approaches utilizing harmonic excitation are the most frequently employed techniques. Mechanical impedance approaches utilizing impact excitation are also used. Recent advances in low-cost PC-based data acquisition hardware and software and signal processing methods have somewhat eased the burden of those researchers seeking to extract LRV bearing and seal rotor dynamic coefficients from laboratory experiments. However, their task remains a considerable challenge because of the inherent factors previously described.

5.3.1 Mechanical Impedance Method with Harmonic Excitation

Impedance approaches are often associated with characterization of an electrical network by a prescribed model circuit of resistances, capacitances, and inductances. With sufficient measured input/output data on an actual system, correlation of *input* (e.g., single-frequency sinusoidal voltage signal) and the resulting *output* (e.g., current signal) leads to a solution of values for all the model's resistances, capacitances, and inductances that would theoretically produce the measured outputs caused by the measured inputs. Such a characterization process is commonly referred to as *system identification*. Quite similar approaches have long been used to characterize mechanical dynamic systems with a suitable linear model in which the values of a discrete model's stiffness, damping, and mass elements are solved by determining what combination of these values produces the "best fit" in correlating measured input and output signals.

For example, suppose a machine is mounted to the floor of a large plant and it is known from experience that if a vibration analysis of the machine assumes the floor to be perfectly rigid, the analysis will be seriously flawed. Common sense dictates that one does not devise an FE model of the entire plant building just to couple it to the vibration model of the machine in question, which occupies only a few square feet of the plant's floor space. If previous experimental data are not deemed applicable, then a mechanical impedance *shaker test* can be performed on the plant floor at the location where the machine will be installed. An alternate technique is to apply an impact force to the floor position in question, measuring simultaneously the impact force signal and the acceleration signal at the floor point of impact. Impact approaches are fairly common and standard *hammer kits* (from small laboratory size up to large sledge-hammer size for power plants machinery) are made for this purpose with the force and motion signals processed through a dual-channel FFT instrument to extract the impact point's mechanical impedance. For very large structures (e.g., plant building) or devices with very high internal damping (e.g., multistage centrifugal pumps) impact techniques may lack sufficient energy input to the structure to achieve adequately high response signal-to-noise ratios to work well. In such cases, multiple impact strikes (several hundred) combined with synchronized signal time averaging has been used to filter out the noncoherent noise, but this is a very specialized procedure. For the sake of the following example, it is assumed that the vertical motion of the floor is significant and that a mechanical shaker is used, as illustrated in Figure 5.9.

If the structure is dynamically linear, then its steady-state response contains only the forcing function frequency, ω. The linearity assumption thus leads to the following equations as the basis for processing measured response to the controlled sinusoidal force input illustrated in Figure 5.9. Here it is convenient to use the complex plane representation for harmonic signals explained in Section 4.2.2 and illustrated in Figures 4.2 and 4.3:

$$(m_s + m_f)\ddot{x}_f + c_f\dot{x}_f + k_f x_f = F_s e^{i\omega t}$$
$$x_f = X e^{i(\omega t + \phi)}$$
(5.10)

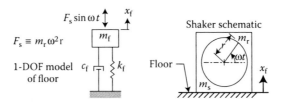

FIGURE 5.9 Vertical shaker test of floor where a machine is to be installed.

Equations 5.10 lead to the following complex algebraic equation:

$$(k_f - \omega^2 m_f - \omega^2 m_s + ic_f\omega)Xe^{i\phi} = F_s \qquad (5.11)$$

For the configuration illustrated in Figure 5.9, the vertical forcing function is equal to the imaginary part of the complex force. The single complex equation of Equation 5.11 is equivalent to two real equations and thus can yield solutions for the two unknowns $(k_f - \omega^2 m_f)$ and c_f at a given frequency. If the excited floor point was in fact an exact 1-DOF system, its response would be that shown in Figure 1.5, and the impedance coefficients k_f, m_f, and c_f would be constants independent of the vibration frequency, ω. However, since an actual structure is likely to be dynamically far more complicated than a 1-DOF model, the "best-fitted" impedance coefficients will be functions of frequency. When it is deemed appropriate or necessary to treat the impedance coefficients as "constants" over some frequency range of intended application, the coefficients are then typically solved using a *least-squares linear regression* fit of measurement data over the applicable frequency range.

For a 2-DOF radial plane motion experiment on a dynamically anisotropic bearing or seal, force and motion signals must be processed in two different radial directions, preferably orthogonal like the standard x–y coordinates. For a concentric fluid annulus, such as typically assumed for radial seals, the dynamic coefficient arrays are formulated to be isotropic, as is consistent with a rotationally symmetric equilibrium flow field. Impedance tests devised strictly for the isotropic model, Equation 2.85, require less data signals than impedance tests for the anisotropic model, Equation 5.9.

There are fundamentally two ways of designating the *inputs* and the *outputs*. In the 1-DOF impedance test schematically illustrated in Figure 5.9, the input is the harmonic force and the output is the resulting harmonic displacement response. However, there is no fundamental reason that prevents these roles from being reversed, since both input and output signals are measured and then processed through Equations 5.10. Likewise, a 2-DOF radial plane motion experiment on a bearing or seal may have the x and y force signals as the controlled inputs with the resulting x and y displacement signals as the outputs, or the converse of this. Both types of tests are used for bearing and seal characterizations.

The apparatus developed by the author at the Rotor Dynamics Laboratory of Case Western Reserve University is the example described here, because it is configured with a maximum of versatility that accommodates the anisotropic model with fluid inertia effects, as embodied in Equation 5.9. Figure 5.10 shows a cross section of the double-spool-shaft spindle assembly of this apparatus. The inner spindle provides the controlled spin speed. The outer spindle, which provides a circumscribing

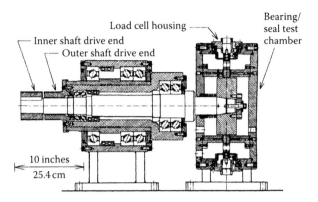

FIGURE 5.10 Double-spool-shaft spindle for bearing and seal test apparatus.

support of the inner spindle, is comprised of two close fitting sleeves that have their mating fit machined slightly eccentric to their centers. These machined-in eccentricities allow the radial eccentricity between the two coaxial spindles to be manually adjusted from zero to 0.030 in. Each spindle is driven independently with its own variable speed motor. The net result of this arrangement is an independently controlled circular journal orbit of adjustable radius (0–0.030 in.) with a controllable frequency and whirl direction independent of the controllable spin speed. The bearing/seal test chamber is designed to be hermetically sealed for testing seals with large leakage through flows and pressure drops or open to ambient as is typical for testing journal bearings. The bearing or seal test specimen can be either very stiffly held by piezoelectric load measuring cells or flexible mounted. Several inductance-type noncontacting proximity probes are installed to measure all x and y radial displacement signals of the journal and the tested bearing or seal. A full description of the complete test facility and data processing steps are given by Adams et al. (1988, 1992) and Sawicki et al. (1997). The journal, attached to the inner spindle on a tapered fit, is precision ground, while the inner spindle is rotated in its high-precision ball bearings (Horattas et al., 1997).

In the fullest application of the apparatus shown in Figure 5.10, all the 12 stiffness, damping, and inertia coefficients shown in Equation 5.9 can be extracted for a rotor dynamically anisotropic bearing or seal. The *inputs* are the x and y radial displacement signals of the journal and the *outputs* are the x and y radial force signals required to rigidly hold the bearing or seal motionless. If the bearing or seal has orbital motion that cannot be neglected, then the *inputs* are the x and y radial displacement signals of the journal relative to the bearing and the inertia effect of the test bearing or seal mass (i.e., D'Alembert force) must be subtracted from the *output* measurements by the x and y load cells that support the test bearing or seal.

Although the apparatus shown in Figure 5.10 produces an orbit that is very close to circular, it is not required that the orbit be assumed perfectly circular because the orbit is precision measured with a multiprobe complement of noncontacting proximity probes. The impedance model postulates that the measured x and y orbit signals (inputs) and force signals (outputs) are harmonic. The processed signals extracted from the measurements can thus be expressed as follows:

$$x = Xe^{i(\Omega t+\phi_x)}, \quad y = Ye^{i(\Omega t+\phi_y)}, \quad f_x = F_x e^{i(\Omega t+\theta_x)}, \quad f_y = F_y e^{i(\Omega t+\theta_y)} \tag{5.12}$$

where Ω is the orbital frequency. (Here, Ω is not necessarily equal to ω, the rotational speed.)

Equations 5.12 are substituted into Equations 5.9 to yield two complex equations. The basic formula $e^{iz} = \cos z + i \sin z$ separates real and imaginary parts of the resulting two complex equations, to yield the following four real equations:

$$\begin{aligned}
F_x \cos \theta_x &= \left[\left(\Omega^2 m_{xx} - k_{xx}\right) \cos \phi_x + c_{xx}\Omega \sin \phi_x\right] X \\
&+ \left[\left(\Omega^2 m_{xy} - k_{xy}\right) \cos \phi_y + c_{xy}\Omega \sin \phi_y\right] Y \\
F_x \sin \theta_x &= \left[\left(\Omega^2 m_{xx} - k_{xx}\right) \sin \phi_x - c_{xx}\Omega \cos \phi_x\right] X \\
&+ \left[\left(\Omega^2 m_{xy} - k_{xy}\right) \sin \phi_y - c_{xy}\Omega \cos \phi_y\right] Y \\
F_y \cos \theta_y &= \left[\left(\Omega^2 m_{yx} - k_{yx}\right) \cos \phi_x + c_{yx}\Omega \sin \phi_x\right] X \\
&+ \left[\left(\Omega^2 m_{yy} - k_{yy}\right) \cos \phi_y + c_{yy}\Omega \sin \phi_y\right] Y \\
F_y \sin \theta_y &= \left[\left(\Omega^2 m_{yx} - k_{yx}\right) \sin \phi_x - c_{yx}\Omega \cos \phi_x\right] X \\
&+ \left[\left(\Omega^2 m_{yy} - k_{yy}\right) \sin \phi_y - c_{yy}\Omega \cos \phi_y\right] Y
\end{aligned} \tag{5.13}$$

Since there are 12 unknowns in these four equations (i.e., stiffness, damping, and inertia coefficients), measured data must be obtained at a minimum of three discrete orbit frequencies for a given equilibrium operating condition. There are several data reduction ("curve fitting") approaches when test data are taken at a multitude of orbit frequencies for a given equilibrium operating condition. For example, a frequency-localized three-frequency fit propagated over a frequency range with several frequency data points will produce frequency-dependent coefficients to the extent that this improves the fitting of the measurements to Equations 5.13 impedance model. However, it is more typical to reduce the measurement data using a least-squares linear-regression fit of all measurement data over a tested frequency range, thereby solving for all the dynamic coefficients as constants independent of frequency.

The apparatus shown in Figure 5.10 has proven to be accurate and very close to "linear" by its performance and repeatability. Data are collected at 50–100 consecutive cycles of orbit excitation (at frequency Ω) and time averaged to remove all noise or other noncoherent signal content (e.g., spin-speed mechanical and proximity-probe electrical run-out at frequency ω). The time-averaged signals from each measurement channel are Fourier series decomposed. Ω components are much larger than the $n\Omega$ harmonics, indicating near linearity of the apparatus. The test results by Sawicki et al. show excellent agreement with theoretical results.

5.3.2 Mechanical Impedance Method with Impact Excitation

As implied in the earlier section, impact techniques are widely used on lightly damped structures with relatively low background structural vibration noise levels. Such structures can be impact excited to yield adequately favorable signal-to-noise ratios. Journal bearings and fluid-annulus seals typically have considerable inherent damping, and that is always an important benefit for controlling rotor vibration to within acceptable residual levels in operating machinery. However, from the point of view of using impact testing to extract dynamic coefficients of bearings and other fluid-annulus elements, their inherent damping capacity typically results in unfavorable signal-to-noise ratios. Nevertheless, the relative ease of conducting impact impedance testing motivated some researchers to pursue the impact impedance approach for various rotor–stator fluid-annuls elements. A notable example is the work of Professor Nordmann of Germany, whose experimental setup is illustrated in Figure 5.11.

Nordmann and Massmann (1984) explored the use of impact testing to extract the stiffness, damping, and inertia coefficients of annular seals. For coefficient extraction, they used the isotropic model presented in Equation 2.85 with the cross-inertial term $m_{xy} = -m_{xy} = 0$, which correctly conforms to the author's axiom (see Section 2.4.4) requiring symmetry of the highest-order coefficient array. Their test setup, illustrated in Figure 5.11, employs two "identical" test seals that are fed from a common central pressurized annular chamber and are axially opposed to cancel axial pressure forces on the quite flexibly supported seal housing. The housing can be impacted at its center of mass in various x–y radial

Bearing and Seal Rotor Dynamics

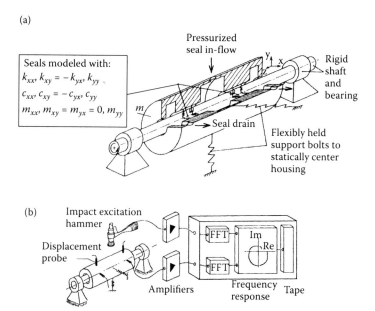

FIGURE 5.11 Experimental setup for impact excitation of radial seals: (a) quarter through-cut schematic illustration of test apparatus and (b) schematic of test measurements and data processing.

directions; so a 2-DOF x–y model is thereby used as a basis for processing the measured signals to extract the five coefficients of the isotropic model. Various algorithms, such as least-squares linear-regression fitting, can be employed to extract the five isotropic-model dynamic coefficients to provide the "best" frequency response fit of the model to the measured time-base signals as transformed into the frequency domain. The 2-DOF model's equations of motion are as follows; factor of "2" is present because the apparatus has two "identical" annular seals:

$$m\ddot{x} + 2(m_{xx}\ddot{x} + c_{xx}\dot{x} + c_{xy}\dot{y} + k_{xx}x + k_{xy}y) = F_x(t)$$
$$m\ddot{y} + 2(m_{yy}\ddot{y} + c_{yy}\dot{y} - c_{xy}\dot{x} + k_{yy}y - k_{xy}x) = F_y(t)$$
(5.14)

Nordmann and Massmann suggest that the questionable quality of their coefficient results may be due to the model needing additional DOFs to adequately correlate with the experimental results. This may very well be a factor, but the author suspects that the more fundamental problem with their results lies with the inherent difficulty of obtaining sufficient energy from an impact hammer into a system that has significant internal damping.

5.3.3 Instability Threshold-Based Approach

As explained in Chapter 1, providing accurate damping inputs to vibration analysis models is possibly the most elusive aspect of making reliable predictions of vibration characteristics for almost any vibratory system. The mass and flexibility characteristics of typical structures are reliably obtained with modern computational techniques such as FE procedures, and thus natural frequencies can be predicted with good reliability in most circumstances. Referring to Figure 1.5 on the other hand, predicting the vibration *amplitude* of a *forced resonance* at a natural frequency is not such a sure thing, because of the inherent uncertainties in damping inputs to the computation model. Similarly, predictions of instability thresholds also suffer from lack of high reliability for the same reason, that is, inherent uncertainties in damping inputs. Motivated by this fundamental reality in vibration analysis as it affects important rotor vibration predictions, Adams and Rashidi (1985) devised a novel experimental approach for extracting bearing dynamic coefficients from instability thresholds.

The apparatus illustrated in Figure 5.12 was first proposed by Adams and Rashidi (1985). The fundamental concept behind the approach is to capitalize on the physical requirement for an exact *energy-per-cycle* balance between *positive and negative damping* influences right at an instability threshold speed. This physical requirement is expressed by Equation 2.81 when $E_{cyc} = 0$. Through adjustment of the test bearing mass, by adding or subtracting weights, one can vary the instability-threshold natural frequency of the 2-DOF system and thereby cause an instability threshold

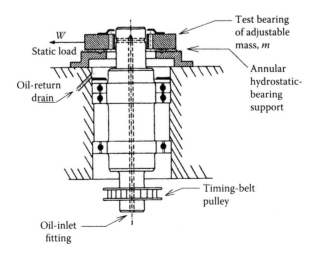

FIGURE 5.12 Vertical spindle rig for controlled instability-threshold-speed tests.

at selected operating conditions spanning a wide range of journal bearing Sommerfeld numbers (dimensionless speed). The controlled test parameters are rotational *speed*, bearing static radial *load*, lubricant *viscosity*, and test bearing *mass*. As with *mechanical impedance approaches*, the experiment mentioned here is correlated with a 2-DOF model given by the following equations:

$$m\ddot{x} + c_{xx}\dot{x} + k_{xx}x + c_{xy}\dot{y} + k_{xy}y = 0$$
$$m\ddot{y} + c_{yy}\dot{y} + k_{yy}y + c_{yx}\dot{x} + k_{yx}x = 0 \quad (5.15)$$
$$c_{xy} \equiv c_{yx}$$

The complete procedure for extracting journal bearing dynamic coefficients at a given Sommerfeld number is summarized by the following sequence:

1. Determine stiffness coefficients using controlled static loading data.
2. Slowly increase spin speed until the instability threshold speed is reached.
3. Capture $x-y$ signals of "linear" instability growth; see Figure 4.11.
4. Invert eigenvalue problem of Equations 5.15 to solve for the damping coefficients.

Basically, this procedure yields a *matched set* of journal bearing stiffness and damping coefficients. Even if the individual stiffness coefficients are somewhat corrupted by experimental measurement errors, they are "matched" to the damping coefficients. That is, by step 4 reproduce the experimentally observed instability threshold *frequency* and *orbit* parameters for the 2-DOF model given by Equations 5.15. Step 4 algorithm uses the following information as inputs (Adams and Rashidi, 1985):

1. Test bearing mass
2. Experimentally or computationally determined bearing stiffness coefficients
3. Eigenvalue, Ω, the frequency of self-excited vibration at threshold
4. Eigenvector information: x-to-y displacement signal amplitudes (X/Y) and the difference in x-to-y phase angles, $\Delta\theta_{xy} \equiv \theta_x - \theta_y$.

Using a standard eigen-solution algorithm, the computation determines the damping coefficient values $(c_{xx}, c_{xy} = c_{yx}, c_{yy})$ that in combination with the *a priori* stiffness coefficient values yield the experimentally observed instability self-excited vibration *threshold speed*, *vibration frequency*, and normalized *orbit ellipse*. A rigorous demonstration of the significant accuracy improvement possible, over standard impedance approaches, with this novel instability threshold experimental approach is presented by Rashidi and Adams (1988).

Equations 5.15 reflect that in contrast to impedance approaches, dynamic force measurements are not needed in this approach, thus eliminating one major source of experimental error. However, the fundamental superiority of this approach lies in its basis that the "matched" stiffness and damping coefficients are consistent with $E_{cyc} = 0$ when the steady operating condition is tuned to its instability threshold (forward-whirl) mode, as described in Section 2.4.1. Using postulated experimental measurement errors, Rashidi and Adams (1988) conclusively show the inherent superiority of this approach over impedance approaches to provide drastically improved prediction accuracy for instability *threshold speeds* and *resonance amplitudes* at critical speeds. In summary, since an instability threshold is inherently most sensitive to nonconservative force effects (i.e., positive or negative damping), it is logical that an instability threshold should be the most sensitive and accurate "measurer" of damping.

5.4 Annular Seals

Developing reasonably accurate rotor dynamic coefficients is even more challenging for radial clearance seals than for journal bearings. That is, the uncertainty in quantifying seal rotor dynamic coefficients for LRV analysis inputs is even greater than the uncertainty for journal bearings. The primary objective here is to identify the major information resources for annular seals and other fluid-annulus rotor dynamic effects. The multistage *BFP* illustrated in Figure 5.13 provides a primary turbo-machinery example where there are several components and locations of fluid dynamical effects that collectively have a dominant influence on the vibration characteristics of the machine.

As described in Section 5.2, journal bearings derive static load-carrying capacity from the hydrodynamic pressure distribution generated between bearing and journal by a viscous lubricant film, which is continuously fed and sheared into the small-clearance converging gap formed between them. At the same time, a journal bearing also reacts to rotor vibration with an important interactive dynamic force that is linearly characterized about the equilibrium position using stiffness and damping elements. On the other hand, the primary function of seals is to control leakage, usually to the lowest flow practical. In the process of performing this primary function, an annular seal also reacts to rotor vibration with an interactive dynamic force that can also be quite significant. The focus here is the LRV

Bearing and Seal Rotor Dynamics

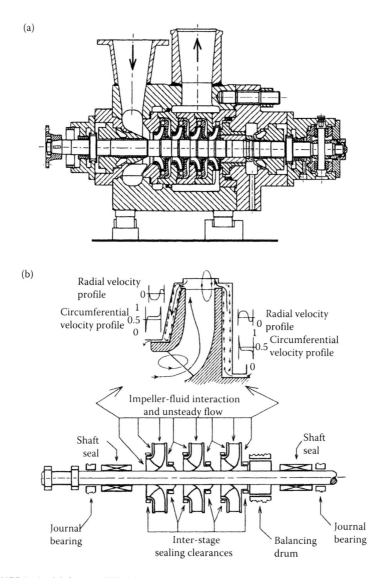

FIGURE 5.13 Multistage BFP: (a) *pump cross section* and (b) *sources of interaction and unsteady-flow rotor forces.*

dynamic characteristics of liquid- and gas-filled small-clearance annular seals and other fluid dynamical effects. Rubbing-type seals are generally not as amenable to linear modeling.

For fixed annular seals like *smooth-bore* and *labyrinth* types, the radial clearance between the close-proximity rotating and nonrotating parts is typically two or more times the clearance of the machine's radial bearings. This makes sense since the seals are not the bearings. On the other hand, it is desirable to have sealing clearances as small as practical, because seal leakage rates increase exponentially with clearance. An exception is *floating ring seals* that may actually have smaller radial clearance than the journal bearings, but this seal type is of minor rotor vibration significance specifically because of its radial float. In many modern high-pressure centrifugal pumps, the potentially beneficial influence of seal radial dynamic interaction effects on rotor vibration are now factored into the design of many pumps. Although it is certainly a good thing to exploit the benefits of seal rotor vibration characteristics, there is a *caveat* to making LRV smooth running too heavily dependent upon sealing clearances. In some sealing components, such as centrifugal pump wear rings, the sealing clearances are likely to enlarge over time from wear, possibly caused by rotor vibration. Thus not only does pump efficiency suffer as this wear progresses, but vibration levels will likely grow as well because of loss of seal rotor dynamic benefits. Often the increased back flow, through the wear ring clearance, to the impeller inlet (eye) causes sufficient disruption to the impeller inlet flow velocity distribution as to also substantially increase the intensity of unsteady flow vibration-causing forces upon the pump rotor. The author has dealt with several such centrifugal pumps, and they require too frequent wear ring replacements to maintain rotor vibration to within safe levels. Such pumps are definitely not of an optimum design (see case study in Section 10.6).

The inputs for annular seals in LRV analyses are handled in the same manner as the stiffness and damping characteristics for journal bearings, except that seals often have fluid inertia effects that are significant and thus need to be included. In further contrast to LRV modeling of journal bearings, the equilibrium position (rotor orbit center) for an annular seal is usually treated as though the rotating part orbits about and relative to the center of its close-proximity nonrotating part. This is done to justify the *isotropic* model shown in Equation 2.85, which is rewritten here as Equation 5.16 with the cross-mass term appropriately set to zero per the axiom given in Section 5.2.4:

$$\begin{Bmatrix} f_x \\ f_y \end{Bmatrix} = -\begin{bmatrix} k_s & k_{ss} \\ -k_{ss} & k_s \end{bmatrix}\begin{Bmatrix} x \\ y \end{Bmatrix} - \begin{bmatrix} c_s & c_{ss} \\ -c_{ss} & c_s \end{bmatrix}\begin{Bmatrix} \dot{x} \\ \dot{y} \end{Bmatrix} - \begin{bmatrix} m_s & 0 \\ 0 & m_s \end{bmatrix}\begin{Bmatrix} \ddot{x} \\ \ddot{y} \end{Bmatrix} \quad (5.16)$$

Test rigs specifically focused on seal rotor vibration characteristics can obviously be simplified by assuming the *isotropic* model. In contrast, the test apparatus shown in Figure 5.10 is by necessity more complicated than most other test rigs for rotor dynamic coefficients because it is designed to allow extraction of the full anisotropic model with fluid inertia, Equation 5.9, which contains 12 coefficients, or 11 coefficients when the symmetry axiom of Section 2.4.4 is imposed upon the inertia coefficient array (i.e., $m_{xy} = m_{yx}$).

5.4.1 Seal Dynamic Data and Resources

The more recent text books on rotor dynamics include information on the LRV characteristics of annular seals. Referring to the Chapter 2 Bibliography, Vance (1988) and Kramer (1993) both provide quite good introductory treatments of seal dynamics. However, the most complete treatment and information resource for seal dynamics is contained in the book by Childs (1993). Childs' book covers a wide spectrum of rotor dynamics topics well, but its coverage of seal dynamics is comparable to the combined coverage for journal bearings provided by Lund et al. (1965) and Someya et al. (1988). It is the single most complete source of computational and experimental data, information and references for seal rotor dynamic characteristics, reflecting the many years that Professor Childs has devoted to this important topic. It is probably safe to suggest that had the untimely death of Professor H. F. Black, Heriot-Watt University, Edinburgh, Scotland not occurred (*ca.* 1980), there would most certainly exist one more major modern resource on the dynamics of seals and other fluid-annulus component effects. Black's work (e.g., 1969, 1971, 1974) provided the major initial impetus for the extensive research and new design information developed on this topic over the last 40 years.

5.4.2 Ungrooved Annular Seals for Liquids

Three commonly used versions of *ungrooved annular seal* geometries are shown in Figure 5.14, with exaggerated clearances for illustrative purposes, as done with the journal bearing illustration in Figure 5.1. Although these ungrooved seals bear some geometric similarity to journal bearings, essential differences distinguish them. First, in most high-pressure applications the fluid being sealed is not a viscous oil but a much lower viscosity liquid like water or other process liquids or gases. The *flow* within the seal

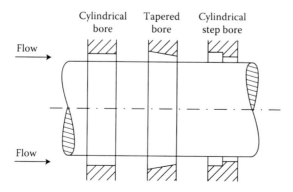

FIGURE 5.14 Ungrooved annular seals (illustrated clearances exaggerated).

clearance is thus usually *turbulent*, in contrast to most oil-film journal bearings that are characterized by the *laminar flow* RLE, Equation 5.1. Second, such seals usually have an axial length much smaller than the diameter (typically $L/D < 0.1$).

The importance of such seals to rotor vibration characteristics is roughly in proportion to the pressure drop across the seal. For high-pressure pumps such as that shown in Figure 5.13, the net effect of the inter-stage sealing clearances, balancing drum, and impeller casing interaction forces is to add considerable radial stiffening and damping to the rotor system. It is relatively easy to calculate that without the liquid inside such a pump; it would likely have one or more lightly damped critical speeds within the operating speed range because the shaft is relatively slender and the two journal bearings are located at opposite ends of the rotor. However, the combined influence of the interstage sealing clearances, balancing drum, and impeller casing interaction forces is to potentially eliminate detectable critical speeds from the operating speed range, at least when all the interstage sealing clearances are not appreciably *worn open*. To maintain good pump efficiency and low vibration levels, a prudent *rule-of-thumb* for such high-pressure pumps is to replace wear rings when the internal sealing clearances wear open to twice the "*as-new*" clearances. Of course, plant machines are like cars in that some owners are quite diligent with maintenance, while others are virtually oblivious to it.

5.4.2.1 Lomakin Effect

The first person to publish about the influence of ungrooved annular seals on rotor vibration was Lomakin (1955, 1958). Figure 5.15 illustrates how a

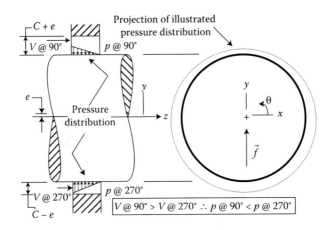

FIGURE 5.15 Lomakin effect pressure distribution in an ungrooved annular seal.

radial-pressure centering force is produced when the rotor and stator of an annular seal are eccentric to each other. Ignoring at this point the effects of shaft rotation and inlet flow preswirl, the entrance pressure loss is highest where the radial gap and thus the inlet flow velocity are largest. Conversely, the entrance pressure loss is lowest where the radial gap and thus the inlet flow velocity are smallest. This effect thus produces a *radial centering force* on the rotor, which increases with eccentricity between seal rotor and stator. That is, the radial displacement causes a skewing of the pressure distribution, producing a *radial stiffness effect* that is called the "Lomakin" effect. The x and y components of the centering force are expressible by directionally integrating the pressure distribution as shown in Equations 5.2 for journal bearings. In this simplest embodiment of the Lomakin effect, with shaft rotation and inlet flow prerotation not included, the centering force vector ($\vec{f}$) is in line with the eccentricity (e) and thus its magnitude is expressible as follows:

$$f = -\int_0^L \int_0^{2\pi} p(\theta, z) R \sin\theta \, d\theta \, dz \qquad (5.17)$$

In precisely the same manner described for journal bearings, the centering force described by Equation 5.17 can be linearized for "small" eccentricities, thus yielding a radial stiffness coefficient as follows:

$$k_r = \frac{f}{e} \approx 0.4 \frac{\Delta p R L}{C} \qquad (5.18)$$

where Δp is the pressure drop, R is the seal radius, L is the seal length, and C is the seal radial clearance.

In this case, k_r is the diagonal stiffness coefficient referenced as k_s in the isotropic model given by Equation 5.16. The centering force stiffness of the *tapered bore* and *cylindrical step bore* ungrooved seals illustrated in Figure 5.15 is explained by the same effect given here for the plain *cylindrical bore* seal. In fact, for the same operating conditions, seal length and minimum clearance, their Lomakin effect is significantly stronger than that of the plain *cylindrical bore* configuration, albeit with accompanying higher leakage flow.

Prior to a wide appreciation of the Lomakin effect by pump designers, their computational predictions for critical speeds based on a "dry pump" model were notoriously unreliable. By accounting for the Lomakin effect, computed predictions and the understanding of rotor vibration characteristics of high-pressure centrifugal pumps are improved over "dry pump" predictions. However, since the Lomakin (1955, 1958) publications it has been conclusively shown, especially in works by Black and Childs, that the additional effects of shaft rotation and preswirl of seal inlet flow are also quite important effects, not only to the radial stiffness (k_s) but also in producing other effects embodied in the coefficients (k_{ss}, c_s, c_{ss}, m_s) contained in the isotropic model of Equation 5.16. The complete modern treatment of annular seal rotor vibration characteristics clearly involves a considerably fuller account of fluid mechanics effects than implicit in the Lomakin effect as well as the classic RLE for laminar oil-film journal bearings.

5.4.2.2 Seal Flow Analysis Models

The predominance of turbulent flow in annular seals has dictated that their proper analysis must incorporate a phenomenological (or semiempirical) aspect to the analysis formulation to account for turbulence. For a limited group of fluid flow problems, top-end super computers are now up to the task of handling simulations of turbulent flow fine structures without the ingredients of semiempirical turbulence models, employing only the Navier–Stokes (N–S) and continuity equations. Such new and highly advanced computational fluid dynamics (CFD) efforts have not yet been applied to many traditional turbulent flow engineering problems, and annular seals are no exception.

For analyses of annular seals, the incorporation of semiempirical turbulence modeling can be inserted into the analysis model in basically two approaches. In the first of these two approaches, the turbulence model is inserted right into the N–S equations. The velocity components are expressed as the sum of their time-averaged and fluctuating parts. The fluctuating parts of the fluid velocity components give rise to the so-called *Reynolds stress* terms, which are handled with a semiempirical turbulence model. In the second approach, a bulk flow model (BFM) is used in a manner similar to that used in traditional calculations for turbulent pipe flow. The fundamental difference between these two approaches is that the BFM approach characterizes the velocity components at any axial and circumferential location by their respective average values at that location. That is, fluid velocity variations across the clearance gap are not considered and thus fluid shear stress variations across the clearance gap are also not considered. Fluid shear stresses in a BFM approach are thus incorporated only at the fluid–solid boundaries (shaft and seal surfaces), employing empirical turbulent friction factors borrowed from the traditional turbulent pipe flow data. The flow path geometric boundaries for ungrooved annular seals are relatively simple. Consequently, the much simpler BFM approach has been the primary approach used by the major technologists who have focused on seal rotor vibration characteristics. A notable exception is the work by Nordamnn and Dietzen (1990) who provide a solution for an ungrooved annular seal using a computational model based on a perturbation of the N–S equations. Professor Childs has stated that the BFM has major shortcomings not previously recognized.

Bearing and Seal Rotor Dynamics

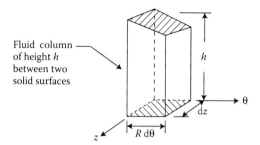

FIGURE 5.16 BFM CV fluid element.

5.4.2.3 Bulk Flow Model Approach

Consistent with the brief description of the RLE provided in Section 5.2.1, the aim here is not to provide all the intricate derivation steps in applying the BFM to annular seals. Instead, the intent here is to facilitate the serious reader's understanding of available derivations of an annular seal BFM, such as that detailed by Childs (1993). To that end, the following perspective is provided.

The BFM employs standard control volume (CV) formulation as covered in fluid mechanics courses of undergraduate mechanical engineering programs. In this application, the CV is a small arbitrary volume of fluid within the seal (Figure 5.16), bounded by seal rotor and stator surfaces, and by infinitesimal differential sides in the axial and circumferential directions. In fact, this is just how Reynolds set up the development of the RLE, except that variation of fluid velocities across the clearance gap are of paramount importance in laminar oil-film bearings and thus are not neglected as they are in the BFM approach.

Fluid flow mass balance for this CV is satisfied by the continuity (mass conservation) equation. Application of Newton's Second Law ($\vec{F} = m\vec{a}$) to this CV leads to two coupled PDEs, one for *circumferential momentum* balance and one for *axial momentum* balance. As Childs (1993) implies, the continuity equation is satisfied by appropriately substituting it into each of the two momentum equations, which are in turn considerably simplified in that derivation step. Employing the coordinate system shown in Figure 5.16, the following two momentum equations for the BFM are thus obtained.

5.4.2.4 Circumferential Momentum Equation

$$-\frac{h}{R}\frac{\partial p}{\partial \theta} = \frac{\rho}{2} u u_s f_s + \frac{\rho}{2}(u - R\omega)u_r f_r + \rho h \left(\frac{\partial u}{\partial t} + \frac{u}{R}\frac{\partial u}{\partial \theta} + w\frac{\partial u}{\partial z} \right) \quad (5.19)$$

5.4.2.5 Axial Momentum Equation

$$-h\frac{\partial p}{\partial z} = \frac{\rho}{2}wu_s f_s + \frac{\rho}{2}wu_r f_r + \rho h\left(\frac{\partial w}{\partial t} + \frac{u}{R}\frac{\partial w}{\partial \theta} + w\frac{\partial w}{\partial z}\right) \quad (5.20)$$

where $u = u(\theta, z)$ is the circumferential velocity, $u_s \equiv (u^2 + w^2)^{1/2}$, $w = w(\theta, z)$ is the axial velocity, $u_r \equiv [(u - R\omega)^2 + w^2]^{1/2}$, $h = h(\theta, z)$ is the film thickness, f_s is the local friction factor for seal, f_r is the local friction factor for rotor, and ρ is the fluid density. f_s and f_r are modeled after empirical friction factors for turbulent pipe flow, thus functions of local Reynolds No. and surface roughness.

Comparison of the RLE with the BFM is informative. First of all, the BFM equations include both temporal and convective inertia terms that are not retained in the RLE. Thus, the BFM includes an accounting of fluid inertia effects for seal rotor vibration characteristics. Second, the BFM has two coupled equations, whereas the classical lubrication model has only one equation, the RLE. This second comparison is interesting in that it shows a fundamental contrast in the developments of the RLE and BFM. The RLE is basically *conservation of mass*, that is, the scalar continuity equation, with *a priori* solutions for axial and circumferential velocity distributions substituted into the continuity equation. On the other hand, the two BFM equations (for the θ and z components of $\vec{F} = m\vec{a}$) have continuity substituted into them. Third, the RLE has the pressure distribution $p(\theta, z)$ as the only unknown field, but the BFM equations have not only the pressure distribution but also circumferential velocity distributions. Thus, while the RLE needs to be accompanied only by pressure boundary conditions, the BFM equations need pressure plus circumferential and axial velocity inlet boundary conditions.

Proper pressure and velocity boundary conditions combined with Equations 5.19 and 5.20 provide a well-posed mathematical problem the solution of which yields BFM simulations for ungrooved annular seal flow. Although this system of two equations is a considerable abridgement from the full N–S equations for this problem, obtaining general solutions is nevertheless still a formidable task, given that the BFM equations are coupled nonlinear PDEs.

However, based on computational solutions plus experiments over a static eccentricity ratio (e/C) range from 0 to 0.9, strong arguments are made that for practical purposes the seal rotor vibration coefficients expressed in Equation 5.16 are nearly constant out to $e/C \approx 0.3$. The mitigating factors that make static eccentricity of less importance to ungrooved annular seals than to journal bearings are the following: (i) seal clearances are typically more than twice the clearances of the journal bearings, which are the primary *enforcers* of rotor centerline static position and (ii) turbulent flow inherently acts to desensitize the circumferential variation of pressure to static eccentricity because of the corresponding reduction of the local Reynolds number where the seal film thickness is smaller and increase where it is larger. This second effect is similar to a (hypothetical) circumferentially cyclic variation in viscosity $\propto h$ with the maximum viscosity at the maximum film thickness and the minimum viscosity at the minimum film thickness. Such a phenomenon in a journal bearing would obviously desensitize it to static eccentricity. It would thus appear that the *isotropic model* of Equation 5.16 is justified for seals much more so than for journal bearings.

Childs (1993) presents in considerable detail the formulation for extracting annular seal rotor vibration characteristic coefficients from the BFM. Perturbation pressure solutions $\Delta p(\theta, z)$ are formulated and obtained for a "small" circular rotor orbital motion of radius ($e \ll C$) about the centered position, that is, about the position for which seal flow is rotationally symmetric. Integration of the perturbation pressure distribution into x and y force components yields the following:

$$f_x(e, \Omega) = -\int_0^L \int_0^{2\pi} \Delta p(\theta, z, e, \Omega) R \cos\theta \, dE \, dz$$

$$f_y(e, \Omega) = -\int_0^L \int_0^{2\pi} \Delta p(\theta, z, e, \Omega) R \sin\theta \, d\theta \, dz$$

(5.21)

Since this perturbation force is a function of orbit frequency, it lends itself to a second-order polynomial curve fit in frequency that directly extracts the *isotropic model* coefficients of Equation 5.16. To that end, expressing the perturbation force by its orthogonal components referenced to the instantaneous *radial* and *tangential* directions of the circular perturbation orbit yields the following expressions (refer Figure 2.13):

$$f_R \cong -(k_s + \Omega c_{ss} - \Omega^2 m_s)e, \quad f_T \cong (k_{ss} - \Omega c_s)e \quad (5.22)$$

Solutions for the *isotropic model* stiffness, damping, and inertia coefficients of Equation 5.16, based on Equations 5.22, are a mathematical curve-fit approximation of the exact frequency-dependent characteristic of the BFM perturbation solution. In a manner similar to data reduction of harmonic-excitation experimental results as covered in Section 5.3.1, Childs uses a least-squares fit of Equations 5.22 over the frequency range of Ω/ω from 0 to 2.

There is a tacit underlying assumption in this whole approach that Adams (1987) calls the *mechanical impedance hypothesis*, which implies that the extracted rotor vibration coefficients are not functions of the *shape* of the imposed harmonic orbit (i.e., circular versus elliptical). This hypothesis is completely consistent with the RLE, but must be assumed for the BFM. Equations 5.22 provide that the BFM-based perturbation solutions are only approximated by the *mechanical impedance hypothesis* and thus the potential influence of *orbit shape* on seal rotor dynamic coefficients should be kept open for discussion. As the more general sample problems in Chapter 4 demonstrate, with typical LRV models that incorporate *anisotropic* journal bearing characteristics, LRV orbits are usually elliptical, not circular.

5.4.2.6 Comparisons between Ungrooved Annular Seals and Journal Bearings

The majority of journal bearings operate with their hydrodynamic films in the laminar flow regime, in which case aligned journal bearings are characterized by two dimensionless parameters, Sommerfeld number (dimensionless speed) and L/D. In some applications, however, the combination of journal surface speed, lubricant viscosity, and bearing clearance place journal bearing hydrodynamic lubricating films into the turbulent regime. Conventional wisdom of the experts is that a quite good approximation for turbulence effects in journal bearings is based on the use of an *apparent viscosity*, which is locally made higher than the actual viscosity as a function of the local Reynolds numbers for journal velocity and localized parameters of pressure gradient and film thickness. This approach is provided by Elrod and Ng (1967). In the Elrod–Ng approach, the RLE Equation 5.1 for laminar lubricant films is still employed, albeit with the local viscosity at each finite-difference grid point modified to its local *apparent viscosity*. There is then an additional dimensionless number (e.g., clearance based Reynolds number) to characterize the journal bearing. The Elrod–Ng approach rests upon a fundamental assumption that temporal and convective inertia terms of the N–S equations are negligible even though it is fluid inertia at the film flow's fine structure level that is an essential ingredient of the turbulence. Thus, even with turbulence effects included, the theory and characterization of hydrodynamic journal bearings is not appreciably different than for laminar hydrodynamic lubrication. In stark contrast, ungrooved annular seals are characterized by several nondimensional parameters, including, but not limited to, the following list of major ones:

$$\text{Pressure drop:} \quad \frac{(p_{in} - p_{out})}{\rho w_0^2} \quad w_0 \equiv \frac{Q}{2\pi C}$$

$Q \equiv$ seal through flow.

Axial and circumferential Reynolds numbers, respectively:

$$R_z = \frac{2w_0 \rho C}{\mu}, \quad R_\theta = \frac{R\omega \rho C}{\mu}$$

Length-to-diameter and clearance-to-radius ratios, respectively: $L/D, C/R$.

Absolute roughness of rotor and stator, respectively: $e_r/2C, e_s/2C$.

Thus it is clear that, unlike journal bearings, annular seals do not lend themselves to the development of nondimensional wide-coverage design charts or tabulations as a practical option. Although the task of developing reasonably accurate journal bearing vibration characteristic input coefficients for rotor vibration analyses can be quite challenging, that task for seals is considerably more challenging than for journal bearings. For applications such as high-pressure centrifugal pumps where rotor vibration performance is dominated by the various fluid-annulus sealing zones (see Figure 5.13), the author recommends that serious analysts use experimentally benchmarked commercially available computer codes such as those from the Turbo-Machinery Laboratory at Texas A&M University.

Section 5.2.6, Subsection "Fundamental Caveat of LRV Analysis," provides a list of significant uncertainty factors affecting journal bearing characteristics. The following comparable list of uncertainty factors for ungrooved annular seals is similar but longer:

- *Clearance uncertainty* via seal rotor and stator diameter mfg. tolerances
- Variations in *fluid viscosity* from fluid temperature variations
- Seal rotor-to-stator *static eccentricity* (assumed zero for isotropic model)
- Seal rotor-to-stator tilt *misalignment*
- Seal *ring distortions* from loads, temperature gradients, wear, and so on
- Basic simplifying *assumptions* leading to the BFM governing equations
- Coefficients for entrance pressure loss and exit pressure recovery
- Entrance circumferential velocity (preswirl)
- Surface *roughness*.

As this list implies, the uncertainty in seal rotor vibration characteristics is no less than that for journal bearings.

5.4.3 Circumferentially Grooved Annular Seals for Liquids

Various fluid-annulus sealing zones, such as those shown in Figure 5.13, are not always ungrooved designs. Circumferential grooves are used in many designs to further reduce leakage flow between stages, through end seals and balancing drum (piston). The number of grooves, their axial spacing, width, and depth are not standardized parameters; different manufacturers have their own variation on the basic idea of circumferentially grooving to improve leakage reduction. The presence of such grooves also provides a more rub-forgiving less seizure-prone rotor–stator combination than without grooves. Grooves are employed on either rotor or stator. Figure 5.17 shows a variety of circumferential groove geometry for annular seals:

a. Labyrinth seals; groove depth much larger than radial tip clearance.
b. Shallow-grooves; groove depth approximately equal to tip clearance.

Published analysis and experimental results are sparse. Those cited by Childs (1993) suggest some trends. First, grooving significantly reduces LRV stiffness and damping effects, possibly as much as 80% reduction with wide deep grooves. Second, having the grooves on the seal stator

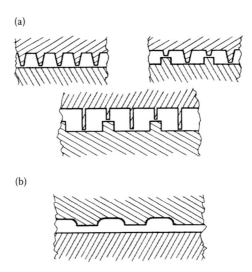

FIGURE 5.17 Examples of circumferentially grooved annular seals: (a) labyrinth seals; groove depth much larger than radial tip clearance and (b) shallow grooves; groove depth approximately equal to tip clearance.

is rotor dynamically more stable than having the grooves on the rotor. A configuration favored by some manufacturers of high-pressure multistage centrifugal pumps employs shallow circumferential grooves that are separated by axially straight lands unlike the sharp saw-tooth or narrow-strip tips in labyrinth seals, as contrasted in Figure 5.17. The advantage of a shallow-groove land-tip configuration is that it retains a significant Lomakin effect (see Figure 5.15). A balancing drum is long (see Figure 5.13). So a shallow groove land-tip geometry thus produces a quite high radial stiffness from a balancing drum because it has the full pressure rise of the pump across it.

From a fundamental fluid mechanics perspective, the flow patterns in a circumferentially grooved fluid annulus are considerably more complicated than in an ungrooved configuration and correspondingly more difficult to analyze. Thus, a BFM approach is unlikely to yield a realistic or accurate characterization of rotor vibration coefficients for circumferentially grooved seals. Nordmann and Dietzen (1990) use a finite-difference solution of the N–S equations, accounting both for turbulence and the geometric complications of circumferential grooves. The comparisons between their computational results and experiments are quite good. At the time of their work, computer costs were a significant factor in obtaining such N–S solutions, but with present work stations and top-end PCs such computational costs are no longer a significant consideration.

5.4.4 Annular Gas Seals

Turbo-machinery with a gas as the working fluid (i.e., compressors and turbines) are quite similar in appearance, function, and principle of operation to turbo-machinery with a liquid as the working fluid (i.e., pumps and hydro turbines). The multistage in-line centrifugal pump illustrated in Figure 5.13 could almost be taken for a centrifugal compressor of similar proportions. In compressible flow turbo-machinery, matters are complicated by the considerable change in process gas density that naturally occurs as the gas progresses through the flow path within the machine. On the other hand, at maximum flow conditions, turbo-machinery for liquids commonly operate with some *cavitation*, particularly at the inlet section of a pump impeller (first stage impeller if a multistage pump) or the exit section of a hydro turbine impeller. Significant amounts of cavitation vapor pockets in a pump act like a hydraulic flexibility (spring) and thereby can significantly contribute to pump flow instability and thus unsteady flow forces exerted upon the rotor.

As with annular seals for liquid-handling machinery, the radial forces developed in annular gas seals are approximately proportional to seal pressure drop and fluid density within the seal. Thus, for comparably sized seals and pressure drop, seal forces developed in gas-handling machines are much less than in liquid-handling machines because of the compressibility and lower density of the gas. Also, the added mass effect, (m_s) in Equation 5.16, is typically negligible and therefore generally not included when dealing with gas seals. The assumption of a seal-force isotropic model, as explained and used for liquid seals, is generally also used for gas seals. With the added-mass terms not included, the isotropic model of Equation 5.16 provides the following linear model that is usually employed for annular gas seals:

$$\begin{Bmatrix} f_x \\ f_y \end{Bmatrix} = -\begin{bmatrix} k_s & k_{ss} \\ -k_{ss} & k_s \end{bmatrix} \begin{Bmatrix} x \\ y \end{Bmatrix} - \begin{bmatrix} c_s & c_{ss} \\ c_{ss} & c_s \end{bmatrix} \begin{Bmatrix} \dot{x} \\ \dot{y} \end{Bmatrix} \quad (5.23)$$

Unlike liquid seals, the gas seal centering stiffness effect (k_s) is usually negligible, often negative. Like liquid seals, the gas seal cross stiffness (k_{ss}) is an important LRV analysis input because it is a destabilizing effect, for example, high-pressure steam turbine blade tip labyrinth seals can cause *steam whirl*.

5.4.4.1 Steam Whirl Compared to Oil Whip

The self-excited rotor vibration in high-pressure steam turbines called *steam whirl* is partially caused by the flow effects in blade tip seals and is embodied in the cross-stiffness coefficient (k_{ss}). Some of the case studies presented in Part 4 of this book deal with the *steam whirl* phenomenon, which is quite similar in its characteristics to the other well-known self-excited rotor vibration phenomenon called *oil whip*. Section 2.4.1 rigorously treats the connection between the *skew-symmetric* part of the stiffness coefficient matrix and such self-excited rotor vibration phenomena. In both *oil whip* and *steam whirl*, the rotor vibrates with a corotational direction orbit, typically at the lowest rotor-system natural frequency, usually near and somewhat below one-half the rotor spin speed frequency. The main difference between *oil whip* and *steam whirl* is in the controlled operating parameters that trigger each of these self-excited rotor vibration phenomena. With *oil whip* there is a rotational speed (threshold speed of instability) above which the self-excited vibration "kicks in." With *steam whirl*, there is a turbine power output level above which the self-excited vibration "kicks in." While both are serious problems requiring solution, *oil whip* is worse because if the *oil whip* threshold speed is encountered below the machine's operating speed, then the machine cannot be safely operated. With *steam whirl*, the turbine can be operated below the power output level where the self-excited vibration "kicks in." Thus, when *oil whip* is encountered, the machine should be shut down and a solution developed. With *steam whirl*, the machine may still be safely operated, albeit at a sufficiently reduced power output level. Thus, the "fix" for a *steam whirl* problem can be made at a later convenient time. Reduced power yields loss of some generating capacity and operation below the machine's best-efficiency power rating (i.e., higher fuel cost/KW-HR).

5.4.4.2 Typical Configurations for Annular Gas Seals

Nearly all annular gas seals can be placed into one of the following four categories:

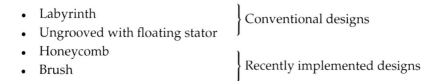

The design features of conventional annular gas seals are not much different than for annular liquid seals. Just like fixed-stator annular liquid seals, gas seals are also not supposed to act as the bearings; thus the radial clearance for fixed-stator annular gas seals is also typically two or more times the clearance of the machine's radial bearings. Consequently, for the much lower viscosity of typical process gases as compared with a typical liquid viscosity, ungrooved annular gas seals with fixed stators are not commonly used because the seal leakage in most cases would be too high with clearances two or more times the bearing clearances. However, annular gas seals with a floating stator (possibly segmented, frequently carbon for nonseizure qualities) can use an ungrooved surface since the stator float makes feasible very small clearances.

For gas handling turbo-machines, labyrinth seals yield considerably lower leakage than ungrooved gas seals of the same axial length and radial clearance because of labyrinth seals' inherent higher resistance to gas leakage flow. Thus, because of the optimum combination of simplicity and relatively low leakage, fixed-stator gas seals with annular grooves are the most common, especially labyrinth seals such as shown in Figure 5.17a. The labyrinth configuration is also inherently more rub-forgiving and less siezure prone than ungrooved seals, a contrast that is even more pronounced with gases than with liquids. The basic labyrinth seal configuration is possibly as old as turbo-machinery, that is, over 100 years old.

In recent times, two relatively new annular gas seal configurations have found their way into some high-performance gas handling turbo-machinery, namely the *honeycomb* seal and the *brush* seal. These two seal types are illustrated in Figure 5.18. The *honeycomb* seal is comprised of deep

(a)

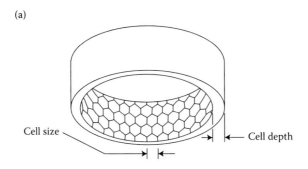

(b)

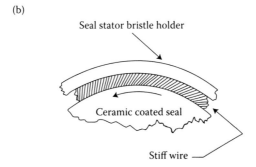

FIGURE 5.18 Recently implemented annular gas seal designs: (a) honeycomb seal and (b) brush seal.

honeycomb-shaped pockets on the seal stator, which reportedly provide lower leakage than comparably sized labyrinth seals of the same clearance and operating conditions. The major improvement provided by the *honeycomb* seal over the labyrinth seal is a significant reduction in tangential flow velocity within the seal, which significantly reduces the destabilizing cross-stiffness effect (k_{ss}). This type of seal has been implemented in centrifugal compressors with back-to-back impellers at the central location sealing the two impeller chambers from each other. Since the axial center of the rotor typically has a large motion participation in the lowest resonant mode shape, the significantly reduced destabilization quality of the *honeycomb* seal provides a considerable increase in the range of compressor operation free of self-excited vibrations, as reported by Childs (1993).

The *brush* seal is illustrated in Figure 5.18b. It uses a tightly packed array of many stiff wire bristles oriented with the direction of rotation, as shown. It reportedly has been determined in recent tests to provide much lower leakage rates than either labyrinth or honeycomb seals. The *brush* seal has also demonstrated favorable rotor vibration characteristics, as reported by Childs (1993). Clearly, the *brush* seal would appear to inherently reduce tangential flow velocity within the seal and thus reduce the destabilizing cross-stiffness effect (k_{ss}). The *brush* seal would also appear to be inherently immune from the potential rotor–stator rub-impact vulnerabilities of other seals since its brush bristles are already in constant contact with the rotor and are relatively compliant. This type of seal is now being used in late model gas turbine jet engines.

Since both the *brush* seal and the *honeycomb* seal are relatively recent developments, their long-term durability and reliability qualities in the field under favorable as well as adverse operating conditions have yet to be firmly established. For example, the author has recently become aware of premature *brush* seal wear-out on the gas turbine jet engines of one major aircraft engine manufacturer. Although not necessarily a safety hazard, significant engine repair costs could readily result. One theory concerning this *brush* seal problem is that bristle motion characterized by circumferentially *traveling waves* occurs in the bristles, leading to their accelerated wear rates.

5.4.4.3 Dealing with Seal LRV-Coefficient Uncertainties

As for annular liquid seals, the most comprehensive single information source on annular gas seal rotor vibration characteristics is Childs' (1993) book. Childs comprehensively shows that the current "ignorance factor" in predictions for seal rotor vibration properties is significantly higher for gas seals than for liquid seals. Also, there is less available laboratory test data on gas seal rotor vibration properties. Unlike journal bearings, annular seals (for liquid and even more so for gas) clearly do not lend themselves to the development of nondimensional wide-coverage design charts or tabulations.

Part 4 of this book provides some of the author's experience in dealing with the inherent uncertainties in seal rotor vibration properties when performing LRV analyses for the purpose of troubleshooting. For example, with adroit use of vibration measurements on a particular vibration-plagued machine, one can often make insight-motivated adjustments to uncertain inputs of the LRV analysis model (e.g., seal LRV coefficients) to improve its correlation with the actual machine's vibration behavior, for example, instability threshold speed or threshold power output, self-excited vibration frequency at the instability threshold, critical speeds, and peak vibration amplitudes at critical speeds. When a model is successfully adjusted to provide a reasonable portrayal of an actual machine's vibration behavior, the author refers to the model as a *calibrated model*. By superimposing promising fixes upon a *calibrated model*, potential corrective actions or retrofits can be thoroughly analyzed and evaluated prior to implementing a specific corrective course of action. This approach brings solid engineering science to bear upon troubleshooting and consequently has a much higher probability of a timely success than randomly trying "something someone heard worked on a different machine at another plant somewhere else."

5.5 Rolling Contact Bearings

Several different configurations of rolling contact bearings (RCB) are used in numerous applications. The most common RCB configurations are *ball bearings*, which can be subdivided into specific categories of *radial contact*, *angular contact*, and *axial contact*. Other commonly used RCB configurations utilize *straight cylindrical*, *crowned-cylindrical*, and *tapered roller elements*. In many applications that employ RCBs, the complete rotor-bearing system is sufficiently stiff (e.g., machine tool spindles) to operate at speeds well below the lowest critical speed, so that the only rotor vibration consideration is proper rotor balancing (Category-1, Table 2.1). In flexible rotor applications, where operating speeds are above one or more critical speeds, the RCBs often have sufficient internal preloading so that in comparison to the other system flexibilities (i.e., shaft and/or support structure) the RCBs act essentially as rigid connections. In applications where the bearings have no internal preload and possibly some clearance (or "play"), the dynamic behavior can be quite nonlinear and thus standard linear analyses are potentially quite misleading (see Section 2.5 of Chapter 2).

RCBs can readily be configured to achieve high stiffness; thus they are frequently used in applications where precision positioning accuracy is important, such as in machine tool spindles. In contrast to their high stiffness potential, RCBs have very little inherent vibration damping capacity, unlike fluid-film journal bearings. Also, unlike a fluid-film journal bearing that will fail catastrophically if its lubricant supply flow is interrupted, an RCB can operate for sustained periods of time when the normal lubrication supply fails, albeit with a probable shortening of the RCB's useable life. Thus, RCBs are usually a safer choice over fluid-film bearings in aerospace applications such as modern aircraft gas turbine engines. In such applications, however, the inability of the RCBs to provide adequate vibration damping capacity to safely pass through critical speeds frequently necessitates the use of *squeeze-film dampers* (SFD) (see Section 5.6) to support one or more of the machine's RCBs. When an SFD is employed, its rotor vibration characteristics are usually the governing factor at the bearing, not the very high stiffness of the RCB in series with the SFD.

Roller bearings inherently possess much *higher load capacity* and Hertzian contact *stiffness* than ball bearings, because a ball's load-supporting *footprint* is conceptually a *point* contact, whereas a roller's load-supporting *footprint* is conceptually a *line* contact. However, in roller bearings each roller has a single axis about which it must spin in proper operation. As rotational speed is increased for a roller bearing, the increased propensity for dynamic skewing of the rollers will impose a maximum useable rotational speed for the bearing. In contrast, a ball's spin may take place about any diameter of the ball. As a consequence, *ball bearings* have much *higher maximum speed* limits than roller bearings, given their inherent absence of dynamic skewing. Given the higher speed capability but inherently lower stiffness of ball bearings compared to roller bearings, it is far more likely that one would possibly need radial stiffness for a ball bearing than for a roller bearing when performing LRV modeling and analyses.

If one focuses on the load paths through an RCB, two important factors become apparent:

1. Each contact between a rolling element and its raceways possesses a *nonlinear* load versus deformation characteristic (F versus δ). That is, since the deformation *footprint* area between rolling element and raceway increases with load, the F versus δ characteristic exhibits a *stiffening* nonlinearity, that is, the slope of F versus δ increases with F.

2. The total bearing load is simultaneously shared, albeit nonuniformly, by a number of rolling elements in compression as illustrated in Figure 5.19. Therefore, the contact forces taken by the rolling elements are *statically indeterminate*, that is, cannot be solved from force and moment equilibrium alone, but must include the flexibility characteristics of all elements.

As first developed by A. B. "Bert" Jones (1946), this combination of *statically indeterminate* and *nonlinear* contact forces requires the use of quite specialized analyses that employ appropriate iterative algorithms to converge on the static equilibrium state of all the rolling elements and raceways

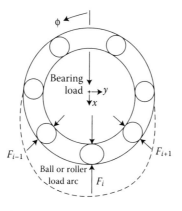

FIGURE 5.19 Typical distribution of contact loads in an RCB.

for a specified combination of externally applied forces and moments. Perturbing the static equilibrium solution yields the RCB stiffness coefficient array. The work of Bert Jones is essentially the foundation of all modern computer codes for RCB load-deflection analyses.

A suitable estimate of RCB radial stiffness for LRV analyses is obtained by assuming that the inner and outer raceways are both perfectly rigid. This simplifying assumption avoids employing the quite formidable and specialized complete static equilibrium-based solution just described, because it geometrically relates all the rolling elements' compressive deflections to a single bearing deflection. The load versus deflection for the ith rolling element is expressible from Hertzian elastic contact theory, such as in the following summary from Kramer (1993) (numbers in B expressions based on steel):

$$F_i = \left(\frac{\delta_i}{B}\right)^n \tag{5.24}$$

where F_i, δ_I are load and deflection of ith rolling element.

Here, $\left. \begin{array}{l} n = 3/2 \\ B = 4.37 \times 10^{-4} d^{-1/3} \end{array} \right\}$ Ball bearings $\quad \left. \begin{array}{l} n = 1/0.9 \\ B = 0.77 \times 10^{-4} L^{-0.8} \end{array} \right\}$ Roller bearings

Units: Ball diameter d mm, roller length L mm, contact force F_j Newtons.

Contact forces occur only when a rolling element is in compression. It is implicit in the approximation here that the bearing has no internal preload and no play (clearance). Then the only source of contact loads is from the applied bearing load and the contact zone will be the 180° arc shown in Figure 5.19. The contact compressive deflection of each rolling element can then be expressed as follows, where x is the relative radial displacement between the raceways:

$$\delta_i = \begin{cases} x \cos \phi_i, & 90° < \phi_i < 270° \\ 0, & -90° < \phi_i < 90° \end{cases} \tag{5.25}$$

Play (clearance) in a bearing tends to make the contact load arc less than 180° and preload tends to make the contact load arc greater than 180°.

With ϕ referenced to the bearing load as shown in Figure 5.19, equilibrating the bearing load by the sum of the components of all the individual contact forces can be expressed as follows:

$$F = \sum_{i=1}^{N} F_i \cos \phi_i = \sum_{i=1}^{N} \left(\frac{\delta_i}{B}\right)^n \cos \phi_i = \sum_{i=1}^{N} \left(\frac{x \cos \phi_i}{B}\right)^n \cos \phi_i = \left(\frac{x}{B}\right)^n S_x$$

$$S_x \equiv \sum_{i=1}^{N} (\cos \phi_i)^{n+1} \tag{5.26}$$

where N is the number of rolling elements within the 180° arc of contact loading.

Rearranging Equation 5.26, the bearing deflection is expressed as follows:

$$x = B\left(\frac{F}{S_x}\right)^{1/n} \tag{5.27}$$

Differentiating the radial bearing force (F) by its corresponding radial bearing deflection (x), bearing x-direction stiffness is obtained in the following equation:

$$k_{xx} = \frac{dF}{dx} = \frac{nx^{n-1}}{B^n}S_x = \frac{n}{x}\left(\frac{x}{B}\right)^n S_x = \frac{n}{x}F \tag{5.28}$$

Visualize each loaded rolling element as a nonlinear radial spring in compression. Each individual rolling element's stiffness in the direction perpendicular to the bearing load can be obtained by projecting a y-direction differential deflection onto its radial direction and projecting its resulting differential radial force back onto the y direction. The radial stiffness of an individual loaded rolling element is obtained by differentiating Equation 5.24, as follows:

$$k_i = \frac{dF_i}{d\delta_i} = \frac{n\delta_i^{n-1}}{B^n} = \frac{n(x\cos\phi_i)^{n-1}}{B^n} \tag{5.29}$$

Projecting the y-direction differential deflection onto rolling element's radial direction and its resulting differential radial force back onto the y direction yields the following:

$$d\delta_i = dy\sin\phi_i \quad \text{and} \quad dF_{iy} = dF_i\sin\phi_i$$

Therefore,

$$dF_{iy} = dF_i\sin\phi_i = k_i\phi\, d\delta_i\sin\phi_i = k_i\, dy(\sin\phi_i)^2$$

The y-direction stiffness for an individual loaded rolling element is thus obtained as follows:

$$k_{iy} = \frac{dF_{iy}}{dy} = k_i(\sin\phi_i)^2 \tag{5.30}$$

Summing all the rolling elements' y-direction stiffness yields the bearing's y-direction stiffness, as follows:

$$k_{yy} = \sum_{i=1}^{N} k_i (\sin \phi_i)^2 = \frac{nx^{n-1}}{B^n} \sum_{i=1}^{N} (\cos \varphi_i)^{n-1} (\sin \phi_i)^2 = \frac{nx^{n-1}}{B^n} S_y$$

$$S_y \equiv \sum_{i=1}^{N} (\cos \phi_i)^{n-1} (\sin \phi_i)^2$$

(5.31)

Combining Equations 5.28 and 5.31 yields the stiffness ratio, as follows:

$$R_k \equiv \frac{k_{yy}}{k_{xx}} = \frac{S_y}{S_x} < 1 \quad (5.32)$$

This easily calculated ratio increases with the number of rolling elements in the bearing. Kramer (1993) provides values for the following example cases.

Number of rolling elements in bearing = 8, 12, 16 gives the following:

Ball bearing, $R_k = 0.46, 0.64, 0.73$, Roller bearing, $R_k = 0.49, 0.66, 0.74$

The bearing stiffness coefficients given by Equations 5.28 and 5.31 are derived as though neither raceway is rotating. There are three cases of rotation one could encounter: (1) only the inner raceway rotates (most typical), (2) only the outer raceway rotates, and (3) both raceways rotate (e.g., intershaft bearings for multispool-shaft jet engines). The cage maintains uniform spacing between the rolling elements, and when it rotates the bearing load and resulting deflection are perfectly aligned with each other only when the bearing load is either directly into a rolling element or directly between two rolling elements. At all other instances, bearing load and resulting deflection are very slightly out of alignment. This produces a very slight cyclic variation of the bearing's stiffness coefficients at roller or ball passing frequency, and thus suggests the possibility of what is generically referred to as *parametric excitation*.

As an input into a standard LRV analysis code, such as RDA, Equations 5.28 and 5.31 provide the following bearing interactive force with stiffness only, but no damping:

$$\begin{Bmatrix} f_x \\ f_y \end{Bmatrix} = -\begin{bmatrix} k_{xx} & 0 \\ 0 & k_{yy} \end{bmatrix} \begin{Bmatrix} x \\ y \end{Bmatrix} \quad (5.33)$$

Of course the chosen x–y coordinate system orientation in a given LRV model may not align with the Equation 5.33 principal x–y coordinate system orientation, which is shown in Figure 5.19. However, as described by

Equation 5.7 and Figure 5.2, bearing and seal LRV coefficient arrays are second rank tensors and thus can be easily transformed to any alternate coordinate system orientation. Equation 5.33 transformed to a nonprincipal coordinate system yields nonzero off-diagonal stiffness terms that are equal, that is, the stiffness array is symmetric. Thus, this model for rolling element bearing stiffness does not embody any destabilizing effect, in contrast to journal bearings.

5.6 Squeeze-Film Dampers

Vibration damping capacity of an RCB is extremely small and therefore to measure it is virtually impossible since any test rig for this purpose would have its own damping that would mask that of a tested RCB. As is well known and shown by Figure 1.5, the benefit of damping is in preventing excessively high vibration amplitudes at resonance conditions. Thus, for the many machines running on RCBs that have the maximum running speed well below the lowest critical speed, the absence of any significant RCB damping presents no problem.

Since an RCB can usually operate for sustained periods of time after the normal lubrication supply fails, RCBs are usually a safer choice over fluid-film bearings in aerospace applications such as modern aircraft gas turbine jet engines. In such applications, however, the inability of the RCBs to provide adequate vibration damping capacity to maintain tolerable unbalance vibration levels through critical speeds frequently necessitates the use of SFD. Typically, an SFD is defined by a cylindrical annular oil film within a small radial clearance between the O.D. cylindrical surface of an RCB's outer raceway and the precision cylindrical bore into which it is fitted in a machine. The radial clearance of the SFD is similar to that for a comparable diameter journal bearing, possibly a bit larger as optimized for a specific application. Figure 5.20 shows a configuration that employs *centering springs*.

An SFD is like a journal bearing without journal rotation. Referring to Equation 5.1, the "sliding velocity term" in the RLE is then zero, leaving only the *squeeze-film term* to generate hydrodynamic pressure within the small annular clearance. As a first-order approximation to compute SFD damping coefficients, the perturbation approach given by Equations 5.4 may be used. However, the factors of *film rupture* and *dissolution of air* in the SFD oil film produce considerably more complication and uncertainty of computational predictions for LRV damping coefficients than these factors do in journal bearings. Also, the neglect of fluid inertia effects implicit in the RLE is not as good an assumption for SFDs as it is for journal bearings.

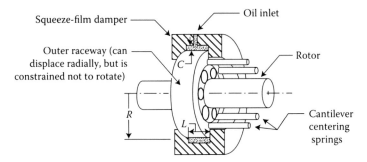

FIGURE 5.20 SFD concept with RCB and centering springs.

5.6.1 Dampers with Centering Springs

The SFD configuration shown in Figure 5.20 employs *centering springs* since there is no active *sliding velocity term* to generate static load-carrying capacity in the hydrodynamic oil film. To create a static equilibrium position about which the vibration occurs and is damped by the SFD, *centering springs* are typically used to negate the bearing static load and maintain damper approximate concentricity. The radial stiffness of the *centering springs* is far less than the radial stiffness of the RCB, as developed in the previous section. Thus, the stiffness coefficient array is essentially the isotropic radial stiffness of the *centering springs*, k_{cs}. Assuming validity of linearization, as postulated for journal bearings in Equation 2.60, the LRV interactive force at a bearing station employing an SFD is then expressible as follows:

$$\begin{Bmatrix} f_x \\ f_y \end{Bmatrix} = -\begin{bmatrix} k_{cs} & 0 \\ 0 & k_{cs} \end{bmatrix} \begin{Bmatrix} x \\ y \end{Bmatrix} - \begin{bmatrix} c_d & 0 \\ 0 & c_d \end{bmatrix} \begin{Bmatrix} \dot{x} \\ \dot{y} \end{Bmatrix} \qquad (5.34)$$

where c_d is the damping coefficient for the concentric damper film.

The SFD's length (L) is typically much smaller than its diameter ($D = 2R$). Consequently, it is customary to consider two cases: (1) SFD does not have end seals, and (2) SFD does have end seals. For Case 1, the supplied oil flow is continuously squeezed out of the two axial boundaries of the damper film and since typically $L/D < 0.2$, solution to Equation 5.1 using the *short bearing* approximation is justified. For Case 2, the use of end seals essentially prevents the significant axial oil flow encountered in Case 1 and thus using the *long bearing* approximation is justified. Also for Case 2, one or more drain holes are put in the damper to maintain a specified oil through-flow to control damper oil temperature.

Postulating a concentric circular orbit for the rotor within the SFD, solution of Equation 5.1 yields an instantaneous radial-plane force vector upon the rotor which can be decomposed into its radial and tangential components. As an example of this, Vance (1988) lists these two force components based on the "short bearing" approximation which is appropriate for the above Case 1 (no end seals) and 180° cavitation zone trailing the orbiting line-of-centers (i.e., minimum film thickness), as follows:

$$\text{Radial component,} \quad F_R = -\frac{2\mu R L^3 \Omega \varepsilon^2}{C^2(1-\varepsilon^2)^2}$$

$$\text{Tangential component,} \quad F_T = -\frac{\pi \mu R L^3 \Omega \varepsilon}{2C^2(1-\varepsilon^2)^{3/2}} \quad (5.35)$$

where μ is the viscosity, R is the damper radius, L is the damper length, Ω is the orbit frequency, C is the damper radial clearance, and ε is the eccentricity ratio e/C.

Radial and tangential force components can also be similarly derived using the "long bearing" solution of Equation 5.1, which is appropriate to the previous Case 2 (with end seals). It should be noted that the force components given by Equations 5.35 are clearly nonlinear functions of the motion. However, they can be linearized for a "small" concentric circular orbit as similarly shown in Equations 5.22 for annular seals. Equations 5.35 can be simplified for $\varepsilon \ll 1$ ($\varepsilon \to 0$) to the following:

$$F_R \cong -\left(\frac{2\mu R L^3 \Omega \varepsilon}{C^2}\right)\varepsilon$$

$$F_T \cong -\frac{\pi \mu R L^3 \Omega \varepsilon}{2C^2} \quad (5.36)$$

Since the radial force approaches zero one order faster than the tangential force (i.e., ε versus ε^2), the only nonzero coefficient retrieved from Equations 5.22 is the diagonal damping coefficient, $c_s \equiv c_d$. Thus, for the "short bearing" approximation with boundary conditions for the 180° cavitation zone trailing the orbiting line-of-centers, the Equation 5.34 damping coefficient for LRV analyses is given as follows:

$$c_d \cong \frac{\pi \mu R L^3}{2C^3} \quad (5.37)$$

For a sufficiently high damper ambient pressure to suppress cavitation, the solution yields a damping coefficient that is twice that given by Equation 5.37.

5.6.2 Dampers without Centering Springs

Eliminating the centering springs makes the SFD mechanically simpler and more compact. The possibility of centering spring fatigue failure does not need to be addressed if there are no centering springs. However, from a rotor vibration point of view, eliminating the centering springs makes the system considerably less simple. The damper now tends to "sit" at the bottom of the clearance gap and it requires some vibration to "lift" it off the bottom. That is a quite nonlinear dynamics problem.

Some modern aircraft engines are fitted with "springless" SFDs while some have spring-centered SFDs. Under NASA sponsorship, Adams et al. (1982) devised methods and software to retrofit algorithms for both types of dampers into the general purpose nonlinear time-transient rotor response computer codes used by the two major U.S. aircraft engine manufactures. Adams et al. (1982) show a family of nonlinear rotor vibration orbits that develop in "springless" SFDs as a rotating unbalance force magnitude is progressively increased. With a static decentering force effect (e.g., rotor weight) and small unbalance magnitudes, the orbit barely lifts off the "bottom" of the SFD, forming a small orbital trajectory that has been likened to a "crescent moon." As unbalance magnitude is progressively increased, it tends to overcome the static decentering force and thus the orbit progresses from the small "crescent moon" trajectory to a distorted ellipse to a nearly concentric circular orbit as the unbalance force overpowers the static decentering force effect.

To provide a reasonable linear approximation to the nonlinear behavior of both "springless" and spring-centered SFDs, Hahn (1984) developed methods and results to approximate SFD dynamic characteristics with equivalent linearized stiffness and damping coefficients compatible with LRV analysis codes like RDA. Such an approach appears to make sense when parametric preliminary design studies are conducted, leaving a full nonlinear analysis to check out a proposed and/or final prototype engine design.

5.6.3 Limitations of Reynolds Equation–Based Solutions

In developing Equations 5.35 through 5.37, a concentric circular orbit trajectory is postulated. If one views the Reynolds equation solution for film pressure distribution in the SFD from a reference frame rotating at the orbit frequency (Ω), the pressure distribution is the same as in an equivalent journal bearing running at static equilibrium with the same eccentricity. In the typical case where cavitation occurs, the respective SFD and journal bearing Reynolds equation solutions are still equivalent. However, there is a quite significant physical difference between the SFD and its equivalent journal bearing. That is, in the journal bearing under static load there is typically an oil inlet groove near where the film gap starts its reduction (or "wedge" effect) and the cavitation zone downstream of the minimum film thickness is fixed in the journal bearing space. On the other hand, in the SFD with a concentric Ω-frequency orbit, the cavitation zone also rotates at Ω around the SFD annulus. Depending on whether end seals are used or not and on the through-flow of oil metered to the SFD, a specific "blob" of oil may be required to pass into and out of cavitation several times at a frequency of Ω during a single residence period within the SFD film. It is reasonable to visualize that as the orbit frequency is progressively increased, the SFD oil *refuses to cooperate* in that manner. Experiments have in fact shown that as orbit frequency is progressively increased, the SFD becomes an oil froth producer and its damping capacity falls far short of Reynolds-equation-based predictions.

Hibner and Bansal (1979) provide the most definitive description on the failure of classical lubrication theory to reasonably predict SFD performance. They show with extensive laboratory testing at speeds and other operating conditions typical of modern aircraft engines that fluid-film lubrication theory greatly over predicts SFD film pressure distributions and damping coefficients. They observed a frothy oil flow out of their test damper. They suggest that the considerable deviation between test and theory stems from *gaseous cavitation*, greatly enhanced by air bubbles being drawn into the SFD to produce a *two-phase* flow that greatly reduces hydrodynamic pressures.

The work of Hibner and other SFD specialists indicates that for low-speed applications, classical hydrodynamic lubrication theory can provide reasonable predictions for SFD performance. But at rotational speeds typical of modern aircraft engines, classical hydrodynamic lubrication theory greatly overestimates SFD damping coefficients and therefore thorough testing of specific SFD configurations is required to reliably determine the actual SFD performance.

5.7 Magnetic Bearings

The generic configuration of an *active* magnetic bearing system is shown in Figure 5.21, which schematically illustrates the essential components.

The main feature of magnetic bearings which has attracted the attention of some rotating machinery designers is that they are *oil-free bearings*. This means for example that with large pipe line compressor rotors supported on oil-free bearings, the elimination of oil precludes the eventual coating of pipeline interior surfaces with lost oil that otherwise must be periodically cleaned out of the pipeline, at considerable service and downtime costs. Interestingly, this feature is not uppermost in the minds of magnetic bearing conceivers, who for the most part are academicians with a particular focus on control theory. They conceived the modern active magnetic bearing as an electromechanical actuator device that utilizes rotor position feedback to a controller in order for the magnetic bearing to provide electromagnetic

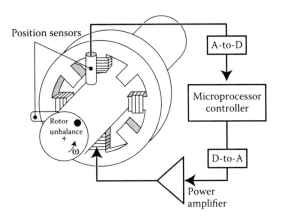

FIGURE 5.21 Active magnetic bearing schematic.

noncontacting rotor levitation with attributes naturally occurring in conventional bearings, that is, static load capacity along with stiffness and damping. Magnetic bearing technologists have focused their story on the fact that the rotor dynamic properties of magnetic bearings are freely prescribed by the control law designed into the feedback control system, and thus can also be programmed to adjust in real time to best suit a machine's current operating needs, such as active tuning "around" critical speeds and "extra damping" to suppress instabilities.

5.7.1 Unique Operating Features of Active Magnetic Bearings

Magnetic bearing systems can routinely be configured with impressive versatility not readily achievable with conventional bearings. In addition to providing real-time controllable load support, stiffness, and damping, they can simultaneously provide feed-forward-based dynamic bearing forces to partially negate rotor vibrations from other inherent sources. They can also employ notch filtering strategies to isolate the machine's stator from specific rotor vibration frequencies such as synchronous forces from residual rotor mass unbalance. In this last feature, notch filtering out the synchronous unbalance forces from the rotor–stator interaction forces seems to be quite nice, but then the rotor wants to spin about its polar inertia principal axis through its mass center, and thus its surfaces will wobble accordingly, meaning that rotor–stator rubs and/or impacts at small rotor–stator radial clearances have an increased likelihood of occurrence.

A natural extension of current magnetic bearing systems is their integration with next-generation condition monitoring strategies, discussed in Chapter 7, Section 7.1. Not only do active magnetic bearing systems possess the displacement sensors inherent in modern conditioning monitoring systems, but they automatically provide the capability of *real-time monitoring of bearing forces*, a long wished-for feature of rotating machinery problem diagnosticians. As alluded to earlier, magnetic bearings being real-time controlled force actuators can also be programmed to impose static and dynamic bearing load signals that can be "intelligently" composed to alleviate (at least partially and temporarily) a wide array of machine operating difficulties such as excessive vibrations and rotor–stator rubbing initiated by transient thermal distortions of the stator or other sources. Clearly, the concept of a so-called *"smart machine"* for next-generation rotating machinery is not difficult to conceptualize when active magnetic bearings are employed for rotor support. For an update on magnetic bearing publications, Allaire and Trumper (1998) provide several papers and Schweitzer (1998) focuses on "smart rotating machinery."

5.7.2 Short Comings of Magnetic Bearings

Magnetic bearing systems are relatively expensive, encompassing a system with position sensors, A-to-D and D-to-A multichannel signal converters, multichannel power amplifiers, and a microprocessor. Also, the lack of basic simplicity with such a multicomponent electromechanical system surely translates into concerns about reliability and thus the need for component redundancy (e.g., sensors).

The most obvious manifestation of the reliability/redundancy factor is that magnetic bearings in actual applications require a backup set of *"catcher" bearings* (typically ball bearings) onto which the rotor drops when the magnetic bearing operation is interrupted, such as by power or primary nonredundant component failure, or when the magnetic bearing is overloaded. The dynamical behavior of the rotor when the catcher bearings take over was initially not properly evaluated by magnetic bearing technologists. But in rigorous application testing, it was found that severe nonlinear rotor vibration can occur when the rotor falls through the catcher bearing clearance gap and hits the catcher bearings.

Fluid-film bearings and RCBs both possess considerable capacities for momentary overloads, for example, shock loads. Since these conventional bearings completely permeate the modern industrial world, their high capacities for momentary overloads are essentially taken for granted since they "do their job" and keep on running. On the other hand, magnetic bearings "saturate" when loads are pushed to their limits and thus provide little capacity for large load increases that momentarily exceed the bearing's design load capacity by substantial amounts. This is a serious limitation for many applications. For static load and lower-frequency stiffness and damping properties, magnetic bearing force capacity is limited

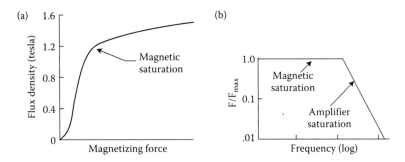

FIGURE 5.22 Magnetic bearing saturation effect load limits (Fleming, 1991).

by the saturation flux density of the magnetic iron, as illustrated in Figure 5.22a. A further limitation is set by the maximum rate at which the control system can change the current in the windings. The magnets have an inherently high inductance which resists a change in current, thus the maximum "slew rate" depends on the voltage available from the power amplifier. In practical terms, the required slew rate is a function of the frequency and amplitude of rotor vibration experienced at the bearing. Figure 5.22b illustrates the combined effects of magnetic saturation and slew-rate limitation on magnetic bearing load limits.

Conventional bearings are not normally feedback-controlled devices, that is, they achieve their load capacity and other natural characteristics through mechanical design features grounded in fundamental mechanics principles. Conversely, the basic operation of active magnetic bearings relies on feedback of rotor position signals to adjust instantaneous bearing forces. As a result, a generic shortcoming of active magnetic bearings stems from this fundamental reliance on feedback control. It is referred to with the terms "spillover" and "collocation error." Feedback control design is traditionally viewed as a compromise between response and stability. Whenever a feedback loop is closed, there is the potential for instability, as is well known.

Specifically for active magnetic bearings, collocation error arises from the sensors not being placed exactly where the bearing force signals are applied to the rotor, and this can produce rotor dynamical instabilities (spillover) that would not otherwise occur. Surely, no longstanding rotor dynamics specialist is enthused about this, since other traditionally recognized rotor dynamical instability mechanisms are always lurking, especially in turbo-machinery.

The magnetic bearing technologists' answer to this fundamental shortcoming is to have programmed into the control law a very accurate dynamics model of the rotor system. However, machines constantly change their dynamic characteristics in response to operating point changes and as a result of normal seasoning and wear over time. This necessitates

Bearing and Seal Rotor Dynamics

continuous automatic real-time recalibration of the dynamics model residing in the programmed control law. Under some well-defined operating modes, rotor dynamical systems can be quite nonlinear, and then having an accurate model in the control law for the actual rotor system becomes a formidable challenge. It prompts one to sarcastically ask "how all those 'stupid' oil-film bearings and ball bearings can do all the things they do?" The answer is, "the designers are smart."

5.8 Compliance Surface Foil Gas Bearings

Gas film bearings of both hydrodynamic and hydrostatic functioning were already being investigated and used in a few novel applications nearly 50 years ago. However, use of those bearings never achieved wide industrial use, primarily because of quite low load capacity at modest rotational speeds and rotor dynamical instability problems at speeds sufficiently high to provide useable static load capacities. Hydrostatic gas bearings utilizing porous media bearing sleeves were also shown to be feasible in laboratory testing and analysis. The foil gas bearing concept achieved success in the predigital-age high-speed tape deck heads by manufacturers such as Ampex. The main modern application of the hydrodynamic air bearing, initially on mainframe computer high-speed flying-head disk readers, has found its present place in PC hard drives. Quite recently, a major Cleveland-based manufacturer of MRI medical scanners has successfully developed and employed hydrostatic air bearings to support the main rotational positioning barrel, advancing the position resolution in this product.

About 25 years ago the gas foil bearing concept evolved into a new family of configuration, namely the *compliance surface foil gas bearing* (Heshmat et al., 1982). Figure 5.23 illustrates two typical compliance surface foil gas

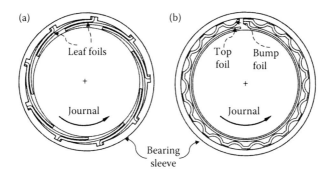

FIGURE 5.23 Two types of compliant surface foil gas journal bearings: (a) leaf-type foil bearing and (b) bump-type foil bearing.

journal bearings. Similar concepts for axial load thrust bearings are also in the mix but not addressed here because of the focus on rotor radial vibration.

The main selling feature touted of the modern compliant surface foil gas bearing has been that it is *oil free*, which can be a definite advantage for a number of applications, not just aerospace applications. Although magnetic bearings (see Section 5.7) were touted by the academically inclined developers primarily for their accurately controllable characteristics like rotor dynamical stiffness and damping, it is actually the *oil-free* nature of magnetic bearings that resulted in their use in a few heavy industry applications such as in large pipeline compressors. However, compliant surface *foil bearings* have the *added benefit* over *magnetic bearings* of use at quite elevated temperatures in excess of 1100°F (593°C).

Most of the development work has been focused on the bump-type foil bearing illustrated in Figure 5.23b. These bearings are assembled with a modest amount of elastic preload, which means that at start-up the journal is in rubbing contact with the inner surface of the *top foil*. Special coatings on the interior surface of top foil are thus required to counter the potential for significant wear accumulation from many starts and stops. As the journal accelerates up to operating speed, the hydrodynamic gas film overcomes the initial preload and thereby separates the journal and top foil. As the bearing radial load comes into play, the top foil and bumper foil elastically deform under the load so as to spread the load-carrying hydrodynamic separating pressure distribution more uniformly over the load-carrying area. This yields a bearing load capacity superior to the original hydrodynamic gas bearings of rigid construction.

An interesting dichotomy between compliant surface *foil bearings* and *magnetic bearings* is the following. Foil bearings are relatively simple in their configuration, with operating properties resulting from the fundamental ambient-gas hydrodynamics of the gas film interacting with the elastically deformable foils. That is, the foil bearings *do not have to be "smart,"* they only have *to let nature take its course*. In stark contrast, operation of the magnetic bearing is anything but simple, involving position sensors, with A-to-D, microprocessor, D-to-A, and power amplifiers (see Figure 5.21). But the magnetic bearing has quite predictable performance, for example, load capacity, dynamic stiffness, and damping coefficients. However, the much simpler configured foil bearing surely presents considerable challenges in predicting its operating performance characteristics, especially rotor dynamical stiffness and damping properties. The elastic deformation of the foils is significantly hardening nonlinear with load, and it is the foil hardening nonlinear deformations that dominate both the static and dynamic bearing properties. The beneficial damping inherent in the dynamic rubbing between the top and bumper foils is also a quite nonlinear mechanism. In consequence, use of foil bearings in a specific application requires significantly more development testing than other alternatives require, at least at this point of time. The main applications thus far for these foil bearings include turbo-chargers and micro gas turbine engines for land-based electric power generation. The *compact oil-free nature* and *high-temperature capability* of the compliant surface foil gas bearing are significant.

One can begin to appreciate the unpredictability of rotor dynamical stiffness and damping coefficients from the work of Howard et al. (2001). Their large scatter of results to fit linear dynamical models leads one to the conclusion that perhaps foil bearings inherently cannot be sufficiently well modeled for rotor vibration predictions using linear vibration models. This assertion is supported by the recent work of San Andres and Kim (2007). Their nonlinear rotor vibration simulations compare amazingly close with their laboratory tests on a small precision 2-bearing high-speed rotor. Their results clearly demonstrate that the rotor vibration characteristics are dominated by the structural nonlinearities of the foils, showing phenomena that could not be predicted by any linear vibration predictive simulation model. For example, they demonstrate that gas foil bearing supported rotors are prone to subsynchronous whirl orbits, albeit at tolerable vibration levels. Specifically, they show subsynchronous orbit frequencies that track subharmonics of the rotor speed (i.e., 1/2, 1/3, etc. of the rotor spin frequency), but more often locking onto a system subsynchronous natural frequency. These subsynchronous orbital motion components may persist over a range of operating speeds, with disappearance then subsequently reappear with further operating changes. Their results also show that adding more rotor unbalance can trigger and worsen the severity of the subsynchronous orbital vibration components. This makes fundamental sense since the addition of rotor unbalance increases the overall vibration level and thereby increases the degree of dynamic nonlinearity, which in turn increases the propensity for the synchronous forcing function to drive energy into harmonics of itself (see Figures 2.17 through 2.19). Their results confirm that the subsynchronous vibration components are a consequence of the structural hardening nonlinearity akin to a Duffing resonator, not a hydrodynamic rotor dynamic instability energized by the gas film.

For a conventional cylindrical journal bearing with a machined journal OD and machined bearing bore ID, the clearance of course has a precise definition. Conversely, the equivalent clearance of a compliant surface foil journal bearing is not as precisely definable because the bearing surface is compliant. If one attempts to progressively increase bearing load to determine at what load magnitude the compliant surface bottoms out on the bearing sleeve (see Figure 5.23), the bearing is likely to fail from overheating before that bottom-out condition is reached. Of course, for what its worth, this test can be done for the journal not rotating. The somewhat nebulous definition of compliant surface foil journal bearing clearance cautions the author concerning the other nonbearing small radial rotor-to-stator radial clearances in an application, for example, radial seal clearances, turbo-machinery blade-tip clearances, balance drum clearance. An important function of any radial bearing is to prevent these other small-clearance components from becoming inadvertent bearings during operation.

5.9 Summary

The specifics of bearing and seal rotor dynamic properties as well as the rotor dynamic effects of turbo-machinery flows (Chapter 6) are primary factors that provide much of the uncertainty in making predictions for rotating machinery vibration. When using rotor vibration predictive analyses for design purposes, one should of course reflect such uncertainties in configuring the design and its prototype test program. However, the use of rotor vibration predictive analyses for *troubleshooting* is a somewhat different endeavor that benefits from having actual vibration measurements made on the machine that is in excessive vibration difficulty. As conveyed by the case studies presented in Part 4 of this book, adroit use of measured vibration characteristics on the actual machine leads to a *calibrated model* that can greatly increase the probability of devising a timely and adequate remedy for the particular rotating machinery vibration problem at hand.

Bibliography

Adams, M. L., "Insights into linearized rotor dynamics, part 2," *Journal of Sound & Vibration*, 112(1):97–110, Academic Press, London, 1987.

Adams, M. L. and Laurich, M. A., "Inside-put pivoted-pad bearing with controllable stiffness," *International Journal of Applied Mechanics and Engineering*, 10(3):395–406, 2005.

Adams, M. L., Padovan, J., and Fertis, D.,"Engine dynamic analysis with general nonlinear finite-element codes. Part 1: Overall approach and development of bearing damper element," *ASME Journal of Engineering for Power*, 104(3):1982.

Adams, M. L. and Rashidi, M., "On the use of rotor-bearing instability thresholds to accurately measure bearing rotordynamic properties," *ASME Journal of Vibration, Stress and Reliability in Design*, 107(4), 1985.

Adams, M. L., Sawicki, J. T., and Capaldi, R. J., *Experimental Determination of Hydrostatic Journal Bearing Rotordynamic Coefficients*, Proceedings of the Fifth IMechE International Conference on Vibration in Rotating Machinery, Bath, England, September 1992, pp. 365–374.

Adams, M. L., Yang, T., and Pace, S. E., *A Seal Test Facility for the Measurement of Isotropic and Anisotropic Linear Rotordynamic Characteristics*, Proceedings,

NASA Sponsored Workshop on Rotordynamic Instability Problems in High Performance Turbomachinery, Texas A&M University, NASA CP-3026, May 1988.

Allaire, P. and Trumper, D. L., (Eds.), "Review of publications on magnetic bearings," *Proceedings of the Sixth International Symposium on Magnetic Bearings*, Boston, August 1998.

Black, H. F., "Effects of hydraulic forces in annular pressure seals on the vibration of centrifugal pump rotors," *Journal of Mechanical Engineering Science*, 11(2):1969.

Black, H. F., *Effects of High Pressure Ring Seals on Pump Rotor Vibration*, ASME paper No. 71-WA/FF-38, 1971.

Black, H. F., "Calculations of forced whirling and stability of centrifugal pump rotor systems," *ASME Journal of Engineering for Industry*, 96:1076-1084, 1974.

Bolleter, U., Wyss, A., Welte, I., and Struchler, R., "Measurement of hydrodynamic matrices of boiler feed pump impellers," *ASME Journal of Vibration, Acoustics, Stress and Reliability in Design*, 109:1987.

Childs, D., *Turbomachinery Rotordynamics—Phenomena, Modeling, & Analysis*, Wiley, New York, 1993, pp. 476.

Elrod, H. G. and Ng, C. W., "A theory for turbulent fluid films and its application to bearings," *ASME Journal of Lubrication Technology*, 3:346–362, 1967.

Fleming, D. P., *Magnetic Bearings-State of the Art*, NASA Technical Memorandum 104465, July 1991.

Guelich, J. F., Bolleter, U., and Simon, A., *Feedpump Operation and Design Guidelines*, EPRI Final Report TR-102102, Research Project 1884-10, 1993.

Hagg, A. C. and Sankey, G. O., "Some dynamic properties of oil-film journal bearings with reference to the unbalance vibration of rotors," *ASME Journal of Applied Mechanics*, 78:302–306, 1956.

Hahn, E. J., *Equivalent Stiffness and Damping Coefficients for Squeeze-Film Dampers*, Proceedings of the Third IMechE International Conference on Vibrations in Rotating Machinery, York, England, 1984.

Heshmat, H., Shapiro, W., and Gray, S., "Development of foil journal bearings for high load capacity and high speed whirl stability," *Transactions of the ASME Journal of Lubrication Technology*, 104(2):149–156, 1982.

Hibner, D. and Bansal, P., *Effects of Fluid Compressibility on Viscous Damper Characteristics*, Proceedings, Conference on the Stability and Dynamic Response of Rotors with Squeeze-Film Bearings, University of Virginia, 1979.

Horattas, G. A., Adams, M. L., and Dimofte, F., "Mechanical and electrical runout removal on a precision rotor-vibration research spindle," *ASME Journal of Acoustics and Vibration*, 119(2): 216–220, April 1997.

Howard, S. A., Valco, M. J., Prahl, J. M., and Heshmat, H., "Dynamic stiffness and damping characteristics of a high-temperature air foil journal bearing," *Transactions of STLE*, 44(4):657–663, 2001.

Jones, A. B., *Analysis of Stresses and Deflections*, New Departure Engineering Data Book, 1946.

Kramer, E., *Dynamics of Rotors and Foundations*, Springer, Berlin, 1993, pp. 383.

Lomakin, A., "Feed pumps of the SWP-220-280 type with ultra-high operating head-capacity performance," *Power and Mechanical Engineering*, 2, 1955 (in Russian).

Lomakin, A., "Calculation of critical speeds and the conditions necessary for dynamic stability of rotors in high-pressure hydraulic machines (pumps) when taking into account forces originating in seals," *Power and Mechanical Engineering*, 4, 1958 (in Russian).

Lund, J. W., Arwas, E. B., Cheng, H. S., Ng, C. W., Pan, C. H. T., and Sternlicht, B., *Rotor-Bearing Dynamics Design Technology, Part-III: Design Handbook for Fluid Film Type Bearings*, Wright-Patterson Air Force Base, Technical Report AFAPL-TR-65-45, 1965, pp. 299.

Morton, P. G., "Measurement of the dynamic characteristics of a large sleeve bearing," *ASME Journal of Lubrication Technology*, 145–155, 1971.

Newkirk, B. L. and Taylor, H. D., "Shaft whipping due to oil action in journal bearings," *General Electric Review*, 28(8):559–569, 1925.

Nordmann, R. and Dietzen, F., A three-dimensional finite-difference method for calculating the dynamic coefficients of seals, in J. Kim and W. Yang (Eds.), *Dynamics of Rotating Machinery*, Hemisphere, 1990, pp. 133–151.

Nordmann, R. and Massmann, H., *Identification of Stiffness, Damping and Mass Coefficients for Annular Seals*, Proceedings of the Third IMechE International Conference on Vibration in Rotating Machinery, York, England, September 1984, pp. 167–181.

Pinkus, O. and Wilcock, D., *COJOUR User's Guide: Dynamic Coefficients for Fluid-Film Journal Bearings*, EPRI, Research Project 1648–1, Final Report CS-4093-CCM, 1985.

Raimondi, A. and Boyd, J., "A Solution for the Finite Journal Bearing and its Application to Analysis and Design," Parts I, II, and III, *Transactions of ASLE*, 1(1):159–209, 1958.

Rashidi, M. and Adams, M. L., "Improvement to Prediction Accuracy of Stability Limits and Resonance Amplitudes Using Instability Threshold-Based Journal Bearing Rotordynamic Coefficients," Proceedings of the Fourth IMechE International Conference on Vibration in Rotating Machinery, Edinburgh, Scotland, September 1988, pp. 235–240.

Reynolds, O., *On the Theory of Lubrication and its Application to Mr. Tower's Experiments*, Philosophical Transactions of the Royal Society, London, England, 177, Part 1, 1886.

San Andres, L., "Analysis of variable fluid properties, turbulent annular seals," *ASME Journal of Tribology*, 684–702, 1991.

San Andres, L. and Kim, Tae-Ho, *Issues on Instability and Forced Nonlinearity in Gas Foil Bearing Supported Rotors*, AIAA Joint Propulsion Conference, Paper No. 5094, Cincinnati, OH, July 8–11, 2007.

Sawicki, J., Capaldi, R. J., and Adams, M. L., "Experimental and theoretical rotordynamic characteristics of a hybrid journal bearing," *ASME Journal of Tribology*, 119(1): 132–142, 1997.

Schweitzer, G., *Magnetic Bearings as a Component of Smart Rotating Machinery*, Proceedings of the IFToMM Fifth International Conference on Rotor Dynamics, Darmstadt, September 1998, pp. 3–15.

Someya, T., Mitsui, J., Esaki, J., Saito, S., Kanemitsu, Y., Iwatsubo, M., Tanaka, M., Hisa, S., Fujikawa, T., and Kanki, H., *Journal-Bearing Databook*, Springer-Verlag, Berlin, 1988, pp. 323.

Sternlicht, B., "Elastic and damping properties of cylindrical journal bearings," *ASME Journal of Basic Engineering*, 81:101–108, 1959.

Bearing and Seal Rotor Dynamics 249

Stodola, A., *Steam and Gas Turbines*, 6th edn, (English translation from German), Vol. 2 of 2, McGraw-Hill, New York, 1927, reprinted in 1945, pp. 1085–1086.
Szeri, A. Z., *Fluid Film Lubrication*, Cambridge University Press, Cambridge, UK, 1998.
Vance, J. M., *Rotordynamics of Turbomachinery*, Wiley, New York, 1988, pp. 388.

PROBLEM EXERCISES

1. Using RDA and assuming a Gaussian (normal) distribution of journal bearing clearance for a high production rotor-bearing product, determine the distribution of the first critical speed and the oil-whip instability threshold speed. Use the single central disk two bearing configuration of Problem 3 of Chapter 1. Use 360° cylindrical oil film journal bearings, 4 in. nominal diameter, 2 in. length ($L/D = 0.5$), nominal radial clearance of 0.004 in., and tolerances of ±0.001 in. on journal OD and bearing bore diameter. Assume an oil viscosity of 0.2×10^{-7} reyns (lb s/in.^{2}).

2. Starting with the three N–S equations and the continuity equation, rigorously develop the RLE, Equation 5.1. Clearly state all assumptions.

3. Concerning the experimental extraction of bearing and seal rotor dynamics coefficient arrays, devise a *tensor filtering* algorithm to research the removal of random signal noise and measurement errors.

4. For a three-pad tilting-pad journal bearing (see Figure 5.6), devise a computer code to determine the static equilibrium journal center x and y eccentricity position coordinates for any specified bearing static radial load magnitude and direction. Use the following assumed functional relationship for individual pad load (see Figures 5.4 and 5.5) and assume no bearing preload:

$$W_p = \frac{W_0 e_p^3}{(c - e_p)},$$

where W_p is the pad load, c is the bearing concentric radial clearance, and e_p is the excentricity of journal into the pivot direction.

5. Repeat the Problem 4 scope with the added option of bearing preload, specified by the ratio of radial pivot-clearance-to-concentric-clearance, $0 < c'/c < 1$.

6. Utilizing the published literature, do a *term paper* on *squeeze film dampers* with particular focus on the effects of film rupture and dissolution of air in the oil, and the influence of fluid inertia.

7. Using the equations provided in Section 5.5, calculate the x and y bearing radial stiffness for a commercially available ball bearing without preload and compare to stiffness data available from the manufacturer's literature.

6

Turbo-Machinery Impeller and Blade Effects

The complete flow fields within turbo-machinery stages, both in radial flow and axial flow machines, are significant influences on rotor vibration. Figure 6.1 illustrates the radial flow fields typical in centrifugal impeller stages. For a typical axial flow machine, Figure 6.2 shows the high-pressure turbine of the large steam-powered multiturbine unit shown in Figure 3.9. Within these machines, flow through the impellers and blade rows interacts considerably with the flow through their respective seals.

6.1 Centrifugal Pumps

Referring to Figure 6.1a, it is not surprising that static and dynamic hydraulic forces are imposed on the rotor of a centrifugal pump by the flow through the pump. These hydraulic rotor forces are dominant factors in determining the vibration behavior of a centrifugal pump, especially high-energy pumps such as those required for boiler feed water service.

6.1.1 Static Radial Hydraulic Impeller Force

A *static radial force* is imposed on a pump impeller because the steady portion of the total pressure distribution over the impeller surface is not of perfect axial symmetry. This static radial hydraulic force is relatively larger in single-tongue volute-casing pumps, and smaller in multitongue volute-casing and diffuser-casing pumps. The combined static radial impeller force from all the impellers of a high-pressure multiimpeller pump, such as shown in Figure 5.13, can easily be much larger than the total weight of the pump rotating element. Thus, the static hydraulic impeller force can readily be the dominant factor in determining journal bearing static loads and thus the LRV stiffness and damping characteristics of the bearings. The static hydraulic radial force on an impeller varies considerably in magnitude and direction with pump flow. Therefore, the rotor dynamic properties of the journal bearings can vary considerably over the

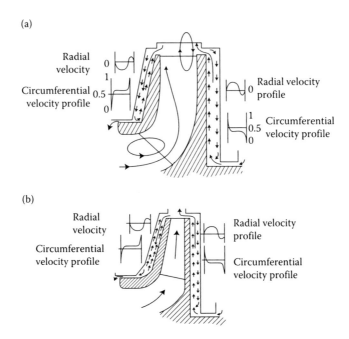

FIGURE 6.1 Centrifugal impeller typical flow patterns: (a) centrifugal pump impeller, radial-plane view and (b) centrifugal compressor impeller, radial-plane view.

operating flow range of a centrifugal pump. For example, if the bearings are unloaded at some pump flow, this can have serious LRV consequences such as oil-whip-induced self-excited large amplitude vibration. Similarly, this variation of the impeller static radial force (and thus the variation of the bearing LRV coefficients) can therefore also shift the location of LRV natural frequencies as a function of pump operating flow.

In the early development period of centrifugal pumps, as speeds and output pressures were being continually increased, it was learned that a significant radial static impeller force was the main reason for high cyclic shaft bending stresses resulting in material fatigue-initiated shaft failures. Stepanoff (1957) was among the first to report on the static radial impeller force in single-tongue volute-type centrifugal pumps, providing the following equation from dimensional analysis calibrated by test results (Figure 6.3):

$$P_s = \frac{K_s H D_2 B_2}{2.31} \qquad (6.1)$$

where P_s is the static force (pounds), H is the pump head (ft), D_2 is the impeller outer diameter (in.), B_2 is the impeller discharge width including

Turbo-Machinery Impeller and Blade Effects

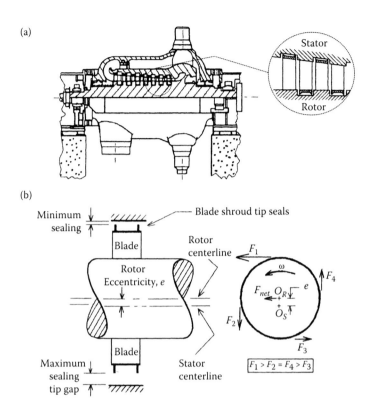

FIGURE 6.2 Contribution to *steam whirl* from the "Thomas–Alford" effect: (a) sectional view of a single-flow high-pressure steam turbine and (b) nonuniform torque distribution resulting from eccentricity.

impeller side plates (in.), and K_s is the empirical coefficient that changes with pump flow approximately as follows:

$$K_s = 0.36\left[1 - \left(\frac{Q}{Q_{BEP}}\right)^2\right], \quad \begin{aligned} Q &= \text{operating pump flow} \\ Q_{BEP} &= \text{best efficient point pump flow} \end{aligned}$$

(6.2)

As Equation 6.2 shows, there is a strong correlation between impeller static radial force and the ratio of the pump's operating flow to its best efficiency flow. This is because the nearly constant average velocity and pressure in centrifugal pump volutes occur only near the best efficiency operating flow. Equation 6.2 is a simple *curve fit* of many test results that show static radial impeller forces to be minimum near the best efficiency flow and maximum at the shutoff (zero flow) condition. The maximum value of K_s depends on various hydraulic design features, with Stepanoff

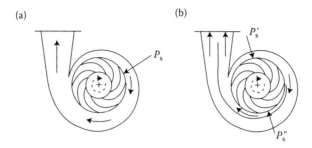

FIGURE 6.3 Static radial hydraulic force on volute-pump impellers: (a) single-volute pump and (b) double-volute pump, $P_s = P'_s - P''_s$.

reporting values for some single-volute pumps as high as 0.6 at shutoff operation.

The well-known *double-volute* (two tongues) configuration, as shown in Figure 6.3b, was devised to divide the pump volute into two equal 180° flow sections, with the intent that each section's static radial impeller force cancels the other's. The double volute does not completely accomplish that objective, but it does yield a drastic force reduction from that of a single-tongue volute. The author is familiar with centrifugal pump designs employing the *tri-volute* (three 120°-arc sections) and the *quad-volute* (four 90°-arc sections). Of course, if one further increases the number of volute tongues, the volute then resembles a diffuser.

Guelich et al. (1987) use the following less confusing form of Equation 6.1, which applies in any consistent system of units and explicitly shows density:

$$K_s = \frac{P_s}{\rho g H D_2 B_2} \tag{6.3}$$

where ρ is the mass density of the pumped liquid and g is the gravitational constant.

In addition to an increased propensity for fatigue-initiated shaft failure from excessive static radial impeller force, the accompanying shaft radial deflections can likely result in rubbing with accelerated wear rates at the close running concentric annular sealing clearances. The work reported by Agostinelli et al. (1959) is probably the most comprehensive source of experimental information on the static radial hydraulic impeller force. Their experimental results for single-tongue volute pumps are approximated well by Equation 6.2. Their results also show measured K_s values at shut-off for both the double-volute and the diffuser-casing pumps that are as low as 20% of the shut-off values measured for pumps using single-tongue volutes, and varying to a far lesser degree over the pump operating flow range than that indicated by Equation 6.2.

6.1.2 Dynamic Radial Hydraulic Impeller Forces

Time-varying (*dynamic*) *hydraulic forces* (both radial and axial) are also imposed on a centrifugal pump impeller. These dynamic hydraulic forces are quite significant and are separable into two categories, as follows:

- Strictly time-dependent *unsteady flow forces*
- *Interaction forces* produced in response to LRV orbital motions

The radial components of these two delineated types of dynamic forces can be incorporated into standard LRV analyses. The *interaction forces* that dynamically *connect* the rotor to the stator can be modeled by bearing-like radial stiffness, damping, and inertia (added mass) coefficients obtained from laboratory tests. The LRV importance of *unsteady flow forces* in a particular pump configuration can be assessed based on the model resonance sensitivity to the dominant frequency force components obtained from tests.

It has been recognized for many years that dynamic hydraulic *unsteady flow forces* on centrifugal pump impellers can be quite significant contributors to overall pump vibration levels as well as pump component failures. This is especially true for high-energy pumps. Furthermore, in contrast to most gas handling turbo-machines, the process fluid's dynamic *interaction forces* in centrifugal pumps have a major influence on LRV natural frequencies, mode shapes, and modal damping. As illustrated in Figure 5.13, interaction forces on a centrifugal pump rotor originate from journal bearings, annular seals, balancing drums, and impeller flow fields.

Most of the relevant research on impeller *dynamic radial forces* of centrifugal pumps is a product of the last 30 years, and most of what is available in the open literature comes from two sources: (1) California Institute of Technology and (2) Sulzer Co. (Pump Division, Winterthur, Switzerland). The work at Cal Tech has been funded primarily by NASA as part of the development of the high-energy pumps for space shuttle main engines. Cal Tech's work focuses primarily on obtaining the bearing-like stiffness, damping, and added mass coefficients for impeller *interaction forces*. The Sulzer work was funded by the Electric Power Research Institute (EPRI) as part of a $10 million multiyear EPRI research project on improving the reliability of *boiler feed water pumps*, and began in the mid-1980s. The Sulzer work covers the *interaction force* bearing-like impeller stiffness, damping, and added mass coefficients as well as the time-dependent impeller *unsteady flow forces*.

6.1.2.1 Unsteady Flow Dynamic Impeller Forces

Impeller *unsteady flow dynamic radial forces* are normalized using the same parameters shown in Equations 6.1 and 6.3 for impeller static radial force.

TABLE 6.1

Normalized (rms) Impeller Hydraulic Dynamic Force Factor (K_d)

Q/Q_{BEP}	$\Omega/\omega = 0.02-0.2$	$\Omega/\omega = 0.2-1.25$	$\Omega/\omega = 1$	Ω_v
0.2	0.02–0.07	0.02–0.05	0.01–0.12	0.2–0.12
0.5	0.01–0.04	0.01–0.02	0.01–0.12	0.1–0.08
1.0	0.002–0.015	0.005	0.01–0.13	0.1–0.06
1.5	0.005–0.03	0.01–0.02	0.01–0.15	0.2–0.10

Ω is the force frequency, ω is the speed, Ω_v is the Vane No. x ω, and K_d values for $\Omega/\omega = 0.2$–1.25 have $\Omega/\omega = 1$ component filtered out.

The corresponding *dynamic force* coefficient K_d is given by Equation 6.4 with values listed in Table 6.1, which are extracted from experimental results reported by Guelich et al. (1993). Good quality hydraulic flow-passage design procedures combined with precision cast or precision milled impellers should yield the low end of the ranges for K_d given in Table 6.1:

$$K_d(\text{rms}) = \frac{P_d(\text{rms})}{\rho g H D_2 B_2} \quad (6.4)$$

where P_d is the dynamic force (rms).

Conversely, poor hydraulic design quality and especially poor impeller dimensional control, such as with cheap low-quality sand cast impellers, will tend to yield the high end of the ranges for K_d given in Table 6.1. The K_d ranges shown for the frequency range of $\Omega/\omega = 0.2$–1.25 have the once-per-rev (synchronous) force component filtered out. The synchronous hydraulic component magnitude is shown in a separate column of Table 6.1, because it is primarily a function of impeller precision and less dependent on the percentage of BEP flow.

High levels of synchronous rotor vibration are usually attributed to the rotor being badly out of balance. However, in high-head-per-stage centrifugal pumps, a large synchronous hydraulic dynamic force may be a primary contributing cause of large amplitude synchronous rotor vibration. As one might expect, synchronous hydraulic impeller forces do not completely mimic rotor mass unbalance forces. Mass unbalance produces a corotational force that is "frozen" in the rotor. On the other hand, a large synchronous hydraulic dynamic impeller force will change in phase angle and somewhat in magnitude with pump flow. A clue to the savvy troubleshooter of poor impeller casting dimensional control is when an unacceptably high synchronous vibration cannot be alleviated over the full operating flow range by performing good rotor balancing procedures.

Figure 6.4 is from experimental results reported by Guelich et al. (1993) on a low specific speed high-head impeller rotating at 4000 rpm in a diffuser casing, typical for a boiler feed pump stage. As the flow is throttled below

Turbo-Machinery Impeller and Blade Effects

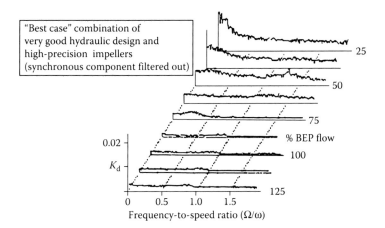

FIGURE 6.4 Spectra (rms) of normalized broadband impeller forces.

the BEP flow, the continuous strong increase in force magnitude results from impeller inlet and exit flow re-circulation (see Figure 6.1a) and flow separation.

6.1.2.2 Interaction Impeller Forces

The handling of impeller LRV *interaction forces* that has evolved over the last 30 years is to *curve fit* experimental data to the same linear *isotropic* LRV model used for most annular seal LRV characterizations. The assumption typically invoked for annular seal LRV coefficient arrays is that the flow field is rotationally symmetric (Chapter 5), and this assumption leads to the *isotropic model* given by Equation 2.85. While this assumption is quite inappropriate for journal bearings, it has been justified for annular seals and yields considerable simplification of both computational and experimental methods to extract LRV coefficient arrays for annular seals. Conversely, the flow field of a centrifugal pump impeller is certainly not rotationally symmetric. Nevertheless, to simplify test rigs and minimize associated costs to extract pump impeller LRV coefficient arrays, initial experiments were based on the *isotropic model* given by Equation 2.85, rewritten as

$$\begin{Bmatrix} f_x \\ f_y \end{Bmatrix} = - \begin{bmatrix} k^s & k^{ss} \\ -k^{ss} & k^s \end{bmatrix} \begin{Bmatrix} x \\ y \end{Bmatrix} - \begin{bmatrix} c^s & c^{ss} \\ -c^{ss} & c^s \end{bmatrix} \begin{Bmatrix} \dot{x} \\ \dot{y} \end{Bmatrix} - \begin{bmatrix} m^s & m^{ss} \\ -m^{ss} & m^s \end{bmatrix} \begin{Bmatrix} \ddot{x} \\ \ddot{y} \end{Bmatrix}$$

(6.5)

To the surprise of some, early test data by Chamieh et al. (1982) suggest that the *isotropic model* is well suited to centrifugal pump impellers. Further

TABLE 6.2

Impeller Dimensionless Stiffness, Damping, and Inertia Coefficients

Source/Type	$\bar{k}^s$	$\bar{k}^{ss}$	$\bar{c}^s$	$\bar{c}^{ss}$	$\bar{m}^s$	$\bar{m}^{ss}$
Cal Tech/volute	−2.5	1.1	3.14	7.91	6.51	−0.58
Cal Tech/diffuser	−2.65	1.04	3.80	8.96	6.60	−0.90
Sulzer/diffuser (2000 rpm)	−5.0	4.4	4.2	17.0	12.0	3.5
Sulzer/diffuser (4000 rpm)	−2.0	7.5	4.2	8.5	7.5	2.0

extensive testing (e.g., Jery et al., 1984; Bolleter et al., 1987), coupled with computational efforts [e.g., Adkins (1985) who did not include impeller-shroud flow effects, and Childs (1999) who included only impeller-shroud flow], led to the realization that the impeller rotor–stator *interaction force* is dominated by the flow field between the casing and impeller shrouds, primarily the inlet-side shroud that has the main radial area projection (refer to Figure 6.1a). Since the flow field between the casing and impeller shrouds can be reasonably viewed to be rotationally symmetric, the experimenters' good fortune with the LRV *isotropic model* is thus understandable. A visualization of the complexity and diversity of such rotationally symmetric flow fields with net inward or outward through-flow is provided by Adams and Szeri (1982), who developed computer solutions of the full nonlinear Navier–Stokes (N–S) equations using Galerkin's method to expand the N–S equations into a truncated set of nonlinear ODEs, which are numerically solved using the method of orthogonal collocation. The simplest of impeller-shroud flow patterns are characterized by a single recirculation cell superimposed upon the through-flow, as shown in Figure 6.1a. Adams and Szeri provide results that show the single recirculation cell evolving into multiple recirculation cells as the rotational Reynolds number is progressively increased.

A summary list of experimentally extracted impeller LRV coefficients for the isotropic model, Equation 6.5, is given in Table 6.2, and the coefficients are made nondimensional as follows:

$$\text{Dimensionless stiffness coefficients, } \bar{k}_{ij} \equiv \frac{k_{ij}}{\pi \rho R_2 B_2 \omega^2}$$

$$\text{Dimensionless damping coefficients, } \bar{c}_{ij} \equiv \frac{c_{ij}}{\pi \rho R_2 B_2 \omega} \quad (6.6)$$

$$\text{Dimensionless inertia coefficients, } \bar{m}_{ij} \equiv \frac{m_{ij}}{\pi \rho R_2 B_2}$$

where ρ is the mass density of pumped liquid, R_2 is the impeller outer (discharge) radius, and B_2 is the impeller discharge width including impeller side plates.

In comparing the Cal Tech and Sulzer results, it should be realized that the Cal Tech test rig used much lower power impellers with different hydraulic design details than the Sulzer test rig impellers. Both the Cal Tech and the Sulzer tests reported that, unlike annular seal inertia coefficients, the cross-coupled inertia coefficient for impellers (i.e., skew-symmetric inertia) is not negligible. In the author's opinion, this is an anomalous conclusion that stems from a lack of due appreciation to the simple fact that the equivalent mechanical-impedance coefficients in Equation 6.5 are just *curve fit* coefficients that are given birth when they are evaluated to provide the best simple curve fit to radial *force* versus *motion* test results over a limited frequency range from a quite complex 3D fluid mechanics flow field. Based on the rigorously argued conclusion in Section 2.4, Equation 2.84, the Cal Tech and Sulzer data *curve fits* should have been done with the inertia matrix constrained to symmetry, that is, $m^{ss} = 0$. Sawicki et al. (1996) show that when the highest order matrix is constrained to symmetry in the data reduction *curve fitting* step, as compared to allowing it to be non-symmetric, all the other coefficients change (adjust) somewhat to provide the best fit possible with the retained coefficients. Sawicki et al. further show that by applying extracted coefficients through Equation 6.5 to the measured displacement signals, the so computed dynamic force signals have comparable accuracy comparisons to the actual measured force signals with or without the physically inconsistent skew-symmetric inertia coefficient. Nevertheless, provided the Cal Tech and the Sulzer coefficient results summarized in Table 6.2 are not used as LRV analysis inputs above the maximum Ω/ω test value (~2), the physically inconsistent curve fit approach should not corrupt analysis answers. Interestingly, the Cal Tech and Sulzer results summarized in Table 6.2 are consistent with each other in magnitudes and signs, except for the skew-symmetric inertia term. Cal Tech's $\bar{m}^{ss}$ values are negative and Sulzer's are positive, yielding opposite physical interpretations and thus supporting the argument inferred from Equation 2.84 that a nonzero $\bar{m}^{ss}$ is a physical inconsistency; that is, it does not make physical sense.

Aside from the anomaly of a nonzero skew-symmetric inertia coefficient shown in Table 6.2, other important general observations are in order. First, it is observed that the radial stiffness term (k^s) is negative, and tends to slightly lower natural frequencies. Second, the skew-symmetric stiffness term (k^{ss}) is positive and significantly complements other similar forward-whirl destabilizing effects such as those from lightly loaded journal bearings and annular seals. Third, the relatively large skew-symmetric term (c^{ss}) is like a gyroscopic effect (see Figure 2.13), which lowers the backward-whirl natural frequencies and raises the forward-whirl natural frequencies and, being rotation-direction biased, reflects convective fluid inertia influences. Finally, the diagonal inertia term (m^s) is like a nonstructural added mass on the rotor and tends to slightly lower natural frequencies.

6.2 Centrifugal Compressors

The radial forces on centrifugal compressor rotors are basically similar to, but considerably less dominant than, those illustrated for centrifugal pumps in Figure 5.13. As stated in the previous section, in contrast to most gas handling turbo-machines, the process fluid's dynamic *interaction forces* in centrifugal pumps have a major influence on LRV natural frequencies, mode shapes, and modal damping. Although the effects of centrifugal compressor aerodynamic forces are an important design consideration, they do not constitute the overwhelming influence on LRV characteristics as the hydraulic forces do in centrifugal pumps. As a consequence, there is far less information in the open literature on centrifugal compressor impeller LRV effects than is summarized in Section 6.1 for centrifugal pumps. No industry-funded research project on centrifugal compressors similar to the $10 million EPRI project on boiler feed water pumps has been launched. Also, the author is not aware of any copiously funded university research projects on centrifugal compressor aerodynamic rotor forces comparable to the Cal Tech work on centrifugal pump impeller forces.

As with most turbo-machinery, the primarily LRV concerns of centrifugal compressor designers are location(s) of *critical speed(s)* within the operating speed range, *unbalance sensitivity*, and *thresholds of instability*. Predictive analyses of centrifugal compressor critical speeds and unbalance sensitivity have traditionally been considered to be sufficiently accurate without accounting for aerodynamic interaction forces. On that basis alone, one might surmise that there is not a single compelling justification to motivate significant research expenditures for laboratory experiments to determine the LRV radial-interaction force coefficients for centrifugal compressor impellers. However, the thresholds of instability are not easy to accurately predict, but are of paramount importance in achieving successful centrifugal compressor designs. Until recently, the absence of data on the LRV coefficients for gas handling annular seals has hampered centrifugal compressor designers in determining what portion of the LRV destabilization from a centrifugal compressor stage originates in the stage's annular sealing clearances and how much comes from the aerodynamic impeller forces.

6.2.1 Overall Stability Criteria

Possibly the definitive publication on centrifugal compressor LRV stability limits is that of Kirk and Donald (1983). The most valuable information in their paper is a chart of *pressure parameter* ($p_2 \Delta p$) versus speed-to-critical-speed ratio (N/N_{cr}) for successfully stable running compressors. Based on several successfully stable compressors and two initially unstable ones that were stabilized through modifications, Kirk and Donald provide an *acceptability* chart redrawn here in Figure 6.5. The two compressor units labeled Unit "A" and Unit "B" are from well-known plants that are identified by name in their paper. They draw attention to compressor configurations. These compressors are most susceptible to LRV instability due to the having *back-to-back* impellers (to minimize rotor axial thrust loads) because the balancing drum seal is then located near the rotor mid-span position. Any destabilizing influence from the mid-span balancing drum seal has the maximum "opportunity" to cause LRV instability because the unstable mode (the first bending mode, Figure 4.13) has its maximum amplitude near the mid-span axial location. In discussion with Professor Childs (1999), the author has learned that at least one manufacturer has quite successfully employed a *honeycomb seal* (Figure 5.18) in the mid-span balancing drum of their *back-to-back* centrifugal compressors. According to Professor Childs, this has extended considerably the margin of rotor dynamic stability on these compressors since the *honeycomb seal* drastically diminishes the effect of seal inlet preswirl and in-seal swirl as well as increases the potentially unstable first mode's natural frequency through increased interactive radial stiffness at the mid-span location. Correspondingly, it is clear from Figure 6.5 that the two compressor units referenced were stabilized by modifications that raised the critical speed, since the

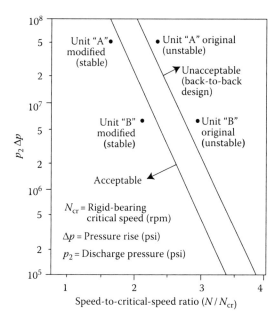

FIGURE 6.5 Criteria for rotor dynamically stable centrifugal compressors.

pressure parameter remained unchanged. That is, the delivered output of the two machines was not lowered as part of the modifications.

The next two sections deal with the Thomas–Alford LRV destabilizing forces in axial flow turbo-machinery. The Kirk and Donald paper shows a calculation approach for the centrifugal compressor LRV destabilizing cross-coupled (skew-symmetric) stiffness coefficient based on a version of the Alford (1965) formulation for axial flow turbo-machinery, modified for centrifugal compressors. They attempt a correlation of their modified Alford calculation with field experience from stable and unstable centrifugal compressors, but this correlation is of questionable meaning because of a potential inconsistency. As shown in Figure 6.2b and further explained in the next two sections, an Alford-type destabilizing force as physically explained is corotational for turbines, but possibly counter-rotational for compressors. However, the subsynchronous LRV instability self-excited vibration on centrifugal compressors occurs in the corotational direction, just like a journal-bearing-induced *oil whip* (explained in Section 2.4). Thus an Alford-force explanation or basis of modeling is potentially inconsistent with the experience from compressors that have exhibited LRV instability. The author has spoken with a number of centrifugal compressor designers, both in the United States and in Europe, and none of them has ever seen a centrifugal compressor experience LRV subsynchronous vibration in the backward whirl orbital direction. Section 2.4 provides a fundamental explanation that covers this.

6.2.2 Utilizing Interactive Force Modeling Similarities with Pumps

Experimental LRV information is summarized in the previous section for centrifugal pumps, taken from two modern well-funded research projects at Cal Tech and Sulzer Pump Division, respectively. Since such intensive experimental research results are not in existence for centrifugal compressors, at least not in the open literature, some assumptions must be made. However, in the author's opinion it makes more sense to assume that strong LRV similarities exist between centrifugal pump and centrifugal compressor impeller *destabilizing interactive forces*, than to invoke the Alford-force approach shown by Kirk and Donald. Specifically, it makes more sense to assume that LRV interactive centrifugal compressor impeller forces (not including annular sealing gaps) are dominated by the flow field between the inlet-side impeller shroud and casing, as shown for centrifugal pump impellers and with destabilization effects likewise corotational. This approach strongly suggests that the primary sources of centrifugal compressor stage LRV destabilization are the annular sealing gaps and the flow field between impeller inlet shroud and casing, assuming the compressor impeller has an inlet-side shroud. As is now known for centrifugal pump impellers, the less the radial projected shroud area,

(a)

(b)

FIGURE 2.15 Photos from the two 1970s catastrophic failures of large 600 MW steam turbine-generator sets. Using nonlinear rotor dynamic response computations, failures could be potentially traced to the large unbalance from loss of one or more large LP turbine blades at running speed, coupled with behavior of fixed-arc journal bearings during large unbalance. (a) LP steam turbine outer casing. (b) Brushless exciter shaft. (c) Generator shaft. (d) LP steam turbine last stage.

(c)

(d)

FIGURE 2.15 Continued.

FIGURE 2.25 Piston and connecting rod of a small reciprocating compressor.

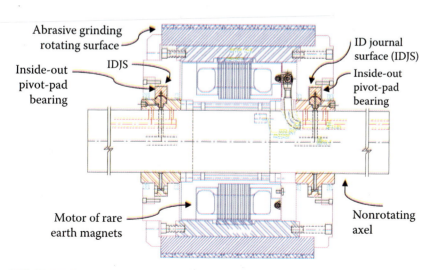

FIGURE 5.7 Next-generation centerless grinder spindle.

FIGURE 5.8 Photo of three-pad inside-out PPJB with three copper pads to facilitate heat removal; three steel thrust sectors.

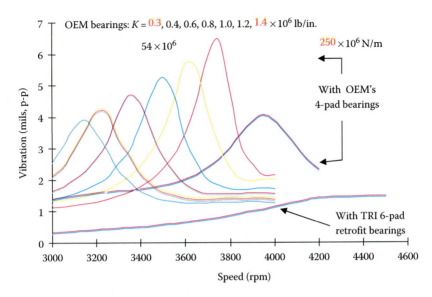

FIGURE 10.4 Predicted shaft unbalance vibration amplitude at bearing no. 1: At the nominal bearing static unit load of 200 psi (13.6 bar), OEM radial bearing stiffness K is computed to be 1.98×10^6 lb/in. (254×10^6 N/m).

FIGURE 10.11 Proxy probe setup.

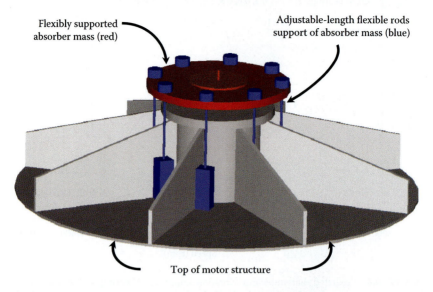

FIGURE 10.19 Vibration absorber (blue and red) atop pump motor.

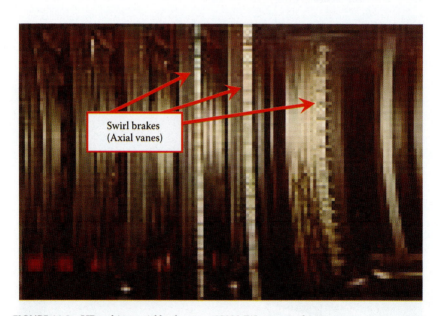

FIGURE 11.3 HP turbine swirl brakes on a 1300 MW steam turbine.

FIGURE 12.7 Air preheater drive/platform with overhang support bars.

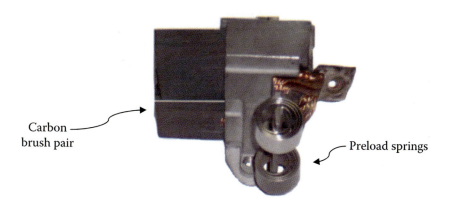

FIGURE 12.10 Carbon brush pair in holder with preload springs.

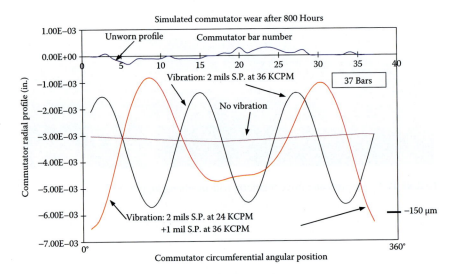

FIGURE 12.12 Simulated commutator radial wear for APU DC starter-generator.

the less the shroud should contribute to LRV interactive forces and the more the annular sealing gaps are the dominant source of potential LRV instability.

For compressor impellers without an inlet-side shroud, the annular sealing gaps are probably the singular dominant source of potential LRV instability. As stated at the end of Chapter 5, in using rotor vibration predictive analyses for *troubleshooting* purposes, as opposed to design purposes, one can adroitly utilize actual vibration measurements made on a troubled machine to adjust uncertain LRV model inputs (such as for compressor stages) to achieve a reasonable agreement between the machine and the LRV model. Such "calibrated models" can greatly increase the probability of devising a timely solution for high levels of subsynchronous instability vibrations.

6.3 High-Pressure Steam Turbines and Gas Turbines

6.3.1 Steam Whirl

As with LRV considerations for most gas handling turbo-machinery, in steam turbines the rotor and journal bearings provide the dominant stiffness effects whereas positive LRV damping comes almost entirely from the journal bearings. However, the journal bearings can also be a troublesome source of LRV instability (oil whip) if journal bearing static loads are insufficient to maintain stable operation, as explained in Section 2.4. In steam turbines, there is an additional destabilizing effect that originates in the turbine stages that can produce subsynchronous forward-whirling rotor vibration quite similar to oil whip. The self-excited rotor vibration caused by this effect is usually referred to as *steam whirl*, and like *oil whip* it can produce large amplitude subsynchronous frequency forward-whirl rotor vibrations. The importance of steam whirl excitation is almost exclusively in the high-pressure turbine section of large steam turbine-generator units, for reasons that are made clear in this section. A number of case studies in Part 4 involve *steam whirl*-induced self-excited rotor vibration and corrective measures.

The operating symptom that distinguishes *steam whirl* from *oil whip* is that it initiates at some *threshold level of turbine power output*, not at some threshold speed, even though the resulting self-excited vibrations from the two respective phenomena are essentially indistinguishable, that is, a forward whirl subsynchronous rotor vibration usually near and slightly below one-half the rotor spin speed frequency. For example, if steam whirl initiates at say 90% of a turbine-generator unit's rated full power output, then operation above 90% rated power will not be possible without the associated subsynchronous rotor vibration. It will likely be necessary to temporarily derate the unit by 10% until a modification or maintenance fix can be implemented to achieve full power output without the occurrence of subsynchronous vibration levels above the recommended safe operating maximum limits.

6.3.1.1 Blade Tip Clearance Contribution

The earliest publication addressing *steam whirl* was that of Thomas (1958). He proposed that observed thresholds of instability correlated to turbine power output could not be properly explained as a journal-bearing-induced instability, which was then already well known to be speed induced, that is, instability threshold speed. Thomas focused on the fact that when a turbine blade row is given a radial displacement eccentricity relative to its closely circumscribing nonrotating casing, blade tip leakage and thus blade row efficiency become circumferentially nonuniform. The essential feature of Thomas' explanation is that the power loading on the turbine blades is correspondingly nonuniform as Figure 6.2b illustrates, with the blades instantaneously passing the position of minimum radial gap (i.e., minimum local leakage) having the highest tangential driving force and the blades instantaneously passing the position of maximum radial gap having the lowest tangential driving force. The resultant sum of all the tangential blade forces upon the rotor then has a net radial force that is perpendicular to the rotor radial eccentricity and in the corotational direction. Assuming that such a net radial force is well approximated using a linear bearing-like LRV coefficient (k_{sw}), the net force shown in Figure 6.2b can then be expressed as follows:

$$\{F_{net}\}_{stage} = \begin{Bmatrix} f_x \\ f_y \end{Bmatrix}_{stage} = -\begin{bmatrix} 0 & k_{sw} \\ -k_{sw} & 0 \end{bmatrix} \begin{Bmatrix} x \\ y \end{Bmatrix} \quad (6.7)$$

$$k_{sw} = \frac{\beta T}{DL} \quad (6.8)$$

where T is the turbine-stage torque, D is the mean diameter of turbine stage blade row, L is the radial length of turbine blades, and β is the linear factor for blade force reduction with radial tip clearance.

$$F_{blade} = F_{blade}^{(0)}\left(1 - \beta\frac{C}{L}\right) \quad (6.9)$$

$F_{blade}^{(0)} \equiv$ Blade tangential force for zero clearance, $C =$ Clearance

Based on numerous published test results, $2 < \beta < 5$ for unshrouded turbines. Although β was originally devised as the change in efficiency per unit change in clearance ratio (C/L), it is now considered to be more an empirical factor to put Equations 6.7 and 6.8 into agreement with laboratory tests and field experience. Consistent with intuition, Equation 6.8 shows that a radial eccentricity of a blade row has a proportionally greater steam whirl effect the shorter the blades, that is, it is inversely proportional to the blade length. Thus, in the higher pressure stages of a steam turbine, although the mean blade diameter is smaller, $1/DL$ is still significantly the largest because the blades are relatively quite short. Furthermore, the torque is significantly higher than in lower pressure stages. Therefore, it is clear that the destabilizing force contribution given by Equation 6.7 is by far the greatest in the high-pressure turbine (see Figure 3.9).

Following Thomas' (1958) explanation in Germany for steam turbines, Alford (1965) proposed the same explanation for gas turbines in the United States, by which time Thomas was already in the early phases of extensive research on test machines to measure the net destabilizing force expressed in Equation 6.7 and shown in Figure 6.2b. In the United States, the net destabilizing force is often referred to as the *Alford* force, but here it is referred to as the *Thomas–Alford* force in recognition that Thomas (1958) provided the first explanation of LRV instability correlated to turbine power output. The major doctoral dissertations under Professor Thomas' direction, by Urlichs (1975) and Wohlrab (1975), contain many of the extensive experimental results on steam whirl developed at the Technical University Munich. A comprehensive English summary of this research is given by Thomas et al. (1976). See also Kostyuk et al. (1984), Pollman et al. (1978), and Urlichs (1976).

6.3.1.2 Blade Shroud Annular Seal Contribution

Early experimental research on steam whirl found that when blade shrouds (see Figure 6.2b) are added to a blade row, the magnitude of the destabilizing force becomes approximately *two or more times* as large as without blade shrouds. Subsequent research on the rotor vibration characteristics of labyrinth annular gas seals (Chapter 5) confirmed that the additional steam whirl destabilizing effect with shrouds is strongly driven by the corotational preswirl of high-pressure steam entering the labyrinth annular tip seals. As shown by one case study in Part 4, this component of the total steam whirl effect can be greatly attenuated by using axially aligned flow-straightening vanes (called *swirl brakes*) just upstream of the annular seals. Without swirl brakes, the total steam whirl force is approximately *two to three times* the value that would be predicted using Equation 6.7 with Equation 6.8. The most significant precision experiments for the blade shroud annular seal contribution to whirl forces were conducted and reported by Wright (1983). He devised a vertical rotor test rig with a two-strip labyrinth air seal in which precisely controlled electromagnetic dampers were tuned to produce neutral stability (instability threshold condition) for a given operating seal pressure drop, rotational speed, seal geometry, and preswirl velocity. Wright's objective was to provide high-precision experimental results to which proposed computational approaches for labyrinth seal destabilizing forces could be compared and evaluated. An explanation is provided here for the labyrinth seal destabilizing effect. Figure 6.6 consists of four circumferential pressure distributions corresponding to four cases, respectively.

The four illustrated cases in Figure 6.6 are summarized as follows:

a. *Journal bearing* operating with a liquid lubricant, an atmospheric ambient pressure, and thus cavitation formed slightly downstream of the minimum film thickness; pressure field governed by fluid viscosity (i.e., Reynolds Lubrication Equation).

b. *Journal bearing* operating with very high ambient pressure (e.g., pressurized water reactor (PWR) reactor coolant pump lower

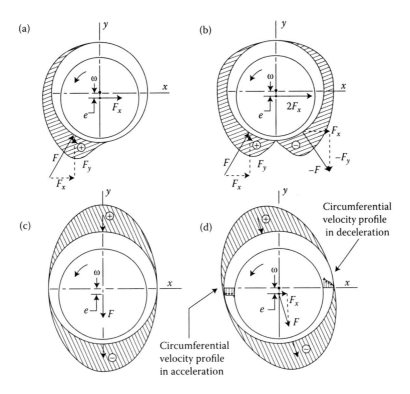

FIGURE 6.6 Circumferential pressure distributions relative to ambient: (a) journal bearing operating with atmospheric ambient pressure (cavitation); (b) journal bearing operating with high ambient pressure (no cavitation); (c) and (d) high rotational Reynolds number fluid annulus with high ambient pressure.

bearing; see Figure 12.1) and thus no cavitation; pressure field also governed by fluid viscosity.

c. *High rotational Reynolds number fluid annulus* in which the major inner core of fluid has a nearly "flat" circumferential velocity profile that is joined to the cylindrical boundaries through very thin boundary layers; pressure field governed by inertia of inner core of fluid.

d. A slightly modified version of (c), which is the basis of the author's own explanation of the labyrinth seal contribution to steam whirl forces.

Figure 6.6a and b are well understood by bearing specialists, and in both of these cases the fluid film force upon the journal has a destabilizing component influence on forward whirling LRV orbits (i.e., perpendicular to journal eccentricity). In (a), with a sufficiently high static bearing load, the squeeze-film damping controls the energy input from the destabilizing force component and the rotor-bearing system is stable. In (b), the entire fluid film force is perpendicular to the journal eccentricity, and thus subsynchronous instability vibration is much more likely. The case illustrated in Figure 6.6c is well known to designers of canned-motor pumps since the fluid-inertia-dominated pressure field acts in contrast to the behavior of a bearing, that is, the circumferential pressure distribution pushes the rotating cylinder in the direction of eccentricity (a decentering force or negative radial stiffness). Cases (a), (b), and (c) are reviewed here as a backdrop for the author's explanation, which focuses on Figure 6.6d.

To simplify the explanation, axial flow is ignored and thus it is assumed that the circumferential mass flow across the radial gap thickness (h) is the same at all angular locations, which is expressed as follows:

$$\text{Circumferential flow/unit axial length, } Q_\theta = \int_0^h V(r,\theta)\, dr = \text{constant}$$

(6.10)

A high rotational Reynolds number fluid annulus has its pressure field controlled by the fluid inertia in the inner core of circulating fluid. Thus, the clearance gap can be thought of as a Venturi meter wrapped around on itself and operating on the Bernoulli equation principle of conservation of energy, with maximum pressure occurring at the maximum radial gap and minimum pressure occurring at the minimum radial gap. With elevation and density changes discounted, the Bernoulli equation can be stated as follows:

$$p + \frac{\rho V^2}{2} = \text{constant} \qquad (6.11)$$

where p is the pressure, V is the fluid velocity, and ρ is the fluid mass density.

If the kinetic energy term is based on the average velocity at each cicumferential location (θ), then the pressure distribution in Figure 6.6c illustrates the result, and is based on the minimum possible local kinetic energy term (per unit of axial length), which is achieved with a perfectly flat velocity profile (zero thickness boundary layer), and is expressed as follows:

$$KE_{min} = \frac{\rho(Q_\theta/h)^2}{2} \tag{6.12}$$

However, the flow in the converging 180° arc is in acceleration whereas the flow in the diverging 180° arc is in deceleration. Therefore, the velocity profile will be "flatter" (thinner boundary layers) in the converging portion than in the diverging portion, as illustrated by the two representative fluid velocity profiles shown in Figure 6.6d. If instead of using the more approximate average-velocity approach at a location (θ), the kinetic energy term is determined by integrating V^2 radially across the fluid gap (h), the so determined kinetic energy will then be slightly greater in the diverging arc than in the converging arc. For any circumferential velocity profile (other than "perfectly flat") at the same radial gap thickness (same average velocity), the kinetic energy term obtained by integrating across the gap thickness (h) is larger than KE_{min}, as expressed in the following equation:

$$KE = \frac{\rho}{2h} \int_0^h [V(r,\theta)]^2 \, dr > KE_{min} \tag{6.13}$$

The Bernoulli equation argument of this explanation is that a higher kinetic energy produces a lower pressure. Consistent with this argument, the pressure distribution in Figure 6.6d is shown to be slightly skewed in comparison to that of Figure 6.6c. Such a skewing of the pressure distribution will produce a destabilizing component from the total fluid force upon the rotating cylinder, as shown in Figure 6.6d. Note the two velocity profiles in Figure 6.6d. The pressure is slightly higher where the profile is more flat than where the profile is less flat. Consistent with the experiments, Equation 6.11 implies the shroud contribution to steam whirl to be proportional to fluid density, and thus like the Thomas–Alford effect, also greatest in the high-pressure turbine.

The high corotation preswirl gas velocity entering such seals naturally contributes strongly to the total circumferential circulation flow within the annular space between two labyrinth seal strips. Without such preswirl, the gas must be circumferentially accelerated (by the boundary layer attached to the rotating boundary) after it enters the space between two labyrinth seal strips, and thus would not have nearly as much of the destabilizing force effect illustrated in Figure 6.6d as with a high preswirl inlet velocity. It is thus quite understandable that the annular labyrinth seal contribution to the total steam whirl effect can be greatly attenuated by using axially aligned flow-straightening vanes (called *swirl brakes*) just upstream of the seal.

From laboratory tests and analyses, it is now known by many LRV specialists that having the grooves of grooved annular seals located on the stator produces less LRV destabilizing effect than locating the grooves on the rotor. Consistent with the above discussion, this is easily understandable because of the difference in the amount of *rotating boundary* area between the two configuration options. Referring to Figure 6.2b, in which the tip clearance is greatly exaggerated, the gas-filled annular chamber between the two sealing strips has four sides, one is stationary and the other three are rotating, because the strips shown are rotor mounted. In the alternate configuration (not shown) in which the sealing strips are stator mounted, the gas-filled rectangular chamber between the two sealing strips has three sides stationary and only one side rotating. Clearly, there is proportionally less boundary layer area available to circumferentially accelerate incoming gas (or liquid) when the sealing strips are stator mounted, and thus less total circumferential circulation flow velocity and therefore less LRV destabilizing effect than with rotor-mounted strips (or grooves).

6.3.2 Partial Admission in Steam Turbine Impulse Stages

Typical fossil-fuel-fired boilers for steam turbines in electric power generating plants in the United States are designed to operate with controlled variable steam flow output. In contrast, European fossil-fuel-fired boilers are typically designed to operate with controlled variable steam pressure output. Thus, it is usual that large steam turbines in the U.S. power plants have impulse stages at the first stage of the high-pressure turbine because the turbine flow and the power output can then be efficiently regulated by throttling impulse stage nozzles.

An impulse stage for a large steam turbine typically incorporates a number (e.g., six) of equally spaced nozzles (Figure 6.7) that are fully open at full power output. To regulate the power below full output, one or more nozzles are throttled. Thus the term "control stage" is sometimes used to identify such controlled-nozzle impulse steam turbine stages. The nozzles are not uniformly throttled, but more typically only one (possibly two) is (are) operated in the partially open setting. This mode of operation is commonly referred to as *partial admission*, and it produces a significant net static radial force on the turbine rotor due to the nonuniform distribution of jet forces on the impulse turbine blade row. So that this static

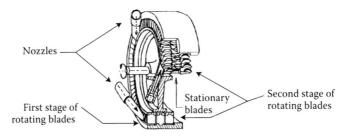

FIGURE 6.7 Laboratory impulse steam turbine stage with three nozzles.

radial force does not add to rotor weight static loads already carried by the high-pressure turbine's journal bearings, the partial admission is configured so that its net static radial force is directed approximately opposite (i.e., up) the weight. This makes sense for bearing static load but it can cause rotor vibration problems. Specifically, the attendant reduction in journal bearing load increases the possibility for subsynchronous instability rotor vibration. The ability of a journal bearing to damp forward whirl subsynchronous LRV modes is reduced as bearing static loads are reduced. Therefore, the additive action of oil whip and steam whirl destabilizing effects can combine to produce large amplitude subsynchronous vibration of the high-pressure turbine at partial admission operation. One of the case studies in Part 4 is concerned with this type of vibration problem.

6.3.3 Combustion Gas Turbines

Multistage axial flow gas turbines are most commonly employed for land-based electric power generation plants and for gas turbine engines of both commercial and military aircraft propulsion systems. Equations 6.7 and 6.9 provide an estimate of Thomas–Alford forces in steam turbines and are in fact the same equations that are similarly applied for combustion gas turbines. The corresponding Thomas–Alford force physical explanations already provided here for steam turbines also apply to gas turbines. It is the author's sense that Thomas–Alford forces are a more important consideration for gas turbine aircraft engines than for gas turbine electric power generating units.

6.4 Axial Flow Compressors

The Thomas–Alford type of destabilizing force described in Section 6.3 and illustrated in Figure 6.2b has been researched concerning its significance to LRV stability of axial compressors. Multistage axial compressors are most commonly employed as an essential portion of modern combustion gas turbines, both for electric power generation plants and for gas turbine engines of both commercial and military aircraft propulsion. Ehrich (1993) reports that the Alford (1965) explanation for destabilizing LRV forces in gas turbines was subsequently extended by Alford to explain LRV instabilities in axial compressors. The direction of the shaft torque that powers a compressor is the same as its direction of rotation, which is of course opposite the torque direction upon a turbine by the shaft it powers. This has led many who have seriously thought about this problem to expect that Thomas–Alford forces in compressors would seek to drive the rotor into backward whirl self-excited rotor vibration, in contrast to the universally accepted forward whirl direction of these forces in turbines. However, according to Ehrich (1993), Alford suggested just the opposite, that is, that these forces also provide energy input to forward whirl modes of compressors just like in turbines. The essence of Alford's argument was that as the rotor displaces radially with respect to its stator, the blades with the instantaneous minimum tip clearance are more efficient and thus more lightly loaded than the blades with the larger clearance.

Current thinking on this reflects the realization that the flow field in such turbo-machinery is quite complex, especially at operating conditions other than at peak efficiency. Although the Thomas–Alford theme is quite important to provide simplified explanations, reality is not so simple. Ehrich (1993) assembled three different sets of experimental Thomas–Alford force results from three different axial compressors. In a quite thorough analysis of all those results, Ehrich concludes that the Thomas–Alford-force coefficient (β), Equation 6.8, is not a simple constant for a specific compressor but is a very strong function of operating condition of the stage (i.e., its throttle coefficient or flow coefficient). Quoting Ehrich: "it is found that the value of β is in the range of $+0.27$ to -0.71 in the vicinity of the stages' nominal operating line and $+0.08$ to -1.25 in the vicinity of the stages' operation at peak efficiency. The value of β reaches a level of between -1.16 and -3.36 as the compressor is operated near its stalled condition." Consistent with Equation 6.8, positive values for β indicate corotational Thomas–Alford forces as in turbines, and negative values for β indicate counter-rotational Thomas–Alford forces. Alford's explanation for compressors appears to be mostly wrong.

For an aircraft gas turbine jet engine application, the axial compressor and gas turbine on the same shaft have equal magnitude torque. In such applications, Ehrich's results indicate that the compressor has either a negligible influence or may even negate some of the destabilizing effect of the forward whirl tendencies from the Thomas–Alford forces in the turbine. Obviously, for electric power generation gas turbines, the turbine torque is considerably larger than the compressor torque; otherwise no power would be generated. Thus, the compressor's Thomas–Alford force importance is less than for the gas turbine aircraft engine high-compressor rotors.

6.5 Summary

Rankine (1869) presented a seriously flawed rotor vibration analysis in which he cast $F = ma$ in a rotor-imbedded (i.e., noninertial) coordinate system without including the requisite correction factor known as *Coriolis acceleration*. Rankine's results led designers to work for several years under the misconception that rotors could not safely operate at speeds in excess of what is now called the *first critical speed*. Not until G. DeLaval in 1895 experimentally showed a steam turbine operating safely above its first critical speed was Rankine's fallacy debunked, leading to the higher-speed higher-power turbo-machines of the twentieth century. Because of the high concentration of power transferred in modern turbo-machines, and for some types of applications the quite high rotational speeds, the process liquid or the gas in turbo-machinery stages provides a number of identifiable fluid–solid interaction phenomena that can quite significantly influence rotor vibration behavior, especially stability. These phenomena are the focus of this chapter.

Bibliography

Adams, M. L. and Szeri, A., "Incompressible flow between finite disks," *ASME Journal of Applied Mechanics*, 49(1):1–9, 104, 1982.

Adkins, D., *Analysis of Hydrodynamic Forces in Centrifugal Pump Impellers*, PhD Thesis, California Institute of Technology, 1985.

Agostinelli, A., Nobles, D., and Mockridge, C. R., "An experimental investigation of radial thrust in centrifugal pumps," *ASME Transactions*, 1959.

Alford, J., "Protecting turbomachinery from self-excited rotor whirl," *ASME Journal of Engineering for Power*, 87:333–344, 1965.

Bolleter, U., Wyss, A., Welte, I., and Struchler, R., "Measurement of hydrodynamic matrices of boiler feed pump impellers," *ASME Journal of Vibration, Acoustics, Stress and Reliability in Design*, 109, 1987.

Chamieh, D., Acosta, A. J., Franz, and Caughey, T. K., *Experimental Measurements of Hydrodynamic Stiffness Matrices for a Centrifugal Pump Impeller*, Workshop: Rotordynamic Instability Problems in High Performance Turbomachinery, Texas A&M University, NASA CP No. 2250, 1982.

Childs, D., "Private communications with M. L. Adams," February 1999.

Ehrich, F., "Rotor whirl forces induced by the tip clearance effect in axial flow compressors," *ASME Journal of Vibrations and Acoustics*, 115:509–515, 1993.

Guelich, J. F., Bolleter, U., and Simon, A., *Feedpump Operation and Design Guidelines*, EPRI Final Report TR-102102, Research Project 1884-10, 1993.

Guelich, J. F., Jud, S. F., and Hughes, S. F., "Review of parameters influencing hydraulic forces on centrifugal impellers," *Proceedings, IMechE*, 201(A3):163–174, 1987.

Jery, B., Acosta, A., Brennen, C. E., and Caughey, T. H., *Hydrodynamic Impeller Stiffness, Damping and Inertia in the Rotordynamics of Centrifugal Flow Pumps*, Workshop: Rotordynamic Instability Problems in High Performance Turbomachinery, Texas A&M University, NASA CP No. 2133, 1984.

Kirk, R. G. and Donald, G. H., *Design Criteria for Improved Stability of Centrifugal Compressors*, ASME Applied Mechanics Division, Symposium on Rotor Dynamical Instability, M. L. Adams (Ed.), ASME Book AMD-Vol. 55, pp. 59–71, 1983.

Kostyuk, A. G., Keselev, L. E., Serkov, S. A., and Lupolo, O. A., "Influence of designs of seals over the shrouding strips on the efficiency and resistance to vibration of turbomachines," *Thermal Engineering*, 31(4):208–210, 1984; *Teploenergetike*, 31(4):36–39, 1984.

Pollman, E., Schwerdtfeger, H., and Termuehlen, H., "Flow excited vibrations in high-pressure turbines (steam whirl)," *ASME Journal of Engineering for Power*, 100:219–228, 1978.

Rankine, W. A., *On the Centrifugal Force of Rotating Shafts*, Vol. 27, Engineer, London, 1869.

Sawicki, J. T., Adams, M. L., and Capaldi, R. J., "System identification methods for dynamic testing of fluid-film bearings," *International Journal for Rotor Dynamics*, 2(4):237–245, 1996.

Stepanoff, A. J., *Centrifugal and Axial Flow Pumps*, 2nd edn, Wiley, New York, pp. 462, 1957.

Thomas, H. J., "Instabile Eigenschwingungen Turbinenläufern angefacht durch die Spaltströmugen Stopfbuschen und Beschauflungen" ["Unstable Natural vibration of turbine rotors excited by the axial flow in stuffing boxes and blading"], *Bull de L'AIM*, 71(11/12):1039–1063, 1958.

Thomas, H. J., Urlichs, K., and Wohlrab, R., "Rotor instability in thermal turbomachines as a result of gap excitation," *VGB Kraftwerkstechnik*, 56(6):345–352, 1976 (in English).

Urlichs, K., *Durch Spaltströmugen Hervorgerufene Querkräfte an den Läufern Thermischer Turbomaschinen [Shearing Forces Caused by Gap Flow at the Rotors of Thermal Turbomachines]*, Doctoral Dissertation, Technical University Munich, 1975.

Urlichs, K., *Leakage Flow in Thermal Turbo-Machines as the Origin of Vibration-Exciting Lateral Forces*, NASA TT F-17409, Translation to English of *Die Spaltströmugen bei Thermichen Turbo-Maschinen als Ursache für Enstehung Schwingungsanfachender Querkräfte, Engineering Archives*, 45(3):193–208, 1976.

Wohlrab, R., *Experimentelle Ermittlung Spaltsströmungsbedingter Kräfte an Turbinenstufen und Deren Einfluss auf die Laufstabilität Einfacher Rotoren* [*Experimental Determination of Forces Conditioned by Gap Flow and their Influences on Running Stability of Simple Rotors*], Doctoral Dissertation, Technical University Munich, 1975.

Wright, D. V., *Labyrinth Seal Forces on a Whirling Rotor*, ASME Applied Mechanics Division, Symposium on Rotor Dynamical Instability, M. L. Adams (Ed.), ASME Book AMD-Vol. 55, pp. 59–71, 1983.

REVIEWS OF PUBLISHED WORK

1. Review the published literature and report on centrifugal pump hydraulic forces imposed on the rotating element.
2. Review the published literature and report on Thomas–Alford forces in axial flow turbo-machinery.
3. Review the published literature and report on balancing drums and their rotor dynamical effects.
4. Review the published literature and report on rotor dynamics stiffness, damping, and inertia arrays for centrifugal pump impellers.

Part III

Monitoring and Diagnostics

7

Rotor Vibration Measurement and Acquisition

7.1 Introduction to Monitoring and Diagnostics

Vibration is the most regularly measured condition parameter in modern rotating machinery, and it is now continuously monitored in many important applications. Bearing temperature is also quite often a continuously monitored condition parameter as is rotor axial position. Some types of rotating machinery vibration problems can be expeditiously diagnosed by correlating vibration level and other such simultaneously monitored parameters, as covered in Part 4 of this book.

Modern vibration monitoring has its genesis in the mid-1950s with the development and application of basic vibration sensors, which are the heart of modern computerized condition monitoring systems. Figure 7.1 shows the traditional fundamental use of vibration monitoring in rotating machinery, that is, to provide warning of gradually approached or suddenly encountered excessively *high vibration* levels that could potentially damage the machinery. *Trending* a machine's vibration levels over an extended period of time can potentially provide early warning of impending excessive vibration levels and/or other problems, and thus provide plant operators with valuable information for critical decision making in order to schedule a timely shutdown of a problem machine for corrective action, for example, rebalancing the rotor.

In recent years, there has been a concerted effort to utilize vibration monitoring in a more extended role, most notably in what is now commonly called *predictive maintenance*, which is an extension and/or replacement of traditional *preventive maintenance*. As illustrated in Figure 7.2 for one proposed version of predictive maintenance, each machine of a given group is provided with specific maintenance actions based on the machine's monitored condition instead of a fixed-time maintenance cycle. In principle this makes a lot of sense, but as most practitioners know, "the devil is in the details." This effort is primarily driven by the current trend in industry and government organizations to drastically reduce maintenance costs, primarily by making large reductions in maintenance and technical support personnel. This prevailing "bean counter" mentality has created new

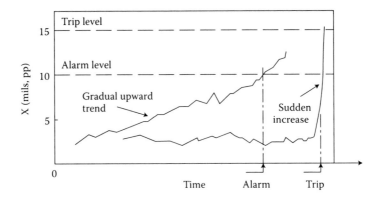

FIGURE 7.1 Tracking of a representative vibration peak amplitude over time.

business opportunities for suppliers of machinery condition monitoring systems and impetus for new approaches to glean increased diagnostic information from already continuously monitored machinery vibration signals.

The invention of the *Fast Fourier Transform* (*FFT*) algorithm in the mid-1960s was developed as an effective means for quickly mimicking the frequently changed radar signal spectrum of enemy ground based antiaircraft missile targeting systems, so that multiple decoy signals could not be distinguished from authentic reflections. The FFT algorithm has subsequently become a primary signal analysis tool, and has been the major modern advancement in rotating machinery vibration signal analysis. The quest of researchers for creative new approaches has been facilitated by the FFT's success and the reductions in maintenance and support personnel. For example, Adams and Abu-Mahfouz (1994) explored *chaos* and

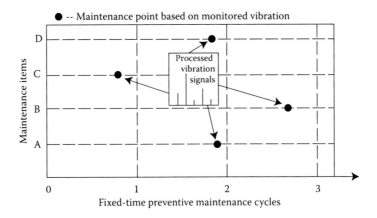

FIGURE 7.2 *Predictive* maintenance contrasted to *preventive* maintenance.

routes to chaos in rotor vibration signals as diagnostic markers for providing improved early detection and diagnosis of impending problems or needed maintenance actions. The development of new machinery vibration signal analysis techniques gleaned from modern chaos theory is predicated on the inherent *nonlinear dynamical character* of many incipient failure modes and wear mechanisms (see Section 2.5 of Chapter 2). Other nontraditional signal analysis methods are also finding their way into machinery vibration diagnosis, such as the signal processing technique called *wavelets* or *wavelet transforms* (WTs).

Over the last 20 years or so the term *expert system* has gained notoriety in those industry and government organizations heavily concerned with rotating machinery. It is a fact that these rotating machinery user sectors have drastically reduced the number of maintenance personnel. Original equipment manufacturers (OEMs) have also undergone similar major contractions involving mergers, downsizing, and the like, with considerable reductions of in-house technology development, and an almost nonexistent development of the next generation of *true specialists* and *experts*. Thus, the so-called *expert systems* struck a welcomed theme among both rotating machinery OEMs and users alike. Naturally, expert systems are at best as good as the information and data stored in them, and glitzy additives such as *fuzzy logic* and *neural networks* have not significantly changed that because they entail a "learning period" that requires a large number of unwanted events to occur. Figure 7.3 illustrates a so-called expert system, which is computer software that contains a programmed knowledge base and a set of rules that key on that knowledge base, as reviewed by Bently and Muszynska (1996) concerning "expert system" application to rotating machinery condition monitoring.

If future major applications of rotating machinery are to be economically successful in an environment of greatly *reduced maintenance personnel* and very few available *true experts*, then new yet-to-be-introduced *machinery management systems* will be required. Development of such new systems was a topic of extensive ongoing research in the author's laboratory at CWRU in the 1990s. The CWRU team developed model-based monitoring-diagnostic and prognostic software, incorporating an array

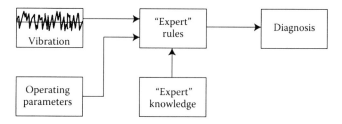

FIGURE 7.3 Flow chart of a rule-based "expert system."

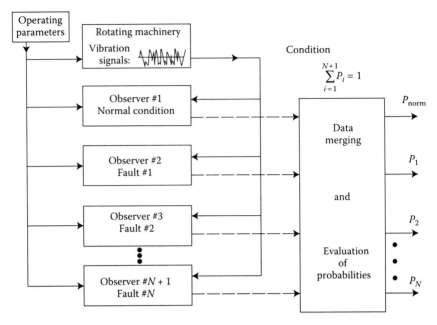

FIGURE 7.4 Real-time probabilities for defined faults and severity levels from statistical correlation of monitored and model-predicted vibration signals.

of machine-specific vibration simulation computer models, specific to an extensive array of operating modes as well as fault types and severity levels. As illustrated in Figure 7.4, each model (called an "observer") is run in real time and its simulated vibration signals are continuously combined with the machine's actual monitored vibration signals and correlated through a novel set of statistical algorithms and model-based filters, as summarized by Loparo and Adams (1998). Probabilities are generated for each fault type and severity level potentially in progress. The vibration models in the observers also remove signal "noise," which does not statistically correlate with the models. In contrast to conventional signal noise filtering techniques, such model-based statistical-correlation filtering allows retention of physical-model correlated low-level and fine-structure signal components, such as in signal chaos content, for on-line or off-line analysis.

One of the many interesting findings by the CWRU team is that the various fault and fault-level specific observer vibration models do not have to be as "nearly perfect" as one might suspect if thinking in a time signal domain and/or frequency domain framework. Because the sum of probabilities is constrained to = 1, a model (observer) only has to be representative enough of its respective operating mode to "win the probability race" among all the "observers" when its fault (or fault combination) type and severity level are in fact the dominant condition. Compared to the rule-based approach inherent in the so-called expert systems, this *physical model-based* statistical approach is fundamentally much more open to correct and early diagnosis, especially of infrequently encountered failure and maintenance-related phenomena and especially of conditions not readily covered within a rule-based "expert system."

An additional benefit of a model-based diagnostics approach is the ability to combine measured vibration signals with vibration computer model outputs to make real-time determinations of rotor vibration signals at locations where no sensors are installed. Typically, vibration sensors are installed at or near the bearings where sensor access to the rotor and survivability of sensors dictate. However, the mid-span locations between the bearings is where operators would like to measure vibration levels the most, but cannot because of inaccessibility and hostile environment for vibration sensors. Thus, the model-based approach provides "virtual sensors" at inaccessible rotor locations.

The field of modern condition monitoring for rotating machinery is now over 50 years into its development and thus is truly a matured technical subject. However, it continues to evolve and advance in response to new requirements to further reduce machinery downtime and drastically reduce maintenance costs.

7.2 Measured Vibration Signals and Associated Sensors

The commonly monitored vibration signals are *displacement, velocity,* and *acceleration*. The respective sensor operating principles are presented in this section. Commercial suppliers of vibration measurement systems provide specific information on their vibration measurement products. For a more formal treatment of machinery vibration measurement, the author refers to the books by Mitchell (1981) and Bently and Hatch (1999), and the paper by Muszynska (1995).

7.2.1 Accelerometers

An *accelerometer* is comprised of an internal *mass* compressed in contact with a relatively stiff force-measuring *load cell* (usually a piezoelectric crystal) by a relatively soft preload *spring*, as illustrated in Figure 7.5. The functioning of an accelerometer is thus derived from the 1-DOF system shown in Figure 1.1. For an accelerometer, the system damping is a negligible effect and thus for explanation purposes the damping is assumed here to be zero. Referring to the free-body diagram shown in Figure 7.5,

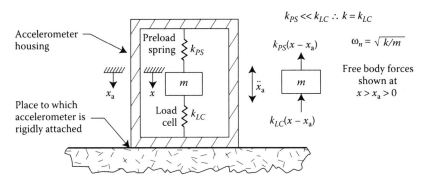

FIGURE 7.5 Elementary schematic for unidirectional accelerometer.

the equation of motion for the mass is obtained as follows:

$$m\ddot{x} + kx = kx_a \tag{7.1}$$

For a sinusoidal motion of the accelerometer housing ($x_a = X_a \sin \omega t$) and the measurement place to which it is rigidly attached, the motion equation for the internal mass is Equation 1.6 with zero damping, as follows:

$$m\ddot{x} + kx = kX_a \sin \omega t \tag{7.2}$$

Thus, for the steady-state solution of Equation 7.2, the normalized response equation shown in Figure 1.5 is applicable, and for zero damping it provides the following ratio for peak acceleration of the internal mass to the housing:

$$\left|\frac{\ddot{X}}{\ddot{X}_a}\right| = \left|\frac{\omega^2 X}{\omega^2 X_a}\right| = \left|\frac{X}{X_a}\right| = \frac{1}{1 - (\omega/\omega_n)^2} \tag{7.3}$$

For a frequency at 10% of the accelerometer's natural frequency (ω_n), Equation 7.3 shows the acceleration of the internal mass to be 1% higher than housing acceleration, at 20% it is 4% higher, at 30% it is 10% higher, and so on.

The accelerometer load cell is usually a piezoelectric crystal and thus only registers compressive loads, necessitating a preload spring to keep it in compression. However, the piezoelectric crystal is inherently quite stiff in comparison with the preload spring. Therefore, the load cell essentially registers all the dynamic force (ala $F = ma \rightarrow a = F/m$) required to accelerate the internal mass. Equation 7.3 shows that for the load cell electrical output to be highly linear with housing acceleration, an accelerometer must be selected with an internal *natural frequency* at least five times *higher* than the maximum end of its intended *frequency* range of use. Consequently, an accelerometer for a relatively high frequency application has a relatively smaller internal mass than an accelerometer for a relatively low frequency range of application. Since a smaller internal mass produces a proportionally smaller peak load-cell force for a given acceleration peak, there is clearly a compromise between sensitivity and frequency range. That is, the higher the accelerometer's internal resonance frequency, the lower its sensitivity. Accelerometer sensitivity is proportional to its internal mass (m), but its internal natural frequency is only proportional to $1/\sqrt{m}$. Consequently, for a given load cell stiffness, the sensitivity varies as $1/\omega_n^2$, that is, a penalty to sensitivity for better linearity. But accelerometers are still the best transducer for high frequencies because of the inherent frequency squared multiplier.

Piezoelectric load cells produce a self-generated electrical output in response to dynamic loading, but at very high impedance. Accelerometers are therefore usually constructed with internal electronics to convert the load cell's signal to a low impedance output suitable for conventional plugs, cables, and data acquisition systems. Mitchell (1981) provides many practical considerations including an explanation on remote location of the electronics for high-temperature measurement places where the piezoelectric load crystal can survive but the signal-conditioning electronics cannot.

7.2.2 Velocity Transducers

The velocity transducer is comprised of a mass (permanent magnet) suspended in very soft springs and surrounded by an electrical coil, as illustrated in Figure 7.6. Also explained by the 1-DOF model, a velocity transducer operates above its natural frequency, in contrast to an accelerometer. Its springs are configured to produce a very low natural frequency so that the permanent magnet typically remains stationary at

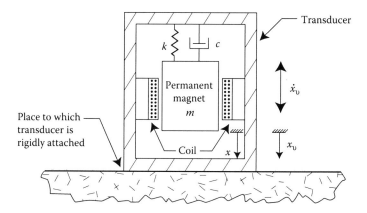

FIGURE 7.6 Elementary schematic for a velocity transducer.

frequencies above 10 Hz. Typically, an internal fluid provides the critical damping of the natural frequency and the roll-off of response below 10 Hz.

With the magnet essentially stationary in the transducer's frequency range of use (typically 10–1500 Hz), vibration of the electrical coil rigidly attached to the housing causes the magnetic flux lines to induce a voltage in the coil *proportional to velocity* of housing vibration. Thus, a velocity transducer produces a self-generated low-impedance velocity-proportional electrical signal that can be fed to monitoring and data acquisition systems without additional signal conditioning. Velocity transducers could therefore be popular in many rotating machinery applications. However, because a velocity transducer has internal moving parts, it is less popular in hostile environments where a relatively higher ruggedness is demanded, as more inherent with an accelerometer. Thus, rugged sensors that are marketed for measurement of velocity are actually accelerometers with built-in integration circuits to output the velocity signal.

7.2.3 Displacement Transducers

7.2.3.1 Background

The internals of many types of rotating machinery, especially turbomachinery, have a number of quite small annular radial clearance gaps between the rotor and the stator, for example, journal bearings, annular seals, balance drums, and blade-tip clearances. Therefore, one obvious potential consequence of excessive rotor vibration is rotor–stator rubbing contact or worse, impacting. Both accelerometers and velocity transducers measure the vibration of nonrotating parts of a machine and thus cannot provide any direct information on rotor motion relative to the stator.

The importance of rotor motion relative to stator motion led to the development of transducers to provide continuous instantaneous rotor-to-stator position measurements, typically at each journal bearing. The earliest rotor-to-stator position measurement device widely applied is commonly referred to as a *shaft rider*, and it is similar to a typical spring loaded IC engine valve tracking its cam profile. That is, a shaft rider is essentially a radial stick that is spring loaded against the journal to track the journal radial motion relative to a fixed point on the nonrotating part of the machine, for example, bearing housing. Shaft riders utilize a position sensing transducer to provide an electrical output linear with shaft rider instantaneous radial position. Some older power plant turbines still use OEM supplied shaft riders. However, shaft riders have two major shortcomings: (1) their mass inertia limits their frequency range and (2) their rubbing contact on the journals is a wear point. Copious lubrication is not a solution to the wear problem, because the uncertainty of contact oil-film thickness is of the same order of magnitude as the rotor relative position changes continuously measured. Therefore, a shaft rider journal contact point is typically a wearable carbon material. The effect of this slowly wearing contact point is to produce a continuous DC drift in position measurements, thus detracting somewhat from the main intent of rotor-to-stator continuous position measurement.

The significant shortcomings of shaft riders led to the development of noncontacting position sensing transducers. Two types of noncontacting transducers that emerged in the 1950s are the *capacitance* type and the *inductance* type. The *capacitance*-type displacement transducer works on the principal of measuring the electrical capacitance of the gap between the transducer tip and the *target* whose position is measured. The *capacitance* method is well suited for highly precise laboratory measurements, but its high sensitivity to material (e.g., oil) variations/contaminants within the clearance gap would make it a calibration "nightmare" for industrial applications. In contrast, *inductance*-type displacement transducer systems have proven to be the optimum rotor-to-stator position measurement method, and are now installed on nearly all major rotating machines in power plants, petrochemical and process plants, naval vessel propulsion drive systems, and many others. It is also the primary rotor position sensor for laboratory test rigs as well.

7.2.3.2 Inductance (Eddy-Current) Noncontacting Position Sensing Systems

Unlike accelerometers and velocity transducers, which are mechanical vibratory systems in their own right, *inductance*-type displacement transducer systems function entirely on electrical principles. As illustrated in Figure 7.7, the system includes a target (shaft), a proximity probe, cables, and an oscillator demodulator (called *proximeter*). A proximity probe is typically made with a fine machine thread on its outer cylindrical surface for precision positioning and houses a helical wound wire coil encased in a plastic or ceramic material. The oscillator demodulator excites the probe's coil with a radio-frequency carrier signal of 1.5 mHz (typical), causing a magnetic field to radiate from the probe's tip. When the probe tip is in *proximity* to an electrically conductive material (target), the induced *eddy currents* in the target absorb electrical energy from the probe coil's excitation and thus attenuate its carrier signal. Within the oscillator demodulator, a DC voltage output is produced from the modulated envelope of the carrier signal, as schematically shown in Figure 7.7.

Figure 7.8 illustrates a typical DC voltage output versus gap. As shown, the DC voltage output calibrates quite linearly over a large gap range between the probe tip and the target. For typical systems now used for monitoring rotor vibration, the linear range is normally from 10 to 100 mils (0.25–2.5 mm). Setting the mean probe-to-target gap at the midpoint of the

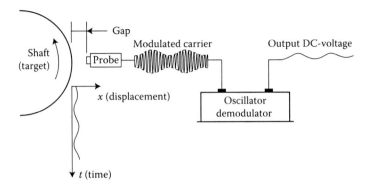

FIGURE 7.7 Inductance (eddy-current) position sensing system.

linear range provides substantially more vibration magnitude operating range than needed for virtually all rotor vibration monitoring applications. In rare catastrophic failures (e.g., Figure 2.15), dynamic motions can quite readily exceed the usable gap range, but this is irrelevant since the vibration monitoring proximity probes are probably destroyed along with the machine.

It is important to point out that the combination of the proximity probe, oscillator demodulator, and their cables form a tuned resonant electrical circuit. Thus, in order to obtain a specified voltage-to-gap calibration factor, the cables must be properly matched to the probe and oscillator demodulator. Adherence to the manufacturer's cable type and length will therefore maintain the system's vibration calibration accuracy with component interchangeability. It is also important to know that the calibration factor is a

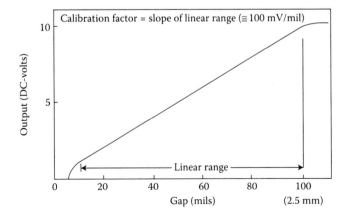

FIGURE 7.8 Typical inductance probe displacement calibration plot.

strong function of the target's material. Therefore, if the manufacturer's supplied calibration factor is in doubt regarding the target material, the system should be carefully recalibrated using the actual target material. Large variations in the probe's ambient temperature and/or pressure may produce variations in the calibration factor that are significant, at least for high-precision laboratory measurements on research rotor test rigs. Mitchell (1981) discusses the influence of probe diameter and excitation voltage on system sensitivity as well as other design and application considerations such as proper probe mounting. With the quite high carrier frequency used to excite the probe's inductance coil, the oscillator demodulator can readily track gap variations linearly at frequencies well over 10,000 Hz, which is considerably higher than needed for virtually any rotor vibration measurement purposes.

Obviously, any residual mechanical run-out of the target portion of the shaft is added to the vibration signal detected by the proximity probe. The periphery of a rotating shaft presents a target that moves laterally across the proximity probe's magnetic field, as illustrated in Figure 7.7. As a consequence, the system's output not only reflects the shaft vibration plus mechanical run-out, but also the superimposed effects of circumferential variations in shaft surface conditions as well as electrical conductivity and permeability variations just below the shaft surface. Except for mechanical run-out, these nonvibration electromagnetic additions to the output signal were not widely recognized until the early 1970s when several apparent excessive rotor vibration problems in plants were diagnosed correctly as "false trips" caused by the nonmechanical electromagnetic signal distortions. That is, the superimposed nonvibration output signal components (commonly called "electrical run-out"), when added to the signal portion representative of actual rotor vibration, indicated fictitiously high vibration levels, triggering automatic machine shutdowns or "trips."

In the years since the nonvibration component in proximity probe system output was first widely recognized, these vibration monitoring systems have been refined to substantially remove nonvibration sources from the output. On the mechanical side, every effort must be taken to provide a smooth shaft target surface free of scratches and with a tight concentricity tolerance to the journal. Mitchell (1981) describes various measures to minimize the electromagnet sources of "electrical run-out." It has become the standard procedure in plants to take the output signal for each probe while the machine is slowly rotated on turning-gear mode or on coast-down near stopping, and to process that data to extract the once-per-rev component (amplitude and phase angle) which is then stored and automatically subtracted in real time from the raw monitored signal. The ultimate precision in journal vibration measurements was demonstrated by Horattas et al. (1997) on the laboratory spindle shown in Figure 5.10 with maximum precision preloaded ball bearings. They mounted a precision grinder/slide on the front of the test rig to remove a test journal's mechanical run-out as achievable with the spindle bearings, that is, less than 0.5 μm TIR residual run-out after grinding. They then processed the remaining slow-speed "electrical run-out" (approximately 0.5 and 0.7 mil pp on x and y probes, respectively) and recorded the outputs as high sampling rate digital signals. By subtracting the entire "electrical run-out" digital signal (not just the once-per-rev component) for each probe from its raw signal at running speed, Horattas et al. (1997) demonstrated journal vibration measurements with an accuracy approaching 0.02 mil (0.5 μm).

Proximity probes are usually installed in pairs at each journal bearing, with their measurement axes at an angular position of 90° with respect to their partner, as illustrated in Figure 7.9a. In this manner, the rotor vibration orbit can be readily viewed in real time by feeding the two signals into the x and y amplifiers of a dual channel oscilloscope. Rotor vibration orbital trajectories are illustrated in earlier Figures, that is, 2.10, 2.11, 2.13, 2.17, and 4.5 through 4.9. Figure 7.9a also illustrates the typical angular orientation of a pair of proximity probes at 45° and 135°. Referring to Figure 2.10, the reason for this is that in most cases the major axis of the orbit ellipse is close to the 45° axis due to the journal bearing oil film being stiffest into the minimum film thickness, that is, along the line of centers. Thus, with one of the two probes located at 45°, its channel yields close to the largest vibration signal of the orbit. The 45° channel is therefore normally selected as the vibration channel used for rotor balancing in the field, that is, it has the highest signal level if it closely aligns with the rotor vibration orbit's ellipse major axis.

By intent, proximity probes measure rotor motion relative to stator motion, and thus do not provide total rotor motion. When measurement of total rotor vibration motion is needed, the combination of a proximity displacement probe mounted with a seismic transducer (accelerometer or velocity transducer) may be employed as illustrated in Figure 7.9b.

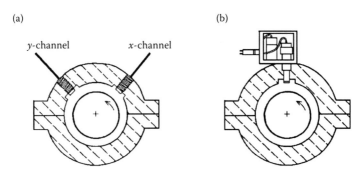

FIGURE 7.9 Proximity probe example mountings: (a) typical 2-probe @90° placement and (b) probe with seismic sensor.

The total rotor displacement signal is then obtained by adding the conditioned outputs of the integrated seismic transducer measurement and the proximity probe displacement measurement. An alternate approach not always feasible in plants but commonly used in laboratory rotor test rigs is simply to mount the proximity probes to an essentially nonvibrating fixture.

7.3 Vibration Data Acquisition

There is a considerable variety of extent and methods used to acquire and log vibration and other diagnostic monitored machinery parameters. The methods and corresponding products available to accomplish data acquisition tasks comprise a constantly changing field that parallels the rapid and perpetual advancements in PCs and Workstations. This section provides neither a historical perspective nor a forecast of future trends for machinery monitoring data acquisition technology. The intent of this section is to present the fundamental steps in data acquisition and a summary of up-to-date methods and devices appropriate for different application categories.

7.3.1 Continuously Monitored Large Multibearing Machines

The main steam turbine-generator sets of large electric power generating plants are a prime example of large machines where the need for constant condition monitoring is driven both by the monetary replacement cost of a machine if seriously damaged (well over $100 million) and the lost generating revenues accrued in the event of an unscheduled outage of a single large steam turbine generator (as high as $500 thousand/24 h day, more in a nuclear plant). A complete rigidly coupled drive line, including a high-pressure turbine (HP), an intermediate-pressure turbine (IP), two low-pressure turbines (LP-1 and LP-2), an AC synchronous generator (2-pole for fossil units, 4-pole for nuclear units), and its exciter (EX), is illustrated in Figure 7.10. The generating unit shown in Figure 7.10 has eight journal bearings and is equipped with x and y noncontacting proximity probes as well as vertical and horizontal accelerometers at each journal bearing, for a total of 32 vibration data channels.

Rotor vibration time-based signals are phase referenced to a single fixed angular position on the rotor, referred to as the *keyphasor*, as shown in Figure 7.11. A *keyphasor* signal must be a very sharply changing signal so as to trigger a time marker for the designated fixed angular position on the rotor. It can be produced by a proximity probe targeting a pronounced shaft surface interruption such as a key way. It can also be produced by a light sensitive optical pickup targeting a piece of reflective tape on the shaft

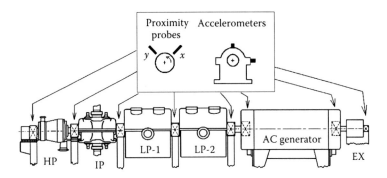

FIGURE 7.10 Vibration monitoring channels for an electric power turbine.

surface. An important use of the *keyphasor* is in prescribing phase angles of the once-per-rev components of all vibration signals for rotor balancing purposes. For large generating units typified by Figure 7.10, continuously updated vibration peak amplitudes (or RMS values) at the bearings are displayed in the plant control room, such as illustrated in Figure 7.12.

For power plants and large process plants, the traditional control room is being replaced by a few computer monitors (i.e., a virtual control room) each having several operations, control and condition monitoring menus. Such *virtual control rooms* need not be located at the plant site. These new virtual control rooms are accompanied by super high-capacity computer data storage units, which make it possible to digitally store all monitored machinery vibration time-base signals on a continuous basis, for any subsequent analysis purposes. This facilitates introducing the next generation of condition monitoring systems developed on a modeled-based

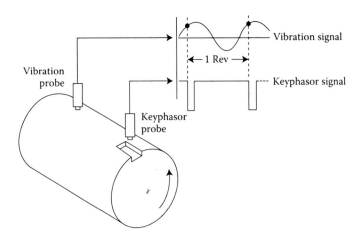

FIGURE 7.11 Vibration signals all referenced to a single keyphasor.

Rotor Vibration Measurement and Acquisition

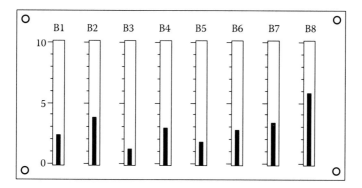

FIGURE 7.12 Control room display of current vibration levels at each bearing.

evaluation of fault and fault-level probabilities, as illustrated in Figure 7.4. For machinery vibration monitoring, the virtual control room monitor has various operator selected menus, such as that illustrated in Figure 7.13.

7.3.2 Monitoring Several Machines at Regular Intervals

Many types of rotating machines are much smaller and more numerous than the electric power generating unit illustrated in Figure 7.10. Unlike modern power plants that are typically dominated by a relatively few large machines, many types of process plants employ several relatively smaller machines too numerous to bear the costs of continuous vibration monitoring systems for every machine.

The lower cost alternative to continuous vibration monitoring is to take vibration data from machines at designated regular intervals. All vibration monitoring system suppliers now market *over-the-shoulder hand-held vibration analyzers* that display on-the-spot vibration analysis outputs such as amplitudes (peak, filtered, RMS, etc.) and FFT spectra. The typical over-the-shoulder unit employs an accelerometer vibration pickup that a maintenance person can securely touch against designated vibration monitoring points on each machine that is routinely checked. Many of

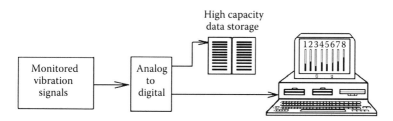

FIGURE 7.13 Viewing vibration levels at bearings in the virtual control room.

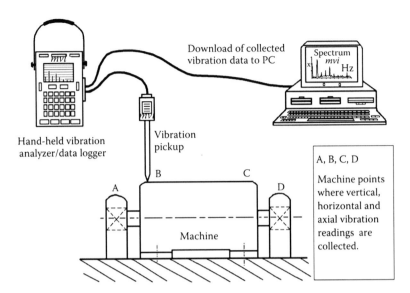

FIGURE 7.14 Portable machinery vibration analyzer and data logger.

these hand-held vibration analyzers are also made to digitally record and store vibration signals that a single maintenance person can acquire from several machines in a single pass through an entire plant, for example, for subsequent downloading into a PC for further analysis and permanent data storage. Figure 7.14 schematically illustrates a typical hand-held vibration analysis and data logger unit.

7.3.3 Research Laboratory and Shop Test Applications

For laboratory and shop test applications, high sampling rate multichannel data acquisition is now universally done quite inexpensively by installing one or more analog-to-digital (A-to-D) expansion boards into a standard desktop PC. Major suppliers of such PC expansion boards also market quite versatile PC software to capture, store and reduce measured signals, and to perform user programmed control operations based on measured signals. Current PC-based high sampling rate multichannel data acquisition setups are quite superior to top-end systems of about 25 years ago and are about 1/20th their cost.

7.4 Signal Conditioning

Raw vibration signals always contain some contamination ("noise") and frequently some actual components that may partially obscure other actual components that comprise the important part of the signal being sought by the measurements taken. Thus, the most frequent signal conditioning operation is *filtering*. A-to-D signal conversion is unfortunately often the first step in data acquisition, with filtering then performed computationally from the digitized signal. However, low-pass analog filtering should be inserted ahead of the A-to-D converter to avoid *aliasing*, which is the "reflection" into the lower end of the spectrum of high-frequency content above the sampling-rate capability of the A-to-D converter. Other frequently performed signal conditioning operations include *integration* (i.e., to extract displacement from measured acceleration or velocity signals, or velocity from measured acceleration signals) and *signal amplitude* conversion.

7.4.1 Filters

Filters that are most often used with vibration signals include *low-pass*, *high-pass*, *band-pass*, *notch*, and *tracking* filters. Filtering is now routinely performed digitally after A-to-D conversion, but the initial signal must first be passed through an analog low-pass filter with cut-off frequency sufficiently below the Nyquist frequency (1/2 sampling rate) to eliminate *aliasing* (false peaks in the FFT amplitude). The analog filter's cut-off frequency must be substantially below the Nyquist frequency because no analog filter has a perfect frequency cut-off, that is, it has its roll-off above the cut-off frequency.

The *low-pass* filter is probably the most frequently employed signal conditioning operation in handling machinery vibration measurement signals. For routine rotating machinery vibration assessments, frequency components above 10 times spin speed are usually not of interest, be they "noise" or true signal. The *low-pass* filter is intended to remove signal content above the designated cut-off frequency and thus *passes* through the remaining portion of the signal that is below the designated cut-off frequency. If using a digital low-pass filter, it is assumed that the original analog signal has already been passed through an analog low-pass filter to avoid aliasing, as described above. It is important to caution here that the typical A-to-D expansion board for PCs does not have an analog low-pass filter to avoid aliasing. However, modern digital tape recorders do (i.e., Sony, TIAC).

The *high-pass* filter is the converse of the *low-pass* filter, removing signal content below the designated cut-off frequency and thus it *passes* through the remaining portion of the signal that is above the designated cut-off frequency. Since routine rotating machinery vibration assessments are usually not focused on frequency components above 10 times spin speed, *high-pass* filtering by itself is not often used in machinery applications. However, the *band-pass* filter, which is a combination of high- and low-pass filtering, is routinely employed in machinery vibration analyses.

A *band-pass* filter is designed to remove signal content outside a designated *frequency band*, and thus is a *low-pass* filter in series with a *high-pass* filter, where the low-pass cut-off frequency is higher than the high-pass cut-off frequency. Again, if filtering digitally, the original analog signal has first been passed through an appropriate analog low-pass filter to avoid aliasing. A *band-pass* filter centered at rotor speed is a standard operation in rotor balancing, since only the synchronous vibration component is processed for rotor balancing purposes. The fundamental basis for this is that balancing procedures are inherently based on the tacit assumption that the vibratory system is linear, and thus only the forcing frequency (once-per-rev) vibration amplitude and phase angle are accommodated in rotor balancing procedures. Synchronous band-pass filtering thus improves balancing accuracy.

The *notch* filter is the opposite of the *band-pass* filter, passing through all the signal content except that which is within a specified *bandwidth*. One interesting application is magnetic bearings, which inherently operate with displacement feedback control, where a *notch*-type filter is frequently used to filter out the once-per-rev bearing force components, so they are not transmitted to the nonrotating structure of the machine, while the bearings continue to provide static load support capacity and damping. The broad band spectra of measured pump impeller hydraulic forces provided in Figure 6.4 have the once-per-rev component removed, that is, *notch* filtered.

A *tracking* filter can employ the functionality of any of the previously described filters, but it has the added feature that its cut-off frequency(s) are made to track a specified signal component. The main application of the *tracking* filter in rotor vibration measurement is to have the center-band frequency of a *band-pass* filter track the once-per-rev frequency tracked by the *keyphaser* signal, illustrated in Figure 7.11. This is a standard feature on rotor vibration signal processing devices as a convenience for tracking synchronous rotor vibration signals as a machine is slowly brought up to operating speed and is coasting down to shut-off or turning-gear condition.

Advanced model-based nonlinear *denoising* filters, which do not remove important low level signal content (e.g., chaos), are inherent in the system schematically illustrated in Figure 7.4, from Loparo and Adams (1998).

7.4.2 Amplitude Conventions

When vibration amplitudes are conveyed, one should also specify which amplitude convention is being used. Although rotating machinery vibration signals always contain frequency components other than just the frequently dominant once-per-rev (synchronous) component, the single-frequency (harmonic) signal is well suited for explaining the different vibration amplitude measurement conventions. The following generic harmonic vibration signal is thus used here:

$$x = X \sin \omega t \tag{7.4}$$

For a vibration signal comprised of only one single harmonic component, there are two obvious choices for conveyance of the vibration amplitude, *single-peak* and *peak-to-peak*, as follows:

Single-peak amplitude (S.P.) = X and peak-to-peak amplitude (P.P.) = $2X$

However, vibration signals frequently contain significant contributions from more than just one harmonic, often several, and thus an *average* amplitude is frequently used to quantify a broad band vibration signal. The two conventional average magnitudes are the *average* absolute value and the RMS average, evaluated over a specified time interval Δt as follows:

$$A = \text{average} = \frac{1}{\Delta t}\int_{t}^{t+\Delta t}|x|\,dt \quad \text{and} \quad \text{RMS average} = \frac{1}{\Delta t}\left[\int_{t}^{t+\Delta t} x^2\,dt\right]^{1/2}$$

For a simple harmonic signal as given in Equation 7.4, these two averages yield the following:

$$A = 0.637X \quad \text{and} \quad \text{RMS} = 0.707X$$

7.5 Summary

The vitally important function of machinery *condition monitoring* rests upon the feasibility of reliable measurement of a machine's "vital life signs," of which vibration is among the most important. The traditional use of rotating machinery vibration monitoring is to provide warning if vibration levels become sufficiently high to potentially damage the machine. While this traditional function of machinery vibration monitoring is of course still of paramount importance, present diagnostic methods now allow a much broader assessment of a machine's condition from its monitored vibration than just saying "the vibration level is too large." *Predictive maintenance* is one example of a capability derived from *condition monitoring*.

Bibliography

Adams, M. L. and Abu-Mahfouz, I., *Exploratory Research on Chaos Concepts as Diagnostic Tools*, Proceedings, IFTOMM 4th International Conference on Rotor Dynamics, Chicago, Illinois, September 6–9, 1994.

Bently, D. E. and Hatch, C. T., *Fundamentals of Rotating Machinery Diagnostics*, Bently Pressurized Bearing Corporation, Minden, NV, 1999.

Bently, D. E. and Muszynska, A., *Vibration Monitoring and Analysis for Rotating Machinery*, Noise & Vibration '95 Conference, Pretoria, South Africa, November 7–9, 1996, Keynote Address Paper, pp. 24.

Horattas, G. A., Adams, M. L., and Dimofte, F., "Mechanical and electrical run-out removal on a precision rotor-vibration research spindle," *ASME Journal of Acoustics and Vibration*, 119(2):216–220, 1997.

Loparo, K. A. and Adams, M. L., *Development of Machinery Monitoring and Diagnostics Methods*, Proceedings of the 52nd Meeting of the Society for Machinery Failure Prevention, Virginia Beach, April 1998.

Mitchell, J. S., *An Introduction to Machinery Analysis & Monitoring*, PennWell, Tulsa, OK, pp. 374, 1981.

Muszynska, A., "Vibrational diagnostics of rotating machinery malfunctions," *International Journal of Rotating Machinery*, 1(3–4):237–266, 1995.

EXERCISES AND REVIEW OF PUBLISHED WORK

1. Formulate the equation(s) and devise an accelerometer users' guide that clearly shows the trade-off between sensitivity and frequency response. Also, utilizing the fundamental principles, determine a practical frequency nonlinearity correctly as one would use an accelerometer approaching its resonance frequency. Comment on the possibility of using an accelerometer in the frequency range significantly above its resonance frequency.

2. When applying an accelerometer with an integrating circuit to output velocity signal, determine the operative criteria for useable frequency range restrictions.

3. Review the available literature on shaft-rider contacting displacement sensors and develop frequency criteria for where inductance-type proximity probes are much more appropriate. Also, find available information to quantify inductance non-contacting proximity probes' performance (sensitivity, linearity) suffers as the used frequency range is increased.

4. Research dynamic system model-based filtering algorithms. Explore potential benefits of employing these advanced signal filtering approaches to rotating machinery vibration monitoring, diagnostics, and prognostics.

8
Vibration Severity Guidelines

8.1 Introduction

Considering the extensive technology development efforts devoted to computing and measuring rotating machinery vibration signals, it has always struck the author as ironic that when all that is "said and done" the fundamental question "at what level does vibration become too much?" is still often left with an uncertain answer, or possibly an answer that is disputed. It parallels the health industry's often changing proclamations on how much of certain "healthy" foods are "enough" and how much of certain "unhealthy" foods are "too much." At the present time, severity criteria for rotating machinery vibration levels are still most heavily governed by "experience." Most industrial rotating machines are not mass produced like consumer products. Therefore, it is not economically feasible to base the experience factor in rotating machinery vibration severity criteria on a rich statistical database stemming from controlled test-to-damage or destruction of machines at various levels of "excessive" vibration, to quantify statistically how long it takes the vibration to damage each machine at each tested vibration level.

There are several new rotating machinery products on the horizon for industrial and consumer applications, such as in the power generation and automotive sectors, that will run at considerably higher rotational speeds than their present forebears (to 100,000 rpm and above). Design solutions for next-generation high-speed rotating machinery will necessitate some fundamental research and development to more accurately quantify just *how much vibration can* be continuously endured by a given machine through its lifetime. The quite approximate *upper limits* provided by contemporary guidelines will probably be unacceptably too conservative or otherwise not applicable to next-generation high-speed rotating machinery.

8.2 Casing and Bearing Cap Vibration Displacement Guidelines

The first rotating machinery vibration severity guidelines widely used in the United States are generally credited to Rathbone (1939). His guidelines grew out of his experience as an insurance inspector on low-speed machines having shaft-to-housing vibration amplitude ratios typically in the range of 2–3. His chart and subsequent versions of it by others are based on *machine casing* or *bearing cap* vibration levels, such as illustrated by the accelerometers on the turbo-generator in Figure 7.10 and the hand-held analyzer in Figure 7.14.

Current severity guidelines bear a strong resemblance to Rathbone's original chart, that is, *as the frequency is higher, the allowable vibration displacement amplitude is less*. Many of the Rathbone-like charts are misleading by subdividing the vibration level into too many zones delineated by too many descriptors such as *destruction imminent, very rough, rough, slightly rough, fair, good, very good, smooth, very smooth,* and *extremely smooth* (in power plant lingo, *"smooth as a baby's a--"*). Such fine striations and descriptors are misleading because they incorrectly imply that the vibration severity guidelines are based on refined engineering science or finely honed experienced-based knowledge. In fact, severity guidelines are based on a collective "voting" by rotating machinery builders, users, and consultants, each having business interests to foster and protect. The most sensible "descendent" of the original Rathbone chart found by the author is provided by Eshleman (1976) based on that given by the German Engineering Society, VDI (1964), reconfigured in Figure 8.1 to show *peak-to-peak vibration displacement amplitude* in both metric and English units.

As labeled in Figure 8.1, the sloping straight lines on this log–log graph are lines of *constant velocity*. Consistent with this, it is widely accepted that between 10 and 1000 Hz (CPS), a given velocity peak value has essentially the same measure of vibration severity. This is a compromise between the *vibration displacement* consideration (e.g., rotor–stator rubbing or impacting caused by excessive vibration displacement) and the *vibration acceleration* consideration (i.e., peak time-varying forces and stresses generated from vibration are proportional to acceleration peak that is proportional to frequency squared times displacement amplitude). In the frequency range of 10–1000 Hz, when specifying vibration level in *displacement*, one needs to know the *frequency* in order to assess the severity, as Figure 8.1 demonstrates. Below 10 Hz the measure of vibration severity is generally characterized by a *displacement value*, whereas above 1000 Hz the measure of vibration severity is generally characterized by an *acceleration value*. This is illustrated in the severity guideline in Figure 8.2, which has been constructed to be numerically consistent with Figure 8.1.

Vibration Severity Guidelines

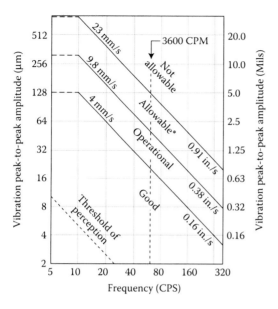

FIGURE 8.1 Bearing cap vibration displacement guideline.

The same vibration severity guideline is embodied in Figures 8.1 and 8.2, and is typical of severity levels now being used for many years to evaluate large turbo-machinery, especially in power plants. Again, one clearly sees the appeal of using *velocity severity levels*, since a particular velocity peak value has the same severity interpretation over the entire frequency range of concern for most plant machinery.

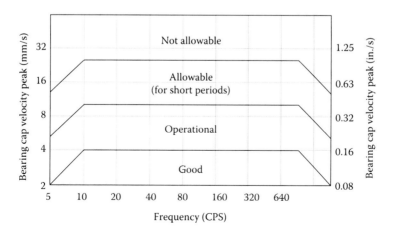

FIGURE 8.2 Bearing cap vibration velocity peak guideline.

8.3 Standards, Guidelines, and Acceptance Criteria

The many standards, guidelines, and acceptance criteria for rotating machinery vibration levels can be a source of confusion for those charged with making plant decisions based on assessing vibration severity in operating machines. The potential for confusion is enhanced by the presence of many governing criteria from several independent groups, as comprehensively surveyed by Eshleman (1976). There are international groups such as the International Standards Organization (ISO) and the International Electrical Commission (IEC). There are nongovernment national organizations such as the American National Standards Institute (ANSI). There are industry trade organizations such as the National Electrical Manufacturers Association (NEMA), the American Petroleum Institute (API), the American Gear Manufacturers Association (AGMA), the Compressed Air and Gas Association, and the Hydraulic Institute. Various engineering societies, such as the American Society of Mechanical Engineers (ASME), also have codes and standards for specific types and classes of rotating machinery that may include vibration criteria. Last but not least, standards and specifications that encompass "acceptable" vibration levels are also mandated by the biggest customer of them all, the U.S. government.

In this last category, the U.S. Navy is noteworthy because its vibration acceptance levels for rotating machinery are significantly lower than those of all the other major standards. Nongovernment user groups of rotating machinery know that to require vibration specification acceptance levels significantly lower than what the well-designed and well-maintained machine will comfortably settle into early in its operating life is a large waste of money. In other words, why pay a significant increase in the purchase price of a new machine so that it can be delivered with tested vibration levels significantly lower than what the machine will comfortably exhibit after a relatively short period of operation. This monetary dichotomy between government and nongovernment groups is of course a "bit" more inclusive than just machinery vibration specification acceptance levels.

Eshleman (1976) provides an excellent survey and comparison of the well-recognized machinery vibration severity-level guidelines and acceptance standards. In many cases, those guidelines and standards have been revised since Eshleman's survey was published. A Bibliography Supplement at the end of this chapter provides a more up-to-date listing, but one should keep in mind that revisions are an ongoing process.

To apply, as intended, a specific rotating machinery vibration criterion (guideline or standard) one must carefully study its documentation, because there are a number of important factors that are not uniformly handled across many guidelines and standards. For example, some criteria are based on the vibration *RMS average* (possibly filtered) and some are based on *single-peak* or *peak-to-peak* values. Some are based on *bearing cap* (or casing) vibration level while others are based on *shaft peak-to-peak displacement* relative to the bearing or *shaft total vibration*. Furthermore, the various criteria usually distinguish between the so-called *flexible supports* and *rigid supports*, that is, whether the support's lowest resonance frequency is below or above the operating speed of the machine. Also, the *relative mass* of the rotor to the stator is an important variable that significantly affects application of some criteria, although this is not always stipulated in various standards and guidelines. Whether from a vendor's or a purchaser's perspective, to help remove potential confusion for one who must apply a given guideline or standard, the article by Eshleman (1976) and the book by Mitchell (1981) provide good complementary reviews.

8.4 Shaft Displacement Criteria

Virtually all major turbo-machines using fluid-film bearings now have continuous monitoring of shaft orbital $x-y$ displacements relative to the bearings (see Figure 7.10). Turbo-machines have quite small radial rotor–stator clearance gaps, for example, at the journal bearings, annular seals, impeller rings, balance drums, and blade-tip clearances. Thus, rotor-to-stator vibration displacement is important for evaluation of turbo-machines' *vibration severity*.

By necessity, displacement transducers are located only near the bearings, because that is where there is access to the rotor and that is where the sensors can survive. Vibration displacement at midspan locations between the bearings would be more informative for vibration severity assessments, that is, small rotor–stator annular radial clearance gaps. But midspan locations are inaccessible and environmentally too hostile for proximity probes and cables to survive. Assessment of vibration severity levels from proximity probe displacement outputs at the bearings should therefore be interpreted with due consideration given to the vibration displacement mode shape of the rotor, such as from a rotor-response simulation (see Chapter 4). The extreme example to demonstrate this point is where rotor flexibility produces a rotor vibration displacement mode shape with nodal points near the bearings. Then the rotor vibration displacement amplitudes at the bearings are relatively small even when the midspan amplitudes are sufficiently high to cause accelerated wear at the small rotor–stator annular radial clearance gaps and to initiate a failure in the machine. Actually, it is a deficient design that operates continuously with vibration nodal points near the fluid-film radial bearings, because the fluid-film bearings are usually the primary source of rotor vibration damping, that is, there is no damping unless the dampers are "exercised." Thus, it is not difficult to visualize two contrasting machines where the one with a substantially higher rotor vibration at the bearings is the significantly "happier" machine than the one with a relatively low rotor vibration at the bearings.

As discussed in Section 7.1 (see Figure 7.4), one of the side benefits of a model-based diagnostics approach is the real-time combination of displacement measurements at the bearings with a simulation model observer to construct rotor vibration displacement signals at midspan locations. That is, the model-based approach provides "virtual sensors" at inaccessible rotor locations. As discussed in Chapter 9, measurement of rotor-to-stator vibration displacement adds considerably to the mix of valuable rotating machinery diagnostics information. However, its interpretation for severity assessment purposes is not as simple as is implied by the use of vibration levels based on bearing cap vibration, for example, Figures 8.1 and 8.2. A meaningful severity interpretation of rotor-to-stator vibration displacement measurements needs to be "calibrated" by information on the machine's vibration displacement shape of the rotor. The proliferation of severity standards and guidelines, as listed here in the Bibliography Supplement, is an attempt by builder and user groups to address this and other machine-specific severity-relevant differences.

One assessment of rotor-to-stator vibration displacement at the bearings is based on *time-varying bearing loads* derived from the measured journal-to-bearing displacement. A severity criterion can thereby be based on the *fatigue strength* of the bearing inner surface material (e.g., babbitt). McHugh (1983) showed a procedure using this approach. This approach, however, does not address the absence of midspan vibration displacement measurements. An experience-based guideline from Eshleman (1999) is tabulated as follows:

Speed (rpm)	Normal	Surveillance	Plan Shutdown	Immediate Shutdown
3600	$R/C < 0.3$	$0.3 < R/C < 0.5$	$0.5 < R/C < 0.7$	$R/C > 0.7$
10,000	$R/C < 0.2$	$0.2 < R/C < 0.4$	$0.4 < R/C < 0.6$	$R/C > 0.6$

Condition: R = peak-to-peak J-to-B displacement, C = dia. bearing clearance.

8.5 Summary

When supplying or purchasing a new machine, the allowable vibration levels mandated by the purchase specification provide definite values, whether based on a conservative or a not-so-conservative standard. However, the application of vibration severity guidelines can become a difficult "call" later on when the machine is out of warranty and its vibration levels have increased above the purchase specification level, but are still below the *alarm* or mandatory *shutdown* (trip) levels such as illustrated in Figure 7.1. The bearing cap vibration severity criteria contained in Figures 8.1 and 8.2 are not conservative compared to many purchase specification acceptance levels, but are realistic for subsequent operating criteria of many machines.

Rotor-to-stator vibration displacement measurements add considerably to the mix of valuable monitoring and diagnostics information, as described in Chapter 9. But its use for severity assessment purposes is not any simpler than severity criteria based on bearing cap vibration levels. A complete vibration severity interpretation of rotor-to-stator vibration displacement measurements needs "calibration" by information on the vibration displacement mode shape of the rotor so that midspan rotor-to-casing vibration displacement amplitudes can be reasonably estimated. For additional information on vibration severity levels in rotating machinery, Eshleman (1988) provides a practical and broad treatment.

Bibliography

Eshleman, R. L., "Vibration standards," in C. M. Harris and C. E. Crede (Eds.), *Shock & Vibration Handbook*, 2nd edn, McGraw-Hill, New York, 1976, pp. 19-1–19-15.
Eshleman, R. L., "Machinery condition analysis," *Vibrations*, 4(2), 3–11, 1988.
Eshleman, R. L., *Machinery Vibration Analysis II, Short Course, Vibration Institute*, 1999.
McHugh, J. D., "Estimating the severity of shaft vibrations within fluid–film journal bearings," *ASME Journal of Lubrication Technology*, Paper No. 82-Lub-1, 1983.
Mitchell, J. S., *An Introduction to Machinery Analysis & Monitoring*, PennWell, Tulsa, OK, pp. 374, 1981.
Rathbone, T. C., "Vibration tolerance," *Power Plant Engineering*, (43), 1939.
VDI, *Beurteilung der Einwerkung Machanischer Schwinggungen auf den Menschen*, VDI Standard, 2056, October 1964.

Bibliography Supplement

The following standards for rotating machinery vibration are from Chapter 19, *Shock & Vibration Handbook*, 4th edn, C. M. Harris (Ed.), 1996.

 1. ISO: *Mechanical Vibration of Machines with Operating Speeds from 10 to 200 rps-Basis for Specifying Standards*, ISO 2372.

2. ISO: *The Measurement & Evaluation of Vibration Severity of Large Rotating Machines in Situ; Operating@Speeds 10 to 200 rps*, ISO 3945.
3. ISO: *Mechanical Vibration of Rotating and Reciprocating Machinery-Requirement for Instruments for Measuring Vibration Severity*, ISO 2954.
4. ISO: *Mechanical Vibrations-Evaluation of Machinery by Measurements on Non-Rotating Parts*, ISO/DIS 10816 Series ISO/DIS 10816/1: "General Guidelines".
5. ISO/DIS 10816/2: *Guidelines for Land-Based Steam Turbine Sets in Excess of 50 MW*.
6. ISO/DIS 10816/3: *Guidelines for Coupled Industrial Machines with Nominal Power above 30 kW and Nominal Speeds between 120 and 15,000 rpm, when Measured In Situ*.
7. ISO/DIS 10816/4: *Guidelines for Gas Turbine Driven Sets Excluding Aircraft Derivatives*.
8. ISO/DIS 10816/5: *Guidelines for Hydraulic Machines with Nominal Speeds between 120 and 1800 rpm, Measured In Situ*.
9. ISO: *Mechanical Vibration of Reciprocating Machines—Measurements on Rotating Shafts and Evaluation*, ISO 7919 Series ISO 7919/1: "General Guidelines."
10. ISO 7919/2: *Guidelines for Large Land-Based Steam Turbine Generating Sets*.
11. ISO 7919/3: *Guidelines for Coupled Industrial Machines*.
12. ISO 7919/4: *Guidelines for Gas Turbines*.
13. ISO 7919/5: *Guidelines for Hydraulic Machine Sets*.
14. ISO: *Rotating Shaft Vibration Measuring Systems, Part 1: Relative and Absolute Signal Sensing of the Radial Vibration from Rotating Shafts*, ISO/CD 10817-1.
15. ISO: *Mechanical Vibration of Certain Rotating Electrical machinery with Shaft Heights Between 80 and 400 mm-Measurement and Evaluation of Vibration Severity*, ISO 2373.
16. NEMA: *Motor and Generators, Part 7—Mechanical Vibration-Measurement, Evaluation and Limits*, MG1–1993, Rev. 1.
17. API: *Form Wound Squirrel Cage Induction Motors-250 hp and Larger*, 3rd edn, API STD 541, 1994.
18. API: *Special Purpose Steam Turbines for Petroleum, Chemical, and Gas Industry Services*, 4th edn, API STD 612, December 1994.
19. API: *Special Purpose Gear Units for Refinery Services*, 4th edn, API STD 613, December 1994.
20. API: *Centrifugal Compressors for General Refinery Service*, 5th edn, API STD 617, April 1994.
21. API: *Rotary-Type Positive Displacement Compressors for General Refinery Services*, 3rd edn, API STD 619, May 1995.
22. API: *Centrifugal Pumps for General Refinery Service*, 8th edn, API STD 610, December 1994.
23. ISO: *Acceptance Code for Gears, Part 2: Determination of Mechanical Vibration of Gear Units During Acceptance Testing*, ISO/WD 8579-2.
24. Compressed Air and Gas Institute, Cleveland, Ohio: *In-Service Standards for Centrifugal Compressors*, 1963.
25. Hydraulic Institute: *Acceptance Field Vibration Limits for Horizontal Pumps*, 14th edn, Centrifugal Pump Applications.

26. ISO: *Mechanical Vibration and Shock-Guidelines for the Overall Evaluation of Vibration in Merchant Ships*, ISO 6954, 1984.
27. ISO: *Guide for the Evaluation of Human Exposure to Whole-Body Vibration*, ISO 2631.
28. ISO: *Code for the Measurement and Reporting of Ship-Board Vibration Data*, ISO 4867.
29. ISO: *Code for the Measurement and Reporting of Ship-Board Local Vibration Data*, ISO 4868.
30. ISO: *Mechanical Vibration-Vibration Testing Requirements for Shipboard Equipment and Military Components*, ISO/DIS 10055.
31. Military Standard: *Mechanical Vibration of Shipboard Equipment (Type 1-Enironmental and Type 2-Internally Excited)*, MIL-STD-167–1 (SHIPS), May 1, 1974.

9
Signal Analysis and Identification of Vibration Causes

9.1 Introduction

The most fundamental assessment of monitored rotating machinery vibration is, of course, provided by the ever important factors illustrated in Figure 7.1. That is, *how large* is the vibration and how is it *trending*. For those who are operating a vibration problem machine on a day-to-day basis, these two pieces of vibration information are often all that the operators use to make operational decisions. However, over the last 40 years the quite advantageous application of the FFT spectrum analysis in troubleshooting rotating machinery vibration problems has sensitized the general community of plant operators to the considerable value of vibration signal analyses in diagnosing the source of vibration problems whose solutions would otherwise be far more elusive. As illustrated in Figure 7.14, portable hand-held vibration analyzer data loggers are now pretty standard maintenance and troubleshooting tools in many types of machinery-intensive plants. There are other less commonly used signal analysis tools that are now beginning to find their way into rotating machinery vibration analysis. This chapter has the dual objectives of (a) introducing the frequently used and presently emerging machinery vibration signal analysis tools and (b) explaining the use of these tools combined with accumulated industry-wide experience in order to help identify specific sources of vibration.

9.2 Vibration Trending and Baselines

Even in the healthiest operating machines, monitored vibration signals may tend to migrate in amplitude and phase angle, even while remaining within a *baseline* "envelope" of acceptable vibration levels. Such benign changes are the normal effects of changes in operating conditions, for example, thermal transients, load changes, river circ-water temperature (seasonal changes), normal wear, and other fluctuations in the machine's overall environment. On the other hand, when the monitored vibration signals begin to grow in amplitude beyond the established *baseline* levels for a given machine, that *trend* should be carefully followed by the plant operators to continually assess the potential need for (a) *temporary changes* of the machine's *operating conditions*, (b) *scheduling an early outage* of the machine for corrective actions, or (c) an *immediate shutdown* dictated by rapidly increasing vibration levels. When a machine's vibration levels begin to grow beyond its established baseline levels, some problem within the machine begins to emerge and growth in vibration levels is often not the only symptom of the underlying problem. Once attention is focused on a machine beginning to show an upward trend in vibration levels, various *vibration signal analysis* tools are then commonly used to seek *identification of the root cause(s)*. Frequently used and presently emerging machinery vibration signal analysis tools are introduced in the next sections of this chapter, followed by a section on the use of these tools to help identify specific sources of vibration.

9.3 FFT Spectrum

The mid-1960s invention of the fast Fourier transformation (FFT) algorithm made feasible the modern real-time *spectrum analyzer*, which transforms time-varying signals from the time domain into the frequency domain, and thereby provides a continuously updated on-the-spot picture of a signal's frequency makeup. In modern times prior to FFT spectrum analyzers, the primary real-time on-the-spot display of vibration signals was in their natural time domain, typically using an oscilloscope.

The mathematical basis for spectrum analysis is the Fourier integral, which was provided by the mathematician Joseph Fourier in the early 1800s, long before the advent of modern rotating machinery. However, in modern times prior to the FFT algorithm, which utilizes modern digital computational methods, the transform of a measured time-base signal into the frequency domain required costly "off-line" processing with a slow turnaround. Specifically, a taped recording of the analog signal was processed through several narrow bandwidth analog filters (Section 7.4 of Chapter 7) with center-band frequencies spanning the relevant frequency range. Pre-FFT spectrum analyzers were cumbersome pieces of electronics equipment to operate successfully, requiring a technician experienced in how to tune and adjust the bandwidth filters to achieve optimum results for a given time-base signal record. Understandably, pre-FFT spectrum analysis was very sparingly used. The mathematical details of Fourier series, Fourier integrals, and FFTs are now standard parts of the mathematics component in college engineering curricula, and are well covered in numerous math and engineering analysis textbooks and handbooks. In the interest of space and brevity, these mathematical details are not covered here. Instead, a more heuristic explanation of spectrum analysis is given here to aid the machinery vibration practitioner in understanding the direct connection between a time-base signal and its frequency spectrum.

The practical underlying idea of the Fourier transform (FT) is that a function (e.g., time-base signal) can be constructed from a summation of sinusoidal functions with a continuous distribution of frequency from zero to a suitable cut-off frequency. For a periodically repeating signal or a defined period, a simpler, more restrictive version of this (the so-called Fourier series) is applicable and sums sinusoidal components only at a discrete set of frequencies that are the integer multiples ($n = 1, 2, \ldots$) of a designated base frequency $\Omega_1 \equiv 2\pi/\tau$, where τ is the duration of one period. Although machinery vibration signatures often contain only a limited number of significant harmonic components, their frequencies are often not all integer multiples of a single base frequency, and therefore the FT, not the Fourier series, is the appropriate tool to map rotating machinery vibration signals from the time domain into the frequency domain.

Figure 9.1, fashioned after a similar illustration in Mitchell (1981), provides a visual connection between a function of time, $X(t)$, and its FT or frequency spectrum. As illustrated, only a few harmonics added together readily produce a time trace from which it is difficult to directly view or identify individual contributing sinusoidal components. By transforming the signal into the frequency domain, the contributing components are readily identified.

Since the development of the FFT algorithm, spectrum analysis of time-base signals has permeated many fields of investigation, especially in diagnosing and troubleshooting vibration problems. In teaching students

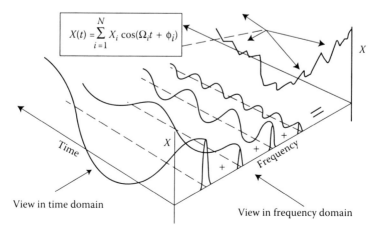

FIGURE 9.1 Illustration of an oscillatory signal's frequency spectrum.

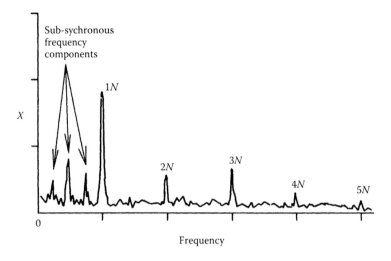

FIGURE 9.2 Example spectrum of a rotating machinery vibration signal.

the practical insight of using FFT signal analysis in vibration problems, the author uses the analogy of the modern paint color mixing apparatus used in retail paint stores. The frequency spectrum of a time-base signal is analogous to a virtually instantaneous process that would identify all the base color components and their respective proportions from a sample of an already mixed paint. Adding up a known ensemble of sinusoidal functions is analogous to adding the prescribed proportions of each base color for a given paint specification, whereas obtaining the frequency spectrum of a multicomponent time-base signal is analogous to figuring out the color components/proportions from the already mixed paint.

The spectrum of a vibration signal measured on a rotating machine is typified by the example in Figure 9.2. The $1N$ (once-per-revolution or synchronous) frequency component is often the largest because of the ever present residual rotor mass unbalance. Harmonic components with frequencies that are integer multiples ($2N, 3N, \ldots$) of the rotational speed are frequently present, usually at relatively small amplitudes. Harmonics at subsynchronous frequencies are also often encountered, from a small percentage of the rotational speed to only slightly less than the $1N$ component.

9.4 Rotor Orbit Trajectories

The example, shown in Figure 9.2, of a frequency spectrum for a vibration signal can be based on *displacement, velocity,* or *acceleration.* As described in the last section of this chapter, there is now a considerable wealth of experience and insight accrued in using such spectra for diagnosing sources of machinery vibration problems. Specifically for *LRV*, rotor orbital vibration *displacement* trajectories provide an additional diagnostic information component for the troubleshooter to analyze when seeking to identify the nature and cause of a rotating machinery vibration problem. Several examples of rotor orbital vibration trajectories are illustrated and described in the earlier chapters of this book. In Chapter 4, which is essentially a user's manual for the RDA code supplied with this book, the discussions pertaining to Figures 4.4 through 4.10 provide a primer for this section. An understanding of the topics covered in Section 2.4 of Chapter 2 further facilitates the adroit use of rotor orbital displacement trajectories in identifying rotor vibration types and sources.

The primary method now widely used to measure rotor orbital displacement trajectories is the *inductance (eddy-current) noncontacting* position sensing system described in Section 7.2 of Chapter 7. Proximity probes for this type of system are commonly installed on major rotating machinery in power plants, petrochemical plants, and others for continuous monitoring and diagnostic purposes. As typified in Figure 7.9a, a pair of proximity probes positioned 90° apart are located at each of a number of accessible axial locations (usually at the radial bearings) such as illustrated in Figure 7.10. By feeding the conditioned output signals from an x–y pair of probes into the x and y amplifiers, respectively, of a dual-channel oscilloscope, the real-time rotor orbital trajectory can be displayed. Thus, one can measure and display in real time the rotor vibration orbits that are computationally simulated from rotor vibration models, as demonstrated in Chapter 4 using the RDA code.

Proximity probes are typically mounted at journal bearings and then measured orbits of the shaft are relative to the bearing (refer Figure 4.4). There is now a considerable wealth of accrued experience and insight in using the geometric properties of rotor vibration orbits for identifying the nature and source of machinery vibration problems. Furthermore, the presence of numerous quite small rotor-to-stator annular radial clearance gaps, such as in turbo-machinery, makes rotor-to-stator relative vibration displacement orbits important information in assessing the well-being of a machine.

In seeking to devise a suitably accurate computer-simulation model for troubleshooting purposes, comparisons between predicted and measured LRV orbits of a troubled machine is of course a proper scientific approach in "fine-tuning" a model for subsequent "what if" studies. However, it is important to realize that the noise-free single-frequency *elliptical orbits* from linear response computer simulations, such as those illustrated in Figures 4.5 through 4.9, are much "cleaner" pictures than is typically obtained for actual measured orbits from unfiltered displacement signals.

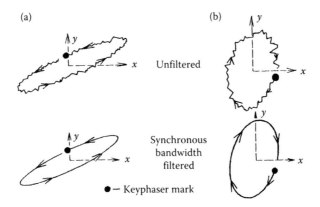

FIGURE 9.3 Illustrated examples of measured rotor synchronous orbits: (a) forward whirl and (b) backward whirl.

Low-pass filtering of raw signals to remove high-frequency components and noise is a first step in "cleaning up" the measured orbit display. Proximity probe signal processing instruments typically have a *tracking filter* option (Section 7.4 of Chapter 7) synchronized by the *keyphaser* signal (Figure 7.11) to track the rotational speed frequency, and thus provide a "clean" synchronous orbit picture, which is comparable to the noise-free single-frequency elliptical orbits from a corresponding computational simulation. Figure 9.3 illustrates two typical measured synchronous rotor vibration orbits, before and after synchronous bandwidth filtering.

When significant nonsynchronous orbit frequency components are present, synchronous bandwidth filtering is ill-advised in general troubleshooting because it removes the subsynchronous and higher harmonic components such as those captured in the Figure 9.2 FFT illustration. On the other hand, for rotor balancing purposes, since only synchronous vibration components are used in the balancing procedure, synchronous bandwidth filtering is naturally applicable. Figure 9.4 illustrates an orbit

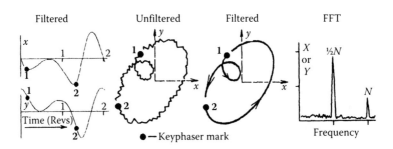

FIGURE 9.4 Measured orbit with synchronous and half synchronous components.

with a period of two revolutions, containing synchronous and half synchronous components, the orbit being typical for cases where the half synchronous component is largest.

Subsynchronous rotor vibrations are often associated with instability self-excited rotor vibrations. The example shown in Figure 9.4 is for the particular case where the subsynchronous component is exactly half the spin speed and thus, being a periodic motion, is easy to illustrate. In contrast, subsynchronous rotor vibrations are more often not at an integer fraction of the spin speed, and thus the motion is not strictly periodic. The real-time orbit display may then look similar to that in Figure 9.4, but will have an additional unsteady "bouncing" motion to it due to its nonperiodic character. An additional unsteadiness is also common because instability self-excited rotor vibrations are themselves often unsteady, as easily observed from a real-time continuously updated FFT that shows the subsynchronous component significantly changing its amplitude (up and down) from sample-to-sample on the spectrum analyzer display.

Referring to the example frequency spectrum in Figure 9.2, a word of caution is in order when observing integer harmonics of the synchronous frequency $(2N, 3N, ...)$ if the signals are proximity probe displacement signals. As described in Section 7.2 of Chapter 7, proximity probe systems produce a fictitious additive vibration component from "electrical run-out" caused by the shaft's circumferential variations in surface conditions, electrical conductivity, and permeability. It is now a standard procedure in plants to take a low-speed output signal for each probe, such as while the machine is slowly rotated on turning-gear mode, and to consider the once-per-rev component of that signal (amplitude and phase angle) as the electrical run-out component, which is then stored and automatically subtracted in real time from the raw signal in normal operation. However, based on the research of Horattas et al. (1997) discussed in Section 7.2 of Chapter 7, proximity probe electrical run-out signals are definitely far from sinusoidal, that is, not a single $1N$ harmonic. Thus, integer harmonics of the synchronous frequency $(2N, 3N, ...)$ are generally contaminated by electrical run-out components even when using the standard procedure, which may remove only the synchronous component of the electrical run-out. As shown by Horattas et al., digitally subtracting out all the electrical run-out removes all its harmonics, but this is not generally done in plants thus far. With present microprocessors it is easy to remove the first 5 or 10 electrical run-out harmonics in real time, and this feature has more recently made its way into the monitoring and diagnostic function now becoming utilized intensively in *predictive maintenance* and *troubleshooting*.

Having reliable vibration baseline data for a machine, subsequent incremental changes to integer-multiple harmonics of spin speed can reasonably be attributed to changes in the vibration spectrum apart from any contamination originating with the "electrical run-out" harmonics. When a machine's vibration levels are within safe conservative levels and running in good condition, the rotor vibration is likely to be well characterized by linear dynamic behavior and integer-multiple harmonics of the spin speed are then more likely to be relatively quite small. When single-frequency dynamic linearity predominates, the rotor orbits are essentially synchronous ellipses. But when a significant nonlinear influence manifests itself, one or more higher harmonics $(2N, 3N, \ldots)$ and one or more subharmonics $(N/2, N/3, \ldots)$ in the synchronous frequency range can become significant. A number of different abnormal conditions can give rise to significant dynamic nonlinearity in the rotor dynamical system, as described in Section 2.5 of Chapter 2. Therefore, the vibration harmonics of spin speed can often be valuable information utilized in troubleshooting rotating machinery problems.

Figure 9.5 illustrates an important example where some progressively worsening influence in the machine causes a progressively increasing static radial force on the rotor (and thus bearings), which leads to increased nonlinear behavior. For example, increased static radial load can develop on a centrifugal pump impeller (Section 6.1 of Chapter 6) as internal stationary vanes and/or impeller vanes become damaged, for example, by cavitation, improper operation of the pump, or poor hydraulic design. Internally generated static radial loads act similar to internal radial misalignment, for example, from casing thermal distortions.

Basically what Figure 9.5 illustrates is what can happen to the normal unbalance forced vibration of the shaft orbital motion due to increased

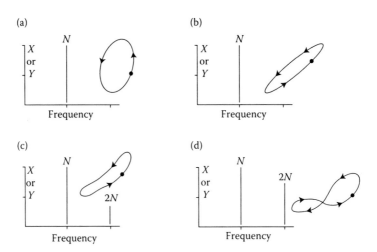

FIGURE 9.5 Filtered orbit and FFT with increasing radial load or misalignment. (a) Nominal radial load; synchronous linear motion. (b) Moderate radial load increase; synchronous linear motion. (c) Substantial radial load; nonlinear motion with some $2N$. (d) Very high radial load; nonlinear motion with high $2N$.

journal bearing dynamic nonlinearity as a progressively worsening static radial load and/or misalignment emerges over time. As the shaft orbit changes from the normal elliptical shape to a "banana" shape to a "figure eight" shape, the progressive increase of the 2N harmonic is reflected. Similar distortions to the normal elliptical orbits can also result from other higher order harmonics (3N, 4N, ...) of spin frequency. Higher journal bearing static loads produce higher journal-to-bearing eccentricity, resulting in increased bearing film dynamic nonlinearity. Therefore, an emerging rich spectrum of higher harmonics can be an indication of excessive radial bearing loads and/or misalignments.

As stated earlier in this section, subharmonics (N/2, N/3, ...) of the spin frequency can be present with rotor dynamical *nonlinearity* and an important example of this is an extension of the "story" already told by the Figure 9.5 filtered signals, which typify worsening conditions as monitored at normal operating speed. What likely happens to orbital vibration of such a machine when it goes through speed coast-down is illustrated in Figure 9.6, fashioned after a case in Bently and Muszynska (1996).

The journal vibration orbit in Figure 9.6a is a slightly different version of the "figure eight" orbit in Figure 9.5d, resulting from a high degree of bearing film dynamic nonlinearity associated with high static radial bearing load and/or misalignment. As this machine coasted down (see Figure 9.10), a significant increase in overall rotor vibration levels was encountered between 4500 and 4100 rpm, as shown by Bently and Muszynska (1996). The orbit and its FFT spectrum at 4264 rpm are a clear "picture" of what is

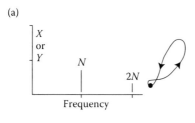

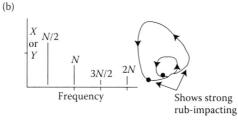

FIGURE 9.6 Changes with high radial load or misalignment during coast-down. (a) Filtered journal orbit and FFT at normal operating speed (5413 rpm). (b) Filtered journal orbit and FFT during coast-down at 4264 rpm.

occurring in the 4500–4100 rpm speed range. At 4264 rpm, the spin speed traverses twice the 2132 cpm unbalance resonance frequency, that is, half the spin speed. Significant bearing dynamic nonlinearity from high radial static load and/or misalignment plus the inherent characteristics of the journal bearings to have low damping for subsynchronous frequencies combine to produce a dominant $\frac{1}{2}N$ subharmonic vibration component in the speed neighborhood of 4264 rpm. As the orbit in Figure 9.6b clearly shows, the significant increase in the overall rotor vibration level near this speed causes strong rotor–stator rub-impacting at the monitored bearing journal or a nearby seal. Once rub-impacting occurs, the dynamic nonlinearity increases even further, synergistically working to maximize the $\frac{1}{2}N$ vibration amplitude through what is tantamount to a so-called *nonlinear jump phenomenon*, similar to that analyzed by Adams and McCloskey (1984) and shown here in Figure 2.18.

Both the orbital vibration illustrated in Figure 9.4 and the field measurement case shown in Figure 9.6b have a period of two revolutions, but are significantly different in a fundamental way. In Figure 9.4, the illustrated motion is the simple summation of two harmonic motions that have exactly a 2:1 frequency ratio. In the Figure 9.6b case however, the motion has a rich spectrum of harmonics of the $N/2$ component, because the sharp redirection of the orbital trajectory in and out of the rub-impacting zone needs several terms in its Fourier series to accurately add up to the orbit shape. The spectrum shown in Figure 9.6b has been truncated beyond the $2N$ component, but the actual spectrum is richer.

It has been well known for over 50 years that large two-pole alternate current (AC) generators with relatively long bearing spans, having significant mid-span static deflection, must have a series of *radial slots* cut along the generator rotor to make its radial static deflection characteristic as close to *isotropic* as is practical (Chapter 12). Otherwise, an intolerably high $2N$ vibration would occur, as inadvertently discovered on early large steam turbine generators in the 1950s. A similar anisotropic rotor stiffness develops when a crack has propagated part way through the shaft, and so proper capture of the $2N$ component of rotor vibration can provide a primary symptom of a cracked rotor. Furthermore, trending the $2N$ component over time can aid in assessing the propagation rate and extent of the crack, for example, how many more hours or days will it take for the shaft to fail. Muszynska (1995) provides insight on this by tracking the $1N$ and $2N$ rotor vibration measurements on rotors with slowly propagating cracks, showing how it is possible in some cases to make an early detection of a slowly propagating material crack through the shaft. Muszynska describes a troubleshooting case study, using *orbits*, combined with the additional tools of *Bode*, *polar*, and *cascade* plots.

Acquiring rotor vibration orbits with permanently installed proximity probe systems for continuous vibration monitoring on major machinery, as typified in Figure 7.10, has a monetary cost not deemed justified for many other rotating machines, although many older machines (e.g., large steam turbine generator units) with valuable remaining operating lives are now retrofitted with proximity probe shaft vibration monitoring systems. More typically, a check of vibration characteristics is routinely collected at regular time intervals using portable hand-held vibration analysis/data logger units as illustrated in Figure 7.14. When vibration problems are detected however, effective troubleshooting can usually be significantly helped by temporarily installing x–y proximity probes at one or more accessible shaft locations. This is frequently implemented by the author and his co-workers in troubleshooting plant machinery excessive vibration problems.

9.5 Bode, Polar, and Spectrum Cascade Plots

The term *Bode diagram* is from the field of feedback control, being a plot of phase angle between harmonic input and output signals versus frequency. Many in rotating machinery vibration have adopted this term to describe steady-state vibration response *amplitude* and *phase angle* versus rotational *speed*. The well-known plot of steady-state vibration amplitude and phase angle versus frequency for a 1-DOF system excited by a sinusoidal force is shown in Figure 1.5, and could be similarly labeled as its Bode diagram. Figure 9.7 illustrates a "Bode plot" of a rotor vibration signal's steady-state response during a gradual roll-up to operating speed or a coast-down from operating speed, as it passes through a critical speed near 1000 rpm. Similar to the steady-state response for the harmonically excited underdamped 1-DOF system (Figure 1.5), passage through critical speeds is typically characterized by a local peak in vibration amplitude and a distinct phase angle shift (180° for the simple 1-DOF system). As covered in Section 1.3, in underdamped natural modes of multi-DOF systems, each mode behaves similar to the 1-DOF system and so the similarity in steady-state response is natural. Rotor unbalances, which excite critical speeds, are in fact like synchronous harmonic excitation forces and differ from the 1-DOF harmonic excitation force only in that the unbalances' force magnitudes vary as the square of the rotational speed (i.e., unbalance force magnitudes $\propto \omega^2$).

On complex machines, phase angle shifts through critical speeds may not be as close to a 180° shift as shown in Figure 9.7. For the 1-DOF system, the phase angle in Figure 1.5 is the phase lag of the steady-state vibration harmonic displacement behind the harmonic excitation force, and thus shifts from 0° to 180° versus frequency through the natural frequency. On the other hand, unbalance synchronous response signals of rotors are time (phase) referenced to a specified angular location on the

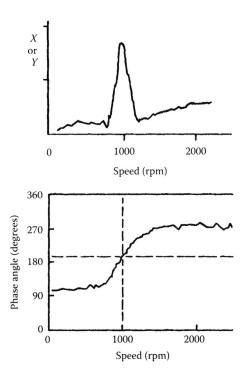

FIGURE 9.7 Bode plot of a rotor vibration signal on speed-up or coast-down.

rotor (keyphaser mark), and thus the phase shift through resonance is not specifically referenced to zero. On very well-balanced rotors, the shift in phase angle through a critical speed is often easier to detect than the speed at which the rotor vibration peaks.

The same information plotted in Figure 9.7 is replotted in polar form in Figure 9.8. The *polar plot* of steady-state vibration is a compact and visually revealing way to present vibration measurements as a function of rotational speed, as a function of time, or as a function of some other parameter in which vibration changes are to be analyzed (see the example in Figure 12.2).

A *cascade plot* (or water fall plot) is a contour-map presentation of vibration *amplitude* (contour elevation) versus *frequency* (horizontal axis) versus spin *speed* (vertical axis), providing an insightful and revealing picture of a machine's rotor vibration characteristics over its entire speed range. Cascade plots can be "busy" when a multitude of vibration frequencies are present. Figure 9.9 shows a *cascade plot* that is interesting but not overly busy.

Figure 9.9 shows a typical case of encroachment upon the oil-whip threshold speed, discussed at length in Section 2.4 of Chapter 2. As is typical, the subsynchronous self-excited rotor vibration mode is the same

Signal Analysis and Identification of Vibration Causes 319

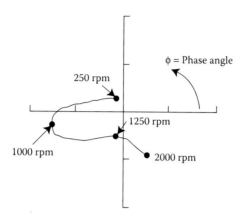

FIGURE 9.8 Polar plot of steady-state vibration amplitude and phase angle.

forward whirling mode as that synchronously excited by unbalance at the first critical speed. The first critical speed mode does not necessarily have exactly the same frequency as when it becomes the self-excited mode at the oil-whip threshold speed, for two reasons. First, the journal bearing effective oil-film stiffness will be somewhat different at the oil-whip threshold

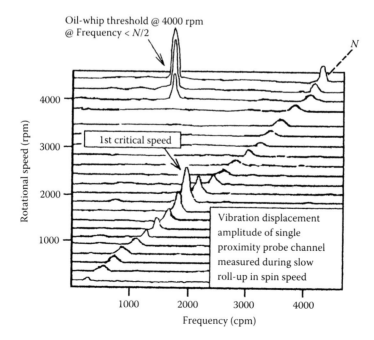

FIGURE 9.9 Cascade of passage through first critical speed to oil whip.

speed than at the critical speed. Second, when oil whip occurs, the typically high orbital vibration amplitudes may produce a frequency increase consistent with achieving a nonlinear limit cycle if the journal orbit fills up most of the bearing clearance circle. As the plot in Figure 9.9 shows, once oil whip is initiated as speed is increased, the oil-whip whirl frequency stays locked onto the self-excited mode's frequency.

A second type of *cascade plot*, fashioned after an example of Bently and Muszynska (1995), is shown in Figure 9.10. It is the same coast-down case as was previously discussed from the rotor orbits and FFTs presented in Figure 9.6. The significant vibration increase between 4500 and 4100 rpm is dominated by the $N/2$ component, as Figure 9.10 also clearly shows.

The derivation that accompanies Figure 4.6 shows that any harmonic orbit (i.e., ellipse) can be composed of a forward-whirl circular orbit and a backward-whirl circular orbit of the same frequency. The cascade plot in Figure 9.10 delineates the forward and backward circular-whirl components for each harmonic, in contrast to Figure 9.9, which is the more common cascade plot that is based on a single time-base signal. By comparing the relative amplitudes of the forward and backward components, the rotor whirl direction for a given harmonic at a specific speed is apparent. Also for a particular harmonic and speed, the major and minor axes of the corresponding orbit ellipse are apparent since the major axis is the sum of the two circular orbit radii and the minor axis is the absolute value of their difference. Thus, the type of cascade plot illustrated in Figure 9.10 is an excellent way to include orbit characteristics, making the "picture"

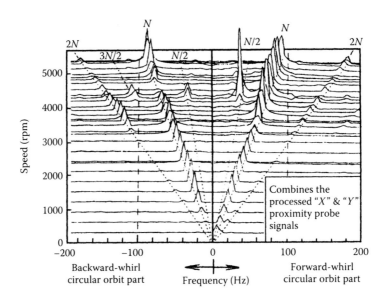

FIGURE 9.10 Cascade for field measurement case shown in Figure 9.6.

complete. A final interesting observation can be made from the actual coast-down case shown in Figure 9.10. That is, when the rub-impacting occurs around 4264 rpm, Figure 9.6b, the $3N/2$ harmonic is essentially present only in the backward-whirl circular orbit component.

The next two sections treat advanced signal analysis methods that have not yet been widely applied in industrial applications, but are described here to show where future machine *condition analysis* innovations lie.

9.6 Wavelet Analysis Tools

Over the last 30 years *wavelets*, which are also called wavelet transforms (*WTs*), have emerged through a confluence of ideas and techniques from such diverse fields as pure mathematics, quantum physics, and electrical engineering. The collection of theory and computational methods now known by the label *wavelets* is now a mature topic in some cutting-edge applications. Some specific applications include (i) computer vision systems that process variations in light intensity at several resolutions levels, similar to how animal and human vision is now postulated to function, (ii) digital data compression of human finger print images, (iii) denoising contaminated time-base signals, (iv) detecting self-similar behavior patterns in time-base signals over a wide range of time scales, (v) sound synthesis, and (vi) photo image enhancements. A number of books on wavelets are now published, but many of these are suitable reading primarily for applied mathematicians and signal analysis specialists. The author has found a few publications that are potentially fathomable by the more mathematically inclined engineers, and these include the article by Graps (1995) and the books by Chui (1997) and Kaiser (1994). Currently marketed machinery condition monitoring systems do not typically utilize *WTs*, but the author's exposure to *WTs* has led him to believe that the capability of next-generation machinery condition monitoring systems would be considerably advanced by their use, once wavelets and their advantages are familiar to machinery vibration engineers. The important implications of *wavelets* for future rotating machinery vibration-based troubleshooting justifies including here a short readable description of *WTs*, to introduce vibration engineers to the topic.

WTs are a powerful extension of the *FT*, the basis for FFT generated spectra. Computationally fast numerical algorithms are readily available for WTs as well, that is, fast wavelet transform (FWT) and are quite similar in their details to FFT algorithms. These two types of transforms are also similar in a fundamental way, and thus the description here of wavelets is keyed to the similarity between the FT and WTs.

When frequency content (spectrum) of a time-base function $x(t)$ is of interest, the first inclination is to compute the FT, which is represented by a complex function of frequency, $X(\omega)$, used to describe the amplitude and phase angle of a sinusoid. This is expressed as follows

$$X(\omega) = \int_{-\infty}^{\infty} x(t)e^{-i\omega t}\, dt \qquad (9.1)$$

Just as the various harmonic frequency components are difficult to see at a glance of a time signal $x(t)$, the time-base information contained in $X(\omega)$ is difficult to see because it is hidden in the phase of $X(\omega)$. The desirability of time localizing spectrum information (*time–frequency localization*) in some signal analysis applications has made the *windowed Fourier transform* (WFT) the primary tool for such needs ever since the development of the FFT algorithm. Time–frequency localization is similar to the music notes written on a sheet of music, which show the musician when (time information) to play which notes (frequency information). A WT is a time–frequency localization, and thus reviewing the WFT is a first stepping stone to understanding wavelets; see Daubechies (1993).

The function $x(t)$ is windowed by multiplying it by a time window function $w(t)$ of a specified time duration t_0, usually with smooth edges. This lifts out the piece of $x(t)$ for the time interval prescribed by the window function, as illustrated in Figure 9.11; similar to Daubechies (1993).

This process is successively repeated to span the specified time range of $x(t)$, with each successive time window shifted by t_0 from the preceding window. Following from Equation 9.1, a WFT $X_{mn}(\omega)$ for each successive nth window is obtained for each of a succession of progressively higher frequency localizations as set in the following equation by the index m (ω_0 and t_0 fixed):

$$X_{mn}(\omega) = \int_{nt_0-\Delta T}^{nt_0+\Delta T} x(t)w(t-nt_0)e^{-im\omega_0 t}\, dt \qquad (9.2)$$

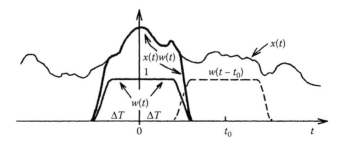

FIGURE 9.11 Time windowing of a function for WFT.

Signal Analysis and Identification of Vibration Causes

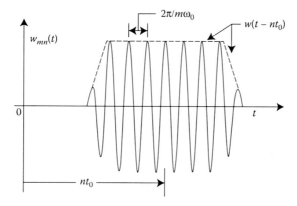

FIGURE 9.12 Illustration of a member of the function family $w_{mn}(t)$.

$\Delta T = t_0/2$, m and n are the real integers $\pm n = 0, 1, 2, \ldots, N$ and $\pm m = 0, 1, 2, \ldots, M$.

The family of WFTs given by Equation 9.2 can be considered as *inner products* of $x(t)$ with the family of functions $w_{mn}(t)$ defined in the following equation:

$$w_{mn}(t) = e^{-im\omega_0 t} w(t - nt_0) \qquad (9.3)$$

As illustrated in Figure 9.12, $w_{mn}(t)$ is a windowed sinusoidal oscillation with a frequency of $m\omega_0$, an amplitude of w, and a time shift of nt_0.

As shown generically in the following equation, a WT W_{mn} is similar to a WFT because it is formed from an inner product of $x(t)$ with a sequence of wavelet functions $\Psi_{mn}(t)$, where m and n likewise indicate frequency and time localizations, respectively:

$$W_{mn} = \int_{t_1}^{t_2} x(t) \Psi_{m,n}(t) \, dt \qquad (9.4)$$

The wavelet functions $\Psi_{mn}(t)$ are localized in time and frequency, similar to the windowed functions $w_{mn}(t)$ in the WFT. Also similar to $w_{mn}(t)$, each $\Psi_{mn}(t)$ integrates to zero over its respective time duration, which means it has at least some oscillations. However, wavelet functions $\Psi_{mn}(t)$ have a basic property that sets them apart from the WFT functions $w_{mn}(t)$. $\Psi_{mn}(t)$ are generated so that the time window interval width is inversely proportional to the localized frequency and thus translated proportionally to its width. In contrast, a given set of WFT $w_{mn}(t)$ functions all have the same width of time interval. This basic property of wavelets makes them ideally suited to analyze signals having highly concentrated (time localized) high-frequency components superimposed on longer lived low-frequency components. This property of wavelets is observed from their basic mathematical specification, given in the following equation and delineated in the next paragraph:

$$\Psi_{mn}(t) = a_0^{-m/2} \Psi(a_0^{-m} t - nb_0) \quad (9.5)$$

Similar to t_0 and ω_0 in Equation 9.2, a_0 and $b_0 > 0$ are fixed, and m and n are as specified for Equation 9.2. For $m > 0$ (higher frequency oscillations in Ψ) the oscillations are packed into a smaller time width, whereas for $m < 0$ (lower frequency oscillations) the oscillations are packed into a larger time width. For a given m, the Ψ_{mn} are time translates of Ψ_{mn} ($n = 0$), with each successive time-shift translation of magnitude $na_0^m b_0$. In the terminology of signal analysis specialists, this generation of a family of wavelet functions Ψ_{mn} is said to be formed by a sequence of *dilations* and *translations* of the *analyzing wave* or *mother wave*, $\Psi(t)$. There are now several recognized "mother waves," usually named for their respective originators. Furthermore, unique mother waves can be formulated for optimum suitability to specific applications, which is an additional advantage of WTs over the FT. Figure 9.13 shows a fairly simple mother wave and two example wavelets generated from it.

The well-known intermittent nature (i.e., nonstationary), often at high frequencies, of rotor vibration signal content symptomatic of a number of specific problems or incipient machine failure phenomena has induced the author to believe that the capability of next-generation machinery condition monitoring systems will be considerably advanced by the use of FWT to augment the present heavy reliance on the FFT. Figure 9.14 presents a visual comparison between *WT* and *WFT*, which clearly delineates the two.

Clearly, WT inherently possesses the simultaneous capabilities of isolating signal discontinuities with high resolution along with detailed frequency analysis over longer time windows. This frequency localization causes many time-base signals to be sparse when transformed into the wavelet domain. This feature yields wavelets quite useful for *feature detection*, signal *noise removal*, and *data compression*. These capabilities are all obvious potential advantages for next-generation machinery condition

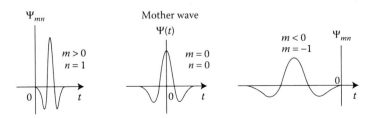

FIGURE 9.13 Two wavelets, which are dilations and translations of $\Psi(t)$.

Signal Analysis and Identification of Vibration Causes

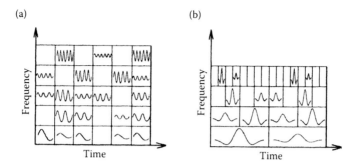

FIGURE 9.14 Time–frequency plane, comparing FT and WT: (a) WFT and (b) WT.

monitoring systems. In contrast to wavelets, the WFT has a single time window interval for all frequencies, and thus the degree of resolution with the WFT is the same at all time–frequency locations.

9.7 Chaos Analysis Tools

WT has now become a signal processing and analysis tool in many applications. Although not yet a standard tool in condition monitoring systems for rotating machinery vibration, wavelets frequently appear in some research and application papers on rotor vibration. In contrast, the use of *chaos* analysis tools in rotor vibration signal analysis is still pretty much the domain of a few researchers in academia, although *chaos* analysis tools are being used in other fields such as in medical research for analyzing monitored heart beat signals. Adams and Abu-Mahfouz (1994) provide an introduction to chaos concepts for analyzing rotating machinery vibration signals. They employ computer simulations to demonstrate how chaos signal analysis techniques can detect some important rotating machinery conditions that are not readily detectable by any of the standard signal analysis tools.

Two of the several models investigated by Adams and Abu-Mahfouz are illustrated in Figure 9.15. The necessary, though not sufficient, requirement for chaotic motion to occur in a dynamical system is *nonlinearity*. Interestingly, many in-process failure mechanisms and various adverse operating conditions in rotating machinery involve significant nonlinear dynamical properties. Section 2.5 of Chapter 2 gives an introduction to nonlinear rotor dynamics. The exploratory work of Adams and Abu-Mahfouz (1994) exposes an abundance of interesting possibilities for machinery condition feature detection using signal mappings that are regularly employed by chaos specialists in their work. Just a few of these are presented here for the two models shown in Figure 9.15.

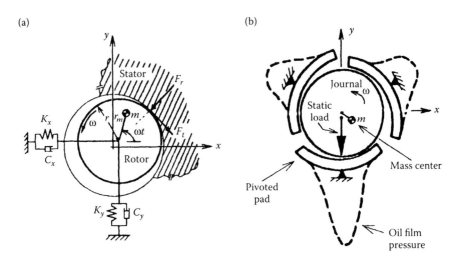

FIGURE 9.15 Simple rotor dynamics models used for chaos studies: (a) unbalance excited rub-impact and (b) three-pad pivoted-pad bearing.

Figure 9.16 presents simulation results for the unbalance-excited rub-impact model illustrated in Figure 9.15a, and shows a confluence of rotor orbital trajectories and their mappings using some typical chaos signal processing tools. The central portion of Figure 9.16 is a *bifurcation diagram*, which here plots the rotor orbit's x-coordinate (with a dot) for each shaft revolution as the reference mark fixed on the rotor passes the same rotational position angle. If the orbital motion is strictly synchronous, only the same dot appears, repeatedly. If a half-synchronous subharmonic component is superimposed, then only the same two dots repeatedly appear. Similarly, the *Poincaré maps* in Figure 9.16 contain a dot deposited for the orbit's (x, y) position at each shaft revolution as the reference mark fixed on the rotor passes the same rotational position angle. The term *quasiperiodic* is used by chaos specialists and others to label nonperiodic signals that are comprised of incommensurate (noninteger related) periodic signals.

Spanning the range of nondimensional unbalance shown in Figure 9.16, the orbital motion goes from quasiperiodic to period-5 motion (five revolutions to complete one period), bifurcates into period-10 motion, becomes chaotic (nonperiodic but nonrandom), and finally emerges from, the chaos zone as period-8 motion. To the author, this is extremely interesting. On the periodic orbits shown, a fat dot is deposited at each keyphaser mark, thus period-5, period-10, and period-8 orbits have 5, 10, and 8 marks, respectively. These dots deposited on the periodic orbits are, by themselves, the Poincaré maps of their respective motion orbits. Thus, for *any periodic motion*, the Poincaré map is a limited number of dots equal in number to the number of revolutions per period of the motion. For a *quasiperiodic motion*,

Signal Analysis and Identification of Vibration Causes

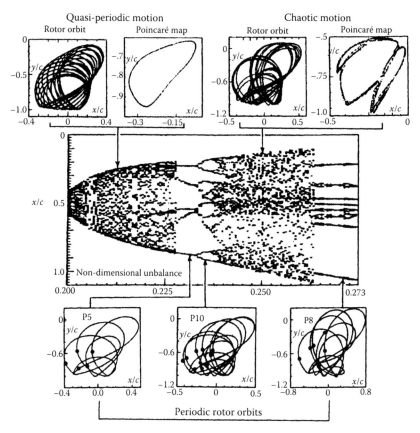

FIGURE 9.16 Rotor orbits and chaos-tool mappings for the unbalance-excited rub-impact simulation model in Figure 9.15a.

the dots on the Poincaré map over time fill in one or more closed loops, with the number of loops equal to the number of superimposed incommensurate periodic components minus one (i.e., $N - 1$). Thus, the quasiperiodic orbit shown in Figure 9.16 has two incommensurate periodic components, thus one loop.

For *chaotic motion*, the Poincaré map has a *fractal* nature to it and therefore has a fuzzy appearance, as displayed for the chaotic orbit in Figure 9.16. There are mathematical algorithms to compute a scalar dimension of such a fractal pattern, as detailed by Abu-Mahfouz (1993). In general, the fuzzier the map, the higher the fractal scalar dimension and the higher the degree of chaos content in the motion. Signal noise content also can produce Poincaré map fuzziness, even without chaos content. Thus, special filtering methods must be employed to remove the noise without removing the chaos content. For this, the author has used in his laboratory model-based observers (see Section 7.1 discussion concerning Figure 7.4) and signal-threshold de-noising in the signal WTs to reconstruct the de-noised signal. Clearly, the example simulation results shown in Figure 9.16 indicate that chaos signal processing tools have a definite potential to significantly enhance the capability of future vibration-based rotating machinery condition monitoring.

Figure 9.17 shows some additional simulation results for the unbalance-excited rub-impact model shown in Figure 9.15a. These results

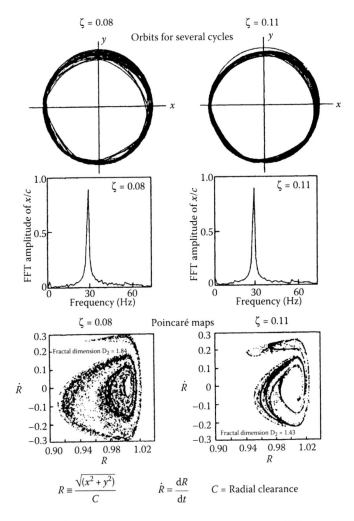

FIGURE 9.17 Poincaré mapping of chaotic response reveals small loss in damping at off-resonance condition (rub-impact model).

were generated to study the detection of small losses in damping capacity at off-resonance conditions. Since the effects of damping on vibration amplitudes are significant primarily at or near a resonance, off-resonance vibration amplitudes are not significantly affected by reduction in damping. As Figure 9.17 results clearly show, the fractal nature of the associated Poincaré maps for 8% and 11% critical damping cases has a quite measurable effect on the degree of chaos in the vibration. However, the FFT signatures for the two compared cases show virtually no difference. It is not difficult to relate the practical implication of early detection of damping loss to rotor vibration. For example, through some progressive deterioration process at a journal bearing, one can readily imagine a slowly progressing loss of damping that would not result in increased vibration levels at operating speed but would result in dangerously high rotor vibration levels on coast-down through critical speed(s).

Figure 9.18 shows one of the several interesting results on chaotic motion with pivoted-pad journal bearings presented by Adams and Abu-Mahfouz (1994). For the three-pad bearing illustrated in Figure 9.15b, with the static load directed into a pivot location (bottom), it is well known that the journal eccentricity will find a stable static equilibrium position on one side of the pivot or the other, but not exactly on the pivot (see Figure 5.3, Chapter 5). This property can be mitigated or even eliminated if the bearing is assembled with *preload* (see Figure 5.6), which is more often not done. The results illustrated in Figure 9.18 show the synchronous unbalance force causing a chaotic orbit with a zero bearing preload, but a more expected small synchronous orbit when a modest amount of preload (15%) is applied by adjusting the pivot clearance to 85% of the bearing's ground

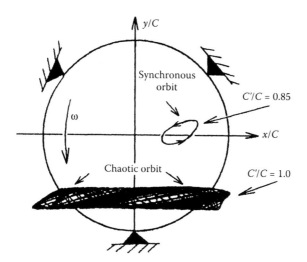

FIGURE 9.18 Chaotic rotor vibration originating in a pivoted-pad bearing.

clearance (C' = pivot clearance, C = bearing pad radius–journal radius). These results suggest some useful applications of chaos-tool signal analysis that can help diagnose such abnormal rotor vibration and other related operating problems. The fuller presentation of these results by Adams and Abu-Mahfouz show chaotic pitching motion of all three bearing pads with the chaotic rotor motion shown in Figure 9.18. They also show that chaotic motion for pivoted pad bearings is possible for other numbers of pads (e.g., four pads) and other operating conditions where the static bearing load is not directed into a pivot.

9.8 Symptoms and Identification of Vibration Causes

Diagnosis of rotating machinery vibration causes has been enormously advanced in modern times through the intensive scrutiny of machinery vibration with modern instrumentation and signal analysis methods. However, identification of vibration causes remains an inexact science, albeit far better now than it was 40 years ago.

9.8.1 Rotor Mass Unbalance Vibration

The most common cause of excessive rotor vibration is *mass unbalance* in the rotor and the primary symptom is of course *excessive once-per-rev (synchronous) vibration*. Excessive vibration is often accompanied by a significant presence of dynamic nonlinearity in a rotor dynamical system (e.g., journal bearing films, rotor–stator rubs). In consequence, integer multiples of the synchronous frequency may also appear with high levels of unbalance-driven vibration (see Figure 9.2). However, vibration components at integer multiples of the synchronous frequency are possible symptoms of other vibration causes as well. Furthermore, in some machinery types, strong synchronous vibration can originate from sources other than rotor mass unbalance, most notably centrifugal pump hydraulic forces (see Section 6.1 of Chapter 6 discussion pertaining to Table 6.1). Therefore, it is readily apparent just from the symptoms associated with excessive rotor unbalance that identification of specific causes of excessive vibration remains an inexact science.

The two most important ramifications of unbalance caused excessive vibration are (1) the long-term abuse that a machine incurs at operating speed if the unbalance situation is not mitigated and (2) the passage through critical speed(s), that is, run-up and coast-down in machines with operating speeds above one or more critical speeds. In a machine where a rotor piece of significant mass detaches from the rotor, coasting the machine down through critical speed(s) without major damage is a primary concern. Many types of rotating machinery typically maintain their quality state of rotor mass balance over long periods of operation, and thus are more likely to exhibit excessive unbalance vibration when something definitive has gone wrong, such as loosening or detachment of a rotor piece. Conversely, some types of rotating machinery are notorious for going *out of balance* in normal operation, for example, *large fans* in power plants, steel mills, and so on, due to uneven accumulation of crud on the fan impeller. In spindles of precision grinders where the grinding wheel diameter is reduced substantially through wear and repeated redressing, nonuniform distribution of grinding wheel density is sufficiently significant in that rotor-mounted automatic balancing devices are a standard spindle attachment that is needed to achieve precision grinding operations.

The rapid growth in the need for larger and higher speed rotating machinery initiated in the early twentieth century quickly clarified the importance and need for well-balanced rotors. Early engineering emphasis was both on (1) developing adequate balancing methods/devices for machinery production and (2) design and construction approaches to make rotors inherently maintain their state of balance in operation. One notable example is the manufacture of large electric generators powered by steam turbines. In the construction of these large generators, it is required that rotors be turned for several hours at their operating speed in shop floor pits at elevated temperatures so as to "season" rotor parts to stable dimensional positions, thereby enabling a stable shop balancing of the rotor.

9.8.2 Self-Excited Instability Vibrations

In forced vibration (like from rotor unbalance), the responsible alternating force is independent of the vibration. In self-excited vibration, the responsible alternating force is controlled by the vibration itself, and vanishes if the vibration ceases. The early twentieth century focus on rotor unbalance vibration quite naturally led early vibration troubleshooters to attribute any excessive rotor vibration to inadequately balanced rotors and/or insufficient ability of rotors to maintain good rotor balance in operation. Such assessments were usually correct. However, in certain landmark cases when repeated rotor rebalancing failed to alleviate an excessive vibration, major discoveries were made of previously unidentified vibration causes. Possibly the most significant example of this is the discovery of *oil whip*, as documented by Newkirk and Taylor (1925). Oil whip is a subsynchronous vibration, triggered when the journal bearings act as negative dampers to the lowest-frequency forward-whirl rotor–bearing vibration mode. Oil whip is scrutinized within the context of a linear model in Section 2.4 of Chapter 2. As Crandall (1983) heuristically explains, oil whip is but one of

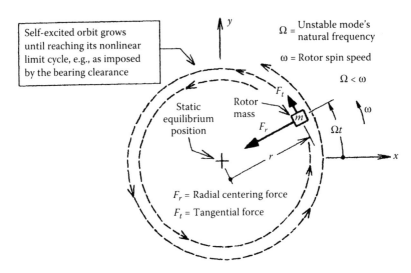

FIGURE 9.19 Growth of rotor orbit for an unstable rotor mode.

the several known self-excited rotor vibration instability mechanisms that share the fundamental characteristic illustrated in Figure 9.19.

The most prominent of this group of instability mechanisms include (1) oil whip, Section 2.4 of Chapter 2; (2) centrifugal pump impeller forces, Section 6.1 of Chapter 6; (3) centrifugal compressor impeller forces, Section 6.2 of Chapter 6; (4) steam whirl, Section 6.3 of Chapter 6; (5) axial flow compressor stages, Section 6.4 of Chapter 6; (6) Coulomb friction, material stress–strain hysteresis, or other rotor-based damping mechanisms rotating synchronously with the rotor; (7) rotor anisotropy (elastic, inertia); and (8) trapped liquid in a hollow rotor. Figure 9.19 illustrates that for any of these instability mechanisms, their reaction force on the rotor in response to a radial displacement from equilibrium has both radial and tangential components. The radial component (F_r) of the reaction force can be either a centering force as in the case of journal bearings, or it can be a decentering force as in the case of a high-Reynolds-number fluid annulus; see Figure 6.6c and d, Chapter 6. However, it is the tangential component (F_t) of the reaction force that supplies the energy to destabilize a potentially self-excited rotor vibration mode. Such destabilizing forces are usually present in most rotating machinery from one or more sources such as itemized earlier in this paragraph. But self-excited vibration occurs only when the dissipative positive damping influences in the system are overpowered by the negative damping influences of the instability mechanism(s) present. An *instability threshold* is an operating condition "demarcation boundary," where the negative damping effects overtake the positive damping effects. In a successfully designed machine, the positive damping effects keep the negative damping effects in-check over the full range of intended operating conditions.

In many modern high-power-density high-speed machines, the supremacy of positive damping effects over the destabilizing negative damping effects is tenuous. Thus, due to subtle differences between so-called identical machines, a particular machine within a group of several of the same configuration may occasionally experience self-excited vibration while the others do not. Similarly, a particular machine may operate free of self-excited vibration for a number of years and then begin to regularly exhibit self-excited vibration, for example, due to accumulated wear or other gradual changes over time or due to hard-to-isolate changes that may occur during overhaul and refurbishment. Such is the nature of oil whip and other similarly manifested instability self-excited rotor vibration causes.

9.8.2.1 Oil Whip

The cascade plot in Figure 9.9 best illustrates the identifying symptoms of oil whip. The machine must typically be rotating above twice the frequency of the potentially unstable mode. As Figure 9.9 suggests, the oil-whip mode is unbalance excited as a critical speed (damped forced resonance) as it is passed through on the way up to operating speed, posing no problem as a critical speed, provided the rotor is adequately well balanced and the mode is adequately damped. As the *oil-whip threshold speed* is encroached upon, the net damping/cycle for this mode transitions from positive to negative and the mode commences vibration at a significant amplitude with its frequency typically below half the rotor spin speed. As Figure 9.9 further shows, with speed increases beyond the oil-whip threshold speed, the subsynchronous rotor vibration does not proportionally track the rotor spin speed but maintains a nearly constant frequency. Also, it does not peak and then attenuate at progressively higher speeds like response through a critical speed does. Thus, oil whip cannot generally be "passed through" as can a critical speed. A further characteristic of oil whip and some other similar self-excited vibrations is that its peak in the frequency spectrum often exhibits significant amplitude fluctuations, for example, as observable from picture-to-picture on a real-time spectrum analyzer display screen. Again, this is the nature of oil whip and other similarly manifested instability self-excited rotor vibrations.

9.8.2.2 Steam Whirl

Steam whirl is described in Section 6.3 of Chapter 6. Its vibration symptoms are quite similar to oil whip, with one notable exception. Its threshold of instability is not rotational speed induced but, instead, power output induced. In power plant jargon, power output is synonymous with "load" on the machine. Steam whirl is a destabilizing mechanism that is of concern primarily in the *high-pressure steam turbine* section of large turbo-generator units. It can produce significantly high subsynchronous forward-whirling rotor vibration, just like oil whip. However, oil whip within the operating speed range usually prevents safe operation, and thus requires an immediate solution. In contrast, since steam whirl is a destabilizing influence whose strength increases with "load" on the machine, not speed, a temporary solution is to derate the machine to a load below which the steam whirl is suppressed by the positive damping, primarily in the oil-film journal bearings. Figure 9.20 illustrates steam whirl vibration symptoms and its threshold power output, and is similar to Figure 9.9, which illustrates the symptoms for oil whip and its threshold speed.

9.8.2.3 Instability Caused by Internal Damping in the Rotor

Damping is a fundamental means both to minimize the amplitude of resonant response and to keep destabilizing influences in check. However, damping mechanisms fixed in the rotor become destabilizing influences in rotors that operate *above a critical speed*. Two widely recognized sources of

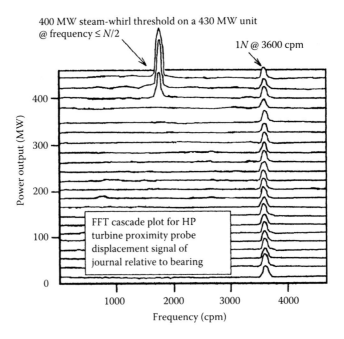

FIGURE 9.20 Steam whirl symptoms on high-pressure steam turbine. Unit should be de-rated to 400 MW pending solution of problem.

rotor-based damping are (1) rotor internal material stress–strain hysteresis-loop energy dissipation and (2) sliding friction between rotor components such as in splines. Both of these rotor-based damping mechanisms are exercised by flexural vibration in the rotor.

Internal rotor damping as a potential source of self-excited vibration was first proposed in a pair of papers by Kimball (1923, 1925), but Smith (1933) provided a simpler insightful explanation. More recently, Crandall (1983) has provided additional clarity to the topic. In essence, irrespective of whether a damping mechanism is linear (e.g., viscous dash pot) or non-linear (e.g., Coulomb sliding friction), there is always an energy dissipating *drag force* at the heart of the damping action. Aerodynamic drag forces are a source of energy dissipation, as airplane and automobile designers well know. However, aerodynamic drag can also impart energy to solid objects such as occurs in strong wind storms, and this essentially explains how rotor internal damping can produce self-excited rotor vibration.

Figure 9.21 illustrates this point using a simple example in which it is assumed that rotor internal damping is the only damping present. As shown, when the rotor spins slower than the orbit natural frequency, rotor internal damping causes an orbital disturbance to decay. Conversely, when

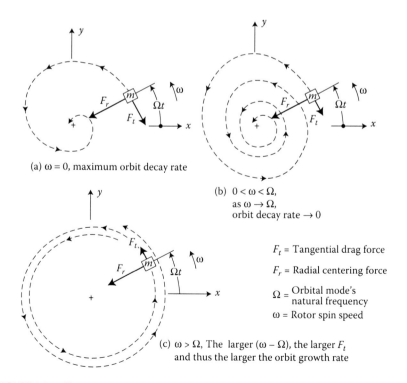

FIGURE 9.21 Transient orbits if rotor-based damping is the only damping present.

the rotor and its internal damping mechanism rotate faster (at a higher frequency) than the orbit natural frequency, the rotor internal damping pulls tangentially in the direction of orbiting and thereby imparts energy to the orbital vibration mode, causing the mode to become self-excited at its natural frequency.

This is similar to an automobile traveling at a speed slower than a high-speed tail wind, which then imparts energy to the vehicle rather than extracting energy from it. Similarly, the strength of the orbit-tangential pulling force from rotor internal damping progressively increases as shaft rotational speed becomes progressively larger than the orbit natural frequency. In actual machines, there are always sources of damping present in the machine's nonrotating portion and thus, for this instability to occur, rotor speed must exceed the orbit natural frequency (i.e., critical speed) by a sufficient margin in order to overpower the positive damping influences present. Thus, when internal rotor damping causes a self-excited rotor vibration, the frequency of the vibration is subsynchronous, similar to oil whip and steam whirl described earlier in this section. Ehrich (1964) provides a comprehensive analytical treatment of self-excited rotor vibration caused by rotor internal damping.

This type of self-excited vibration is not often diagnosed in heavy industrial machines (e.g., in power plants) where the rotors are supported on oil-film journal bearings. But the destabilizing mechanism of rotor internal damping may well be a significant contributing factor in some cases where oil whip or steam whirl are the primary sources of self-excited rotor vibration. On the other hand, rotor internal damping is more likely to be diagnosed as the primary source of self-excited forward-whirling subsynchronous rotor vibration on gas-turbine aircraft jet engines and aero-derivative gas turbines for power generation, where rotors are supported on rolling-element bearings and in some configurations coupled by splines to accommodate differential thermal expansions.

9.8.2.4 Other Instability Mechanisms

Chapter 5 provides a background on bearing and radial-seal rotor vibration characteristics, with attention focused on factors affecting dynamical stability. Chapter 6 provides additional insights into the instability mechanisms for self-excited rotor vibration originating in turbo-machinery stages of pumps, turbines, and compressors. Crandall (1983) gives insightful tutorial descriptions of several rotor instability mechanisms and Ehrich (1976) provides detailed analyses and symptom descriptions.

9.8.3 Rotor–Stator Rub-Impacting

The boiler feed water pump illustrated in Figure 5.13a exemplifies many types of rotating machinery that possess quite small internal annular radial clearance gaps (e.g., at end seals, interstage seals, bearings, blade tips, balance drums, etc.). These small internal radial clearances are essential to the efficient functioning of such machines and are among the most important reasons for the close attention paid to operating vibration levels in rotating machinery. This is because one of the deleterious effects of excessive rotor vibration is the contact between rotor and stator at locations with small rotor–stator radial clearances.

Occurrences of rotor–stator rubs and rub-impacts can be roughly grouped into the following categories: (1) rotor vibration levels become high for any reason (e.g., excessive unbalance, resonance, self-excited instability, etc.), resulting in contact between rotating and nonrotating components, with the rotor–stator contact somewhat passive in its effect on the overall vibration; (2) similar to category (1) except that the rotor–stator contact contributes a strong influence on the ensuing vibration; (3) rotor–stator contact is initiated by excessive static radial rotor forces and/or casing thermal distortions, and excessive vibrations may or may not result; and (4) rotor rub force magnitude is modulated in synchronization with the once-per-rev component of rotor vibration orbit, providing an asymmetric friction-induced heat input to the rotor, causing it to develop a "thermal bow," which initiates a slowly precessing vibration phase angle (forward or backward) because the rotor "high spot" and "hot spot" do not coincide due to thermal inertia and phase lead or lag between the unbalance and the synchronous orbital response it excites. In effect, the "hot spot" tries to catch up to the "high spot," but as the "hot spot" migrates so does the "high spot" in response. This phenomenon is sometimes referred to as "vector turning" because when a representative vibration signal is polar plotted over time (see Figure 9.8), the vibration "vector" slowly rotates (e.g., couple of hours per turn) in one direction, as further explained in the corresponding case study of Chapter 12. When there is insufficient heat-removal capacity available, the turning vector does not stay within acceptable vibration levels, but instead continues to slowly spiral outward.

In brush-type exciters of large AC generators, this "vector turning" phenomenon can be instigated by the brushes' rubbing contact friction forces being synchronously modulated by brush inertia, that is, brushes are spring preloaded to track rotor contact-surface orbital motions. This type of rubbing friction-induced vibration is also known to occur from rotor rubbing of radial seals where the rubbing contact friction force is synchronously modulated by the effective support stiffness of the seal. The field measurement case explained earlier in this chapter pertaining to Figures 9.6 and 9.10 is a prime example of category (3) in which the rub-impacting significantly worsens subsynchronous vibration on coast-down.

Figure 9.22, from tests and corresponding computer simulations, Adams and Loparo (2000), demonstrates transition from category (1) to category (2).

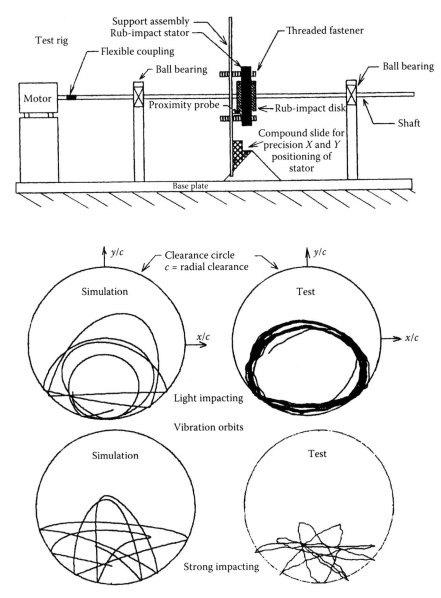

FIGURE 9.22 Unbalance induced rub-impacting on a flexible-rotor test rig.

Referring to the orbital responses in Figure 9.22, the stronger the rub-impacting, the higher the degree of dynamic nonlinearity and thus the greater the amplitude of harmonics of the synchronous frequency $(2N, 3N, 4N, \ldots)$, which will appear in the rotor vibration FFT spectra.

Rub-impact in rotating machines is a group of phenomena involving continuous or dynamically intermittent contact between rotating and stationary (nonrotating) machine components. Rotor–stator contact is of course undesirable but tolerated in most machines for brief periods during initial wear-in and operating transients. Persistent rotor–stator contact considerably accelerates the wearing open of the small rotor–stator radial clearance gaps, significantly reducing machine efficiency and thereby shortening the time cycles between machine repairs and major overhauls. Furthermore, persistent rotor–stator contact puts a machine in jeopardy of failure from the potentially devastating effects of very large amplitude rotor vibration, which can be triggered by severe rotor–stator rub-impact events. In the worst-case scenario, if rub friction is sufficiently intense to impose backward whirl on the rotor vibration orbit, it can provide a mechanism for the ensuing vibration to tap directly into the primary torque-transmitted power through the shaft, leading to an immediate catastrophic failure of the machine.

In modern aircraft gas turbine jet engines, blade tip clearances are understandably quite small, because of inefficiencies associated with back leakage at turbo-machinery blade rows. Therefore, blade tip-on-casing rubs are not unusual in service, given hard landings, ingestion of hail and birds, and tight-maneuver engine rotor gyroscopic effects in military fighter jets. In recent years starting from 2000, the author has been retained by the largest manufacturer of gas turbine jet engines to help develop a test rig and specialized simulation software to study engine core compressor blade tip-rub contact with the engine casing shroud. The test facility that resulted from this collaboration is described by Padova et al. (2005) and the tip-rub-induced blade vibration software development by the author is described by Turner et al. (2005). That work initially focused on gaining new knowledge and insight into the fundamental phenomenon of blade-on-casing rub-induced blade vibration that could potentially lead to shortened blade life expectancy. More recently, this work has focused on determining improved rotor vibration bearing-like model representation of rubbing blade rows to better predict how blade–casing contact changes critical speed locations.

9.8.4 Misalignment

Vance (1988) provides an insightful description of vibration symptoms associated with misalignment. Piotrowski (1995) provides an excellent handbook on shaft alignment methods and procedures. Vance concisely points out the key symptoms that distinguish misalignment caused vibration from rotor unbalance vibration. Specifically, excessive misalignment typically produces a large twice running speed ($2N$) harmonic component of vibration and a high level of axial vibration. He cites bent shafts and improperly seated bearings as special cases of misalignment that yield similar symptoms. Furthermore, for machines that operate below the first critical speed, the misalignment-induced running-speed axial vibration at the two ends of the shaft or across the coupling will be approximately 180° out of phase with each other (like in the 150–210° range). In contrast, these signals will usually be nearly in-phase when rotor unbalance is the primary source of vibration. On this point, Vance cautions that transducer orientations in opposite directions (e.g., at opposite ends of the shaft) will impose an inadvertent 180° error in phase measurement if the readings are not properly interpreted for transducer orientations.

9.8.5 Resonance

With adequacy of rotor balance quality and available damping, passing through critical speeds is a tolerable fact of life for many types of modern rotating machinery, because the critical speed vibration peaks are endured only for the brief time periods while the machine passes through the critical speed(s). However, if by some design flaw, installation error, or component deterioration, a machine's operating speed is quite near a critical speed, excessive unbalance-driven vibration will most likely result. To achieve acceptable vibration levels in such an undesirable circumstance requires a state of rotor balance quality, which is possibly beyond what is practically achievable. Ironically, it is in this very circumstance for which it is most difficult to achieve a high-quality rotor balance. This is because accurate vibration phase angle measurement is an essential ingredient for achieving quality rotor balancing. But the close proximity of a critical speed to the machine's operating speed will cause vibration signal phase angle to be quite unsteady. This is explained by the response of the 1-DOF system to a harmonic excitation force (see Figure 1.5), which clearly shows the steep change in phase angle near the natural frequency that will be caused by continuously occurring small perturbations in natural frequency.

Aside from the critical speed operation, resonances in nonrotating components such as the machine housing or attached components (e.g., piping) are not uncommon, and if not properly diagnosed and corrected can shortly lead to a failure. These types of vibration are usually relatively easy to diagnose and correct, in comparison to vibration problems inherent in the rotor–bearing system. Vibrations of this type may occur at the synchronous frequency ($1N$) and/or its harmonics ($2N, 3N, \ldots$) such as from a vane-passing frequency, and are thus readily identified by their strong dependence on rotor speed. Spectral cascade plots against rotor speed (e.g., Figures 9.9 and 9.10) are therefore quite useful in diagnosing this category of vibration.

9.8.6 Mechanically Loose Connections

Looseness at nonrotating connections such as bearing caps, bearing mounts, or base mounts is likely to result in a vibration problem since the bearings and mounts constrain the shaft to its rotational centerline. The dynamical characteristics precipitated by the looseness of these components will quite likely introduce a significant degree of dynamic nonlinearity into the vibratory system, for example, intermittent in-and-out of hard contacting as components vibrate through dead-band gaps created by the mechanical component looseness. Therefore, the excessive vibration produced by such mechanical looseness usually yields a rich vibration spectrum with several prominent harmonics ($2N, 3N, \ldots$) of the synchronous spin frequency and possibly prominent integer subharmonics ($N/2, N/3, \ldots$) and their integer multiples ($2N/3, 4N/3, \ldots, 3N/2, \ldots$) as well. The singular presence of such a subsynchronous harmonic may lead to a misdiagnosis that the vibration root cause is one of the subsynchronous self-excited vibration types covered earlier in this section. In seeking to differentiate mechanical looseness-caused vibration from other sources (e.g., excessive rotor unbalance, self-excited vibration), taking vibration measurements in different directions and locations on the machine (e.g., see Figure 7.14) can be helpful because looseness-caused vibration tends to be directionally biased as dictated by the specific direction and location of the looseness (see case studies in Section 12.7 of Chapter 12).

Looseness of a rotor-mounted component (e.g., thrust collar, spacer collar, impeller ring, slinger disk, etc.) is also likely to cause a vibration problem. Such a looseness is likely to induce a mass unbalance on the rotor, but not necessarily resulting in a synchronous vibration, although it could. If the looseness combined with other factors involved allows the rotor-loose component to spin at a speed different from the shaft speed, then a nonsynchronous vibration is likely to be present. Prevailing friction conditions, clearances, and fluid or aerodynamic drag forces provide a wide range of possibilities for various steady or unsteady vibrations to result. The loose component could possibly lock its rotational speed into one of the rotor–bearing system's subsynchronous orbital natural frequencies and thereby disguise itself as one of the previously described self-excited vibration types. If the loose component is a driven element like a turbine disk, then a resulting nonsynchronous vibration would be at a frequency above the rotor spin frequency. Otherwise, any resulting nonsynchronous vibration will likely be at a subsynchronous frequency. A rotor-loose component will possibly cause additional symptoms to help it to be identified, for example, axial shuttling of the rotor if the thrust collar is loose.

9.8.7 Cracked Shafts

The initiation and subsequent propagation of a crack through the shaft is of course among the most dreaded failure types in rotating machinery. As a consequence, early diagnosis and careful trending of vibration symptoms for cracked shafts is a well-studied topic within the machinery vibration monitoring field. Muszynska (1995) provides an insightful explanation of cracked-rotor vibration symptoms, employing a simple 2-DOF single-mass rotor dynamics model. Muszynska's model embodies the two prominent symptoms that a rotor crack superimposes on a simple unbalance-only vibration model. These two effects are (1) a bending stiffness reduction aligned with the crack direction and (2) a crack-local shift in the bending neutral axis (the rotor therefore bows) corresponding to the crack direction.

The first of these two effects produces a twice-rotational-speed ($2N$) vibration component. This is similar to what would occur prominently in long two-pole generators were it not for the standard radial slots that are cut in such generator rotors to equilibrate the principle bending stiffnesses. The second of these effects produces a synchronous (N) vibration component that vectorially adds to the preexisting residual unbalance synchronous vibration.

The vibration symptoms for a developing rotor crack are therefore (1) the emergence and growth of a $2N$ vibration component simultaneously with (2) the emergence of a progressive change in synchronous vibration amplitude and phase angle from the rotor-bow-induced unbalance. An additional symptom is apparent when rotor x and y displacements are monitored. That is, the rotor orbit will appear similar to the typical orbit with N and $N/2$ harmonics superimposed (see Figure 9.4) except that the period of the cracked-shaft orbit is one revolution (not two) as immediately detectable from the presence of only one keyphaser mark per period of orbital vibration. There have been some remarkably accurate predictions of how long a rotor can operate before it fails from a sudden through fracture precipitated by a shaft crack propagation. In one well-substantiated case, the supplier of the vibration monitoring system predicted the exact number of operating days remaining for a primary nuclear reactor coolant pump shaft in a PWR commercial electric power generating plant. Although that nuclear plant's operators were unfortunately skeptical of the prediction, after the shaft failed as predicted (to the day) they became converted "believers."

9.8.8 Rolling-Element Bearings, Gears, and Vane/Blade-Passing Effects

Wear and other damage in rolling element bearings yields vibration components that are symptomatically related to specific bearing features such as inner raceway, outer raceway, separator, and rolling element damage.

Similarly, gear sets also produce unique vibration signatures with specific features that can be diagnostically related to specific wear and other damage types and locations. The level of rotor and casing vibration components present for these diagnostic purposes is typically of much lower amplitudes than the overall levels of residual vibration, and thus typically are not of great significance with regard to their potential for vibration-caused damage to a machine. Incipient bearing failure detection is of course important on its own. Adams (Michael) and Loparo (2004) demonstrate the significant capability of combining advanced signal processing with remote sensing of vibration for early detection of impending failure in rolling element bearings.

Similarly, vane-passing and blade-passing frequencies are commonly present in turbo-machinery vibration signatures, but likewise at amplitudes typically much smaller than the overall levels of residual vibration. Vane- and blade-passing vibration components are of diagnostic significance primarily in assessing the respective hydraulic or aerodynamic operating factors of such machinery. For example, the propensity for certain types of internal damage such as to impeller and diffuser vanes in high-power-density centrifugal pumps (e.g., boiler feed water pumps) can be assessed from the strength of vane-passing vibration components. Vane-passing vibration components can also be related to acoustic resonance problems. The highly specialized component and machine-specific natures of the vibration signature components for these categories relegates their fuller treatments to component and machine-specific references. Vance (1988) provides introductory treatments, but an entire book chapter could readily be employed to give comprehensive treatments to each of these topics. Makay (1978) provides the definitive charts of vibration symptom identifications for centrifugal pumps. The Makay charts correlate root-cause symptoms and severity levels to vibration frequency components (normalized by rotational speed frequency) as functions of pump flow (normalized by best-efficiency flow).

9.9 Summary

This chapter identifies and explains the symptoms of known causes of excessive rotating machinery vibration within the context of modern measurement and monitoring-detection technologies. An initial call for vibration diagnosis of a specific machine usually begins with an alert from vibration monitoring that shows the machine's vibration levels are going to exceed or have already exceeded the experience-based "normal" or "allowable" levels for the machine (e.g., see Figures 7.1 and 8.1). In seeking to eliminate the problem, identification of the root cause(s) of the excessive vibration is a far more rewarding approach than the trial-and-error method which may never converge to a good solution. There are many unfortunate cases in which machine owners/operators have lived with vibration-plagued machines for years without a good diagnosis and solution, and thereby have borne the expenses associated with the all-too-frequent costly repairs precipitated by long duration exposure to excessive vibration levels. The spare parts and rebuild business attending such situations can be lucrative indeed, easily exceeding several times the initial profits on the sale of the defectively designed machinery.

As indicated throughout this chapter, the FFT spectrum is presently the major rotating machinery vibration diagnostic tool. A scrutiny of vibration spectral characteristics for various problem root causes, as described and explained in this chapter, shows that many different root causes produce similar looking rich spectra. This shows the need to combine FFT analysis with other diagnostic tools such as rotor orbit measurement and analysis. This also shows the significant value of developing further improvements and better tools in the condition monitoring field. So this chapter also presents new emerging diagnostic approaches that can be major advancements for next-generation condition monitoring products.

Bibliography

Abu-Mahfouz, I., *Routes to Chaos in Rotor Dynamics*, PhD Thesis, Case Western Reserve University, 1993.

Adams, M. L. and Abu-Mahfouz, I., *Exploratory Research on Chaos Concepts as Diagnostic Tools for Assessing Rotating Machinery Vibration Signatures*, Proceedings IFTOMM Fourth International Conference on Rotor Dynamics, Chicago, IL, September 6–9, 1994.

Adams, M. L. and McCloskey, T. H., *Large Unbalance Vibration in Steam Turbine-Generator Sets*, Third IMechE International Conference on Vibrations in Rotating Machinery, York, England, 1984.

Adams, M. L. and Loparo, K. A., *Model-Based Condition Monitoring from Rotating Machinery Vibration*, EPRI Project WO3693-04, Final Report (in press), 2000.

Adams (Michael) and Loparo, K., *Analysis of Rolling Element Bearing Faults in Rotating Machinery—Experiments, Modeling, Fault Detection, and DIagnosis*, Eighth IMechE International Conference on Vibrations in Rotating Machinery, Swansea, Wales, UK, 2004.

Bently, D. E. and Muszynska, A., *Vibration Monitoring And Analysis For Rotating Machinery*, Noise & Vibration '95 Conference, Pretoria, South Africa, November 7–9, 1996, Keynote Address Paper, pp. 24.

Chui, C. K., *Wavelets: A Mathematical Tool for Signal Analysis*, Society for Industrial and Applied Mathematics (SIAM), 1997.

Crandall, S. H., *The Physical Nature of Rotor Instability Mechanisms*, Proceedings ASME Applied Mechanics Division Symposium on Rotor Dynamical Instability (Editor: M. L. Adams), AMD Vol. 55, pp. 1–18, 1983.

Daubechies, I., *Different Perspectives on Wavelets*, American Mathematical Society, ISBN 0-8218-5503-4, Short Course (1993).

Ehrich, F. F., "Shaft whirl induced by rotor internal damping," *ASME Journal of Applied Mechanics*, 1964.

Ehrich, F. F., "Self-excited vibration," in C. M. Harris and C. E. Crede (Eds.) *Shock & Vibration Handbook*, 2nd edn., McGraw Hill, New York, pp. 5-1–5-23, 1976.

Graps, A., "An introduction to wavelets," *IEEE Computational Science and Engineering*, (Summer), 50–61, 1995.

Horattas, G. A., Adams, M. L., and Dimofte, F., "Mechanical and electrical runout removal on a precision rotor-vibration research spindle," *ASME Journal of Acoustics and Vibration*, 119(2):216–220, 1997.

Kaiser, G., *A friendly Guide to Wavelets*, Birkhauser, Boston, 1994.

Kimball, A. L., "Internal friction theory of shaft whirling," *Physics Review*, 21(2):703, 1923.

Kimball, A. L., "Internal friction as a cause of shaft whirling," *Philosophical Magazine*, 49(6):524–727, 1925.

Makay, E., *Survey of Feed Pump Outages*, Electric Power Research Institute, Project-641, Final Report EPRI FP-754, pp. 97, April 1978.

Mitchell, J. S., *"An Introduction to Machinery Analysis & Monitoring,"* Penn Well, Tulsa, Oklahoma, pp. 374, 1981.

Muszynska, A., "Vibrational diagnostics of rotating machinery malfunctions," *International Journal of Rotating Machinery*, 1(3–4):237–266, 1995.

Newkirk, B. L. and Taylor, H. D., "Shaft whipping due to oil action in journal bearings," *General Electric Review*, 28:559–568, 1925.

Padova, C., Barton, J., Dunn, M., Manwaring, S., Young, G., Adams, M. L., and Adams, M. L., "Development of an experimental capability to produce controlled blade tip/shroud rubs at engine speed," *ASME Journal of Turbomachinery*, 127:727–735, 2005.

Piotrowski, J., *Shaft Alignment Handbook*, Marcel Dekker, pp. 574, 1995.

Smith, D. M., "The motion of a rotor carried by a flexible shaft in flexible bearings," *Proceedings of the Royal Society (London) A*, 142:92–118, 1933.

Turner, K., Adams, M. L., and Dunn, M., *Simulation of EngineBlase Tip-Rub Induced Vibration*, Proceedings ASME Torbo Expo Gas Turbine Conference, Reno, NV, June 6–9, 2005.

Vance, J. M., *Rotordynamics of Turbomachinery*, Wiley, New York, pp. 388, 1988.

Part IV

Trouble-Shooting Case Studies

10

Forced Vibration and Critical Speed Case Studies

10.1 Introduction

Vibration excited by residual rotor unbalance is always present in all rotors at all operating speeds, because it is of course impossible to make any rotor perfectly mass balanced. Therefore, the objective concerning unbalance-excited vibration is its minimization, not total elimination. Chapter 8 addresses the fundamental question of whether the residual vibration of a machine is within its acceptable limits or is excessive. When vibration levels are deemed excessive and it has been established that the excitation is unbalance (see Section 9.8), the proper corrective course of action is often simply to rebalance the rotor. This is especially typical in some machinery types that are inherently susceptible to going out-of-balance in normal operation, like large fans in power plants and steel mills where crud collects nonuniformly on fan blades. Many machines are designed with externally accessible rotor balance planes where balance correction weights can be added. Thus, in-service rebalancing of the rotor does not typically require opening up the machine and is considered a relatively routine procedure.

However, the root cause for excessive unbalance-excited vibration can be other than the rotor being too far out-of-balance. As explained in Section 9.8, an inadequately damped resonance condition can cause excessive vibration, even when the excitation force is not large, as Figure 1.5 clearly shows. If for any reason the operating speed is quite near a critical speed, then the vibration levels can readily become excessive for continuous operation. A critical speed near the operating speed can be the result of some design flaw, installation error, component deterioration, or support/foundation changes over time. For similar causes, transient passage through a critical speed may exhibit vibration levels that are dangerously high even for a short duration passage through critical speed, such as in a slow coast-down of the machine.

The case studies presented in this chapter are not of the category where routine rebalancing of the rotor is the solution to the excessive vibration problem. Each case study presented here typifies the more difficult ones where routine rebalancing does not solve the problem. As these cases demonstrate, identification of both root cause(s) and the most cost-effective solution(s) or fixes can be enormously aided by using analysis models.

10.2 HP Steam Turbine Passage through First Critical Speed

Figure 10.1 shows the steam turbine portion (generator not shown) of the HP–LP drive line of a 350 MW cross-compound turbo-generator. Cross-compound units typically have 50% of the power capacity on each of two drive lines. One drive line contains an HP turbine, one or two LP turbines plus an AC generator and possibly a drive-line-mounted exciter. The other drive line similarly contains an IP (or reheat cycle) turbine, one or two LP turbines plus an AC generator and possibly a drive-line-mounted exciter. Cross-compound units are powered by a single fossil-fired boiler, and have main flow steam lines connecting the two drive lines. That is, reheated exhaust from the HP turbine is piped across drive lines to the IP turbine steam inlet and IP turbine exhaust is piped proportionally to all the LP turbines of both the drive lines.

The excessive vibration experienced with the unit in Figure 10.1 occurred primarily in the HP turbine section during coast-down, where it exhibited a clear resonance of the HP turbine, with a 20 mils (0.5 mm) peak-to-peak synchronous vibration of the HP rotor at its bearings. This vibration problem had all the symptoms of an HP turbine critical speed that had become insufficiently damped over a period of several months. Furthermore, the

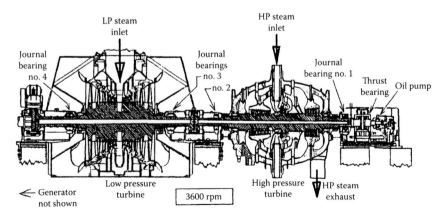

FIGURE 10.1 Cross-compound 350 MW turbo-generator, HP–LP portion.

coast-down speed at which this critical speed vibration peaked varied from coast-down to coast-down, anywhere between 1400 and 2000 rpm.

The author's preliminary diagnosis was that one of the HP rotor's two tilting-pad journal bearings was statically unloaded (at least during coast-down), probably due to bearing alignment shifting. This would explain a significant reduction in bearing damping capacity with the attendant excessively high resonance vibration peaks. This would also explain the nonstationary HP turbine critical speed (i.e., from bearing stiffness variations caused by bearing load variations). From assembly drawings of the machine, the author developed a complete drive-line finite-element-based rotor vibration model for the complete HP–LP drive line (see Chapter 4 for computer modeling examples). Even though the excessive vibration was localized primarily in the HP section of the HP–LP drive line, the model (236 DOF) was configured to include not only the HP turbine but also the LP turbine and generator sections, all connected by rigid couplings and with a total of six journal bearings. The author's approach on this point is not to guess whether, or how much, the LP and generator sections affect the vibration problem, but instead to include the complete drive-line rotor. This approach also provides an analysis model readily available for other future analyses needed for rotor vibration problems anywhere on the same drive line.

A journal bearing's oil-film stiffness and damping properties are strongly influenced by its static load. In the computer model for this case, journal bearing stiffness and damping coefficients were generated for all six journal bearings using the EPRI COJOUR code referenced in Chapter 5. Bearing coefficients were determined for the nominal alignment case with bearing design loads as well as several off-design out-of-alignment bearing load distributions. Since large turbo-generators have more than two journal bearings, the bearing loads are statically indeterminate. This means that bearing static loads are strongly influenced by bearing alignments. Therefore, it is common on large turbo-generator units for vibration characteristics to change significantly as the bearing support structures shift, for example, as a function of operating point and/or from support shifting and settling over time.

As detailed by McCloskey and Adams (1992) for this troubleshooting case, the analysis showed that unloading of bearing no. 2 fully accounted for the excessive vibration symptoms on this unit. They also provide details of an extensive parametric study, which shows that by adding preload to the statically unloaded tilting-pad journal bearing no. 2 (preload of $C'/C = 0.7$), the HP turbine vibration peak at its critical speed is reduced to half the level of that using an unloaded bearing no. 2 with no preload. This change was made to the actual no. 2 bearing with the result that the HP critical speed vibration peak amplitude was in fact approximately halved with no negative side effects of adding the modest amount of preload to no. 2 bearing. The long-term permanent fix by the power company owner of this machine was to replace the two original OEM HP turbine tilting-pad bearings with a superior non-OEM tilting-pad design discussed in Chapter 12 and in the EPRI symposium paper by Giberson (1993).

10.3 HP–IP Turbine Second Critical Speed through Power Cycling

In this case study, the HP–LP steam turbine drive line of a 240 MW 60 Hz (3600 rpm) generating unit is similar to that illustrated for the prior case in Figure 10.1. Excessive rotor vibration was consistently experienced on the unit's HP–IP rotor during power following in the 145–185 MW range. This characteristic was reported to exist since the unit's original commissioning several years earlier. However, it was only after a recent scheduled refurbishment outage that this vibration increased to levels that were deemed sufficiently excessive to require root cause identification and substantial attenuation. The author undertook vibration measurements of the entire drive line employing a 16-channel digital tape recorder. Examination of these measurements suggested the root cause to be the *HP–IP rotor's second critical speed*, as evidenced from the vibration peak accompanied by a sharp change in phase angle measurement, as shown in Figure 10.2. This critical speed shifted up or down across 3600 rpm operating speed as power output of the unit cycled during power demand following. The author's preliminary assessment was as follows. The root cause appeared to be the HP–IP journal bearing static load changes commensurate with the variable partial emission of the impulse turbine control stage during power cycling (see Section 6.3.2 of Chapter 6 with Figure 6.7).

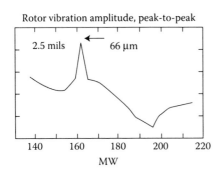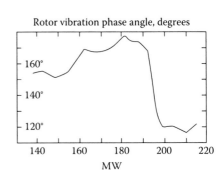

FIGURE 10.2 Measured shaft vibration versus power at HP–IP bearing no. 2.

Forced Vibration and Critical Speed Case Studies

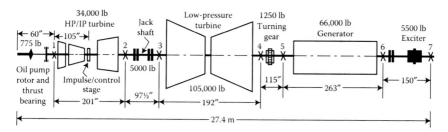

FIGURE 10.3 Rotor vibration model for 240 MW 3600 rpm turbine generator.

Using the RDA code supplied with this book, the author developed a rotor vibration computer model of the unit's entire drive line, as shown in Figure 10.3. This model was used to parametrically study rotor unbalance vibration as influenced by journal bearing hydrodynamic oil-film stiffness variations at 3600 rpm resulting from variations in net impulse turbine nozzle radial force when transitioning through power changes. The computer model simulations verified the initial diagnosis that the root cause was the *HP–IP rotor's shifting second critical speed*. A composite of simulation results are plotted in Figure 10.4, showing the influence of journal oil-film radial stiffness. The OEM's HP bearing configuration (Figure 10.5a) was a four-pad pivoted-pad configuration with dead-weight load between the bottom two pads. As was expected, and as Figure 10.4 shows, the computer model predicted that the second critical speed transitions up through the 3600 rpm operating speed as journal bearing input

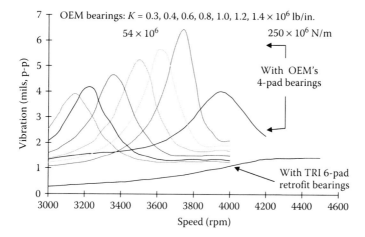

FIGURE 10.4 (See color insert following page 262.) Predicted shaft unbalance vibration amplitude at bearing no. 1: At the nominal bearing static unit load of 200 psi (13.6 bar), OEM radial bearing stiffness K is computed to be 1.98×10^6 lb/in. (254×10^6 N/m).

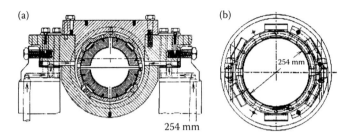

FIGURE 10.5 Turbine pivoted-pad bearing configurations: (a) OEM and (b) TRI.

stiffness is set at progressively higher values, from 0.3 million lb/in. (54 million N/m) to 1.4 million lb/in. (250 million N/m). Figure 10.4 also shows the unbalance vibration response predicted with a six-pad non-OEM journal bearing retrofit at bearings no. 1 and no. 2. This retrofit six-pad pivoted-pad bearing design, with weight centered on the bottom pad, has substantially higher radial stiffness than the OEM design, and its stiffness is much less sensitive to bearing load changes during power following of the unit. Based on the drastic vibration attenuation predicted by the computer model with the six-pad bearing retrofit, the authors recommended this retrofit. It was supplied by Turbo Research Inc. (TRI). A sketch of this bearing configuration is shown in Figure 10.5b and described by Giberson (1993).

This bearing retrofit was purchased and installed at bearings no. 1 and no. 2 shortly thereafter by the power company owner of this unit during the next scheduled outage. Following that scheduled outage and bearing retrofit, the author again took shaft vibration measurements on the entire machine's drive line (Adams and Adams, 2006). Those vibration measurements showed a more than twofold reduction in peak vibration levels. Elimination of the second critical speed from the operating power range was also evidenced from a rotor vibration phase angle gradual change versus unit power.

10.4 Boiler Feed Pumps: Critical Speeds at Operating Speed

10.4.1 Boiler Feed Pump Case Study 1

The rotor sectional view shown in Figure 10.6 is from a four-stage boiler feed water pump (BFP) somewhat similar to that shown in Figure 5.13a. In the power plant of this case study, the BFPs are installed as variable speed units with operating speeds from 3000 to 6000 rpm, each with an

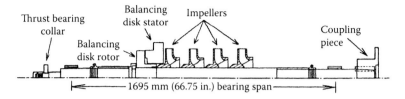

FIGURE 10.6 Rotor sectional view for a four-stage BFP.

induction motor drive through a variable speed fluid coupling. In this plant, the BFPs are all "50%" pumps, which means that when a mainstream turbo-generator is at 100% (full load) power output, two such pumps are operating at their nominal operating condition. This plant (located south of Melbourne, Australia) houses four 500 MW generating units, each having three 50% BFPs installed (i.e., one extra 50% BFP on standby), for a total of 12 boiler feed pumps, all with the same configuration. Full-load operating speed ranges for each 50% BFP is 5250–5975 rpm.

The BFPs at this plant had experienced a long history of failures, with typical operating times between overhauls under 10,000 h, with the attendant significant monetary cost. Based on the operating experience at other power plants (at U.S. power plants) employing the same BFP configuration with quite similar operating ranges, these BFPs should have been running satisfactorily for over 40,000 h between overhauls. Using vibration velocity peak monitored at the outboard bearing bracket, these BFPs were usually taken out of service for overhaul when vibration levels exceeded 15 mm/s (0.6 in./s). To wait longer significantly increased the overhaul rebuild cost, that is, incurred more damage. The dominant vibration frequency was synchronous ($1N$).

The author's preliminary diagnosis was that these pumps were operating quite near a critical speed and that the resonance vibration resulting from this worked to accelerate the wearing open of interstage sealing ring radial clearances. As these inter-stage clearances wear open, the overall vibration damping capacity diminishes significantly, typically leading to a continuous growth of vibration levels. To confirm this preliminary diagnosis, the author developed an RDA finite-element-based computer model (see Chapter 4) for this BFP configuration in order to compute lateral rotor vibration unbalance response versus rpm. The manufacturer (OEM) of the pump provided a nominally dimensioned layout of the assembled pump, including weight and inertia for concentrated masses (impellers, balancing disk rotor, thrust bearing collar, coupling piece, and shaft sleeves). The pump OEM also provided detailed geometric dimensions for the journal bearings, interstage radial seals, and other close-clearance radial annular gaps. This cooperation by the pump OEM greatly expedited the development of the RDA model, eliminating the need to take extensive dimension measurements from one of the BFPs at the plant or repair shop, which were at a considerable distance outside the United States (Australia).

The radial annular gaps have clearance dimensions that are quite small and are formed by the small difference between a bore (inside diameter, ID) and an outside diameter (OD), each with tolerances. The size of each of these small radial clearance gaps is very influential on the respective bearing or seal stiffness, damping, and inertia coefficients (Chapters 5 and 6) and thus very influential on the computed results for rotor vibration response. However, these small radial gaps vary percentage-wise significantly and randomly because of their respective ID and OD manufacturing tolerances plus any wearing open due to in-service use. BFPs are thus one of the most challenging rotating machinery types to accurately model and analyze for rotor vibration. The net result is that even in the easiest of cases, a realistic rotor vibration analysis for troubleshooting purposes (as opposed to design purposes) requires several trial input cases to get the model predictions to reasonably portray the vibration problem the machine is exhibiting. By iterating the model inputs per radial-clearance manufacturing tolerances and allowances for wear, a set of inputs is sought that produce rotor vibration response predictions concurring with the machine's vibration behavior. When a good agreement model is achieved, the author refers to it as the *calibrated model*. Through computer simulations, the calibrated model can then be used to explore the relative benefits of various fix or retrofit scenarios, as was done in the successful steam turbine case studies presented in the previous two sections.

A calibrated model was not initially achieved for this pump vibration problem in that all reasonable model variations for input dimensions failed to produce predicted unbalance responses having a resonance peak below 8000 rpm, which is considerably above the operating speed range. Since the power plant in this case was in Australia, a visit to the plant had not initially been planned. However, given the failure of all initial RDA model variations to replicate or explain the BFP vibration problem, a trip to the plant was undertaken to study the pumps firsthand.

Poor hydraulic conditions in BFPs, such as from inaccurate impeller castings, can produce strong synchronous rotor vibrations (see Section 9.8 and Table 6.1) so several of the impellers were inspected for such inaccuracies. In the course of further searching for the vibration problem root cause, a number of serious deficiencies were uncovered in the local BFP overhaul and repair shop's methods and procedures, all of which collectively might have accounted for the vibration problem. Luckily, on the last day of the planned 1-week visit to the plant, the root cause was discovered, but it could have been easily overlooked. In the process of discussing installation-after-overhaul details with mechanics at the plant, it was revealed that between the inner journal bearing shells and the axially split outer housings there was a clearance of about 0.001 in. (0.025 mm) into which

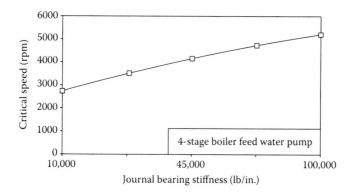

FIGURE 10.7 Computed unbalance-excited critical speed versus bearing stiffness (interposed gasket).

a gasket was interposed and compressed as the two housing halves were tightly bolted together. This use of gaskets had been discontinued many years earlier in most U.S. power plants. The net result of the interposed gasket was to reduce the effective bearing stiffness to a value significantly below the range that had been reasonably assumed in the initial (unsuccessful) attempts to develop a calibrated RDA model. When the relatively soft gasket effect was incorporated into the RDA model bearing stiffness inputs, a resonance peak showed up right in the normal operating speed range.

An analysis study was conducted to compute critical speed (speed at which unbalance-excited vibration response peaks) as a function net bearing stiffness, using a stiffness value range consistent with the interposed gasket. A summary of the results for this analysis is shown in Figure 10.7. A bearing stiffness value of 100,000 lb/in. (12.8×10^6 N/m) places the critical speed right at the normal full load operating speed range. The variability of gasket stiffness also explained the plant's experience with the excessive vibration fading in-and-out over time.

The gasket stiffness is in-series with the bearing oil film's in-parallel stiffness and damping characteristics. Since the gasket is much less stiffer than the journal bearing oil-film stiffness, the gasket also reduces considerably the damping action of the oil films. The use of a gasket between the bearing inner shell and outer housing was clearly the "smoking gun," placing the critical speed near the normal full-load operating speed while depriving the attendant resonance of reasonable damping. All bearings were reinstalled with metal shim strips, at the 45° and 135° angular locations relative to horizontal, providing a bearing pinch of about 0.001 in. (0.025 mm).

A simplified view of the BFP nonplanar critical speed response shape is shown in Figure 10.8, which flattens the actual nonplanar response shape into a plane for plotting purposes. This is helpful in showing the rotor axial locations where residual rotor mass unbalance will have the most

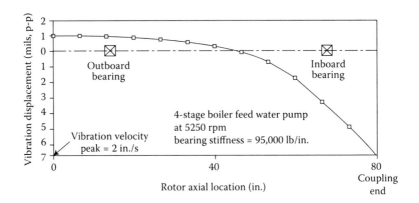

FIGURE 10.8 Critical speed rotor response shape with typical unbalances.

effect in exciting the critical speed resonance vibration. The unbalance vibration response shape here is in fact nonplanar, similar to the example isometric illustration shown in Figure 4.9. The obvious conclusion drawn from the computed unbalance response shape shown in Figure 10.8 was that coupling unbalance probably contributed significantly to this vibration problem, because the repair shop's rotor balancing procedure, as witnessed by the author, was inadequate in several areas, particularly for the coupling. The flexible couplings employed on these BFPs are of the diaphragm type and are well suited to such applications, being more reliable than gear couplings that require maintaining lubrication. With a properly functioning flexible coupling, the BFP is sufficiently isolated from the driver (lateral vibration wise) so that the analysis models can justifiably terminate at the pump half of the coupling. Experience has shown this to be well justified.

10.4.2 Boiler Feed Pump Case Study 2

A second BFP vibration case study presented here involves the BFP shown assembled in Figure 10.9. It is similar in size and capacity to that in the previous BFP case study (Figure 10.6), being a "50%" pump for a 430 MW steam turbo-generator unit. The BFP shown in Figure 10.9 is actually a three-stage pump for boiler feed, but has a small fourth stage (called a "kicker stage"), which is to supply high-pressure injection water at pressures above feed water pressure.

This BFP was observed to have a critical speed at 5150 rpm, although the manufacturer's design analyses did not support this observation. This is a variable speed pump with a maximum operating speed of 6000 rpm. The 5150 rpm critical speed was in the frequently used operating speed range and produced excessive vibration levels, primarily at the inboard

Forced Vibration and Critical Speed Case Studies

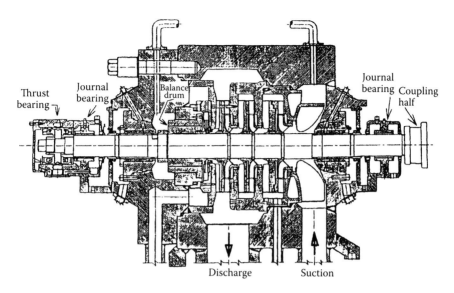

FIGURE 10.9 Assembly view of a BFP.

end of the rotor (i.e., coupling end at suction inlet). A clue was supplied by Dr. Elemer Makay. At a number of power plants employing the same BFP design, he observed BFP inboard journal bearing distress in the top half of the bearing bore. This bearing distress was consistently centered about 10° rotation direction from the top center. The journal bearings were of a design employing a relieved top-half pocket. The specific elbow geometry of the pump inlet piping suggested to Dr. Makay that there was a significant upward hydraulic static force on the suction end (inboard end) of the rotor.

A finite-element-based unbalance response model was developed by the author from detailed OEM information supplied by the electric power company owner of the plant. A lengthy double-nested iteration study was undertaken in which an upward static rotor force was applied on the rotor model at the suction-stage impeller. Through a trial-and-error iteration, this upward static radial force was directed so as to produce an inboard journal eccentricity direction of 10° rotation from the top center, as motivated by the bearing distress observations of Dr. Makay. From each of several values assumed for this force, a set of journal bearing static loads were calculated. A set of bearing stiffness and damping coefficients were in turn calculated for each set of bearing static loads. Each set of bearing stiffness and damping coefficients were then used as inputs to the finite-element-based unbalance response model to compute rotor vibration response versus rpm using a typical set rotor unbalances.

Through several iterations, a force of 3477 pounds yielded journal bearing rotor dynamic coefficients (using the EPRI COJOUR code referenced in Chapter 5) that predicted an unbalance-excited critical speed of 5150 rpm. Furthermore, at this predicted 5150 rpm critical speed, the rotor vibration response shape showed high inboard (coupling end) vibration levels as observed on the BFPs at the plant. In fact, the critical speed rotor vibration response shape was very similar to that shown in Figure 10.8 for the earlier BFP case study presented in this section. The problem was eliminated by retrofitting a different journal bearing configuration that shifted the critical speed considerably above the 6000 rpm maximum operating speed.

10.4.3 Boiler Feed Pump Case Study 3

The boiler feed pump shown in Figure 10.10 experienced a number of forced outages that were accompanied by excessive vibration levels. One of these outages involved a complete through-fracture of the pump shaft just adjacent to the balancing drum runner. As detailed by Adams and Adams (2006), the authors were retained to diagnose the root cause(s) and develop a cost-effective fix. This pump was not equipped with shaft targeting noncontacting displacement proximity probes. So the author's first step was to retrofit X and Y (90° apart) proximity probes near each pump journal bearing, to obtain shaft vibration displacement measurements adequate for successful root cause diagnoses. These four retrofitted proximity probes were installed in parallel with four velocity pickups, to capture any proximity probe mounting motions. Figure 10.11 shows the outboard end of the pump with the author's retrofitted vibration sensors (x and y proximity probes and velocity pick-ups).

A parallel task was undertaken to develop a rotor unbalance vibration response RDA computer model for this pump. Computer model prediction

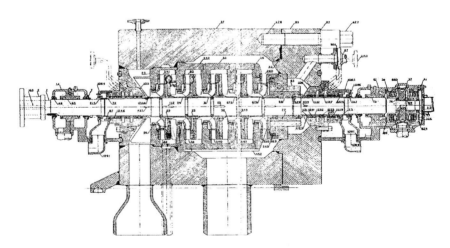

FIGURE 10.10 Variable-speed boiler feed pump of a 600 MW generating unit.

Forced Vibration and Critical Speed Case Studies

FIGURE 10.11 (See color insert following page 262.) Proxy probe setup.

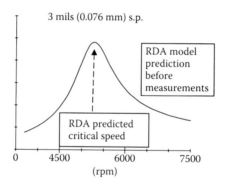

FIGURE 10.12 Journal vibration.

results are shown in Figure 10.12 and predict a critical speed at 5250 rpm, right near the normal full load operating speed. Subsequent to these model predictions, the unit was restarted and all eight channels of the newly installed vibration channels were recorded as a function of pump rotational speed during roll-up. A sample of these vibration measurements is plotted in Figure 10.13. These rotor vibration measurements clearly show a vibration peak at about 5100 rpm, quite close to the predicted 5250 rpm critical speed. This critical speed was judged to be a strong contributing factor to the excessive pump vibrations and associated outages. The author engineered wear-ring surface geometry modifications for this pump to shift the critical speed well above the operating speed range.

10.5 Nuclear Feed Water Pump Cyclic Thermal Rotor Bow

A cross-sectional layout of a PWR nuclear power plant feed water pump configuration is shown in Figure 10.14. This plant houses two 1150 MW

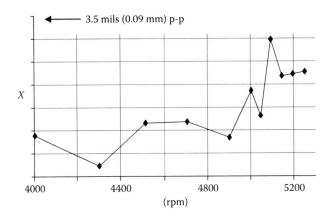

FIGURE 10.13 Boiler feed pump shaft vibration measurements.

PWR generating units. Each unit employs two 50% feed water pumps, for a total of four feed water pumps. All four of these pumps had experienced cyclic rotor vibration spikes that were synchronized with the seal injection water flow control. After several months of unsuccessful in-company vibration measurements and troubleshooting diagnoses of this excessive

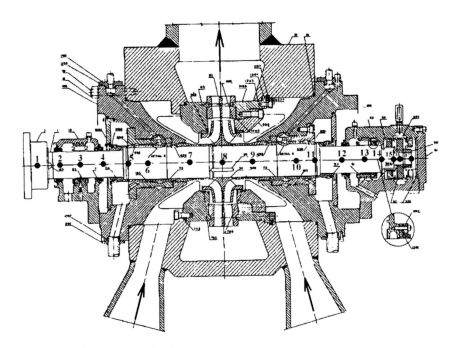

FIGURE 10.14 Nuclear feed water pump.

Forced Vibration and Critical Speed Case Studies 363

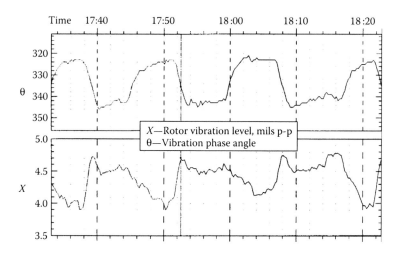

FIGURE 10.15 Fifty-minute vibration record from feed water B-Pump Unit-2.

vibration problem, the author was retained by the power company owner to see if a rotor vibration computer model analysis could identify the root cause of the excessive vibration. A more detailed documentation of this troubleshooting case is reported by Adams and Gates (2002), where plant and pump OEM are identified. The correlation between pump seal injection water control and vibration signals is shown in the 50-min sample of the vibration data in Figure 10.15.

Exhaustive computer rotor vibration analyses of this pump were conducted by the author utilizing the RDA software (rotor mass station numbers shown in Figure 10.14), both for unbalance-forced vibration and instability self-excited rotor vibration root causes. It was no surprise to the author that these analyses eliminated critical speeds and self-excited vibration phenomena as likely root causes, but the plant insisted upon these analyses as the first step.

With those analyses out of the way, the author was free to study if a cyclic thermal bowing of the rotor could be the root cause of the problem that could explain the plant's pump rotor vibration. The 10°F cyclic seal injection differential temperature swings synchronized with the 15 min/cycle rotor vibration characteristic, as shown in Figure 10.15. A close examination of this pump configuration in Figure 10.14 shows a typical arrangement employing shaft sleeves to form the rotating parts of the shaft seals. There are two mating sleeves on each axial side of the impeller. Each of these two-sleeve combinations was modeled by a single hollow cylinder of a nominal length. A calculated 10°F differential thermal expansion for the two-sleeve model (in steel) was computed to be 1.2 mils (0.03 mm), which was calculated to impose a 23,000 pound (10,250 N) compressive force on the sleeves, since the much higher cross-sectional area of the shaft virtually

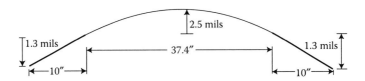

FIGURE 10.16 Computed shaft bow by sleeve-to-shaft differential expansion.

prevents this differential thermal expansion. Under perfect manufacturing and assembly conditions (i.e., no tolerances), the compressive restraining force would be coaxial with the shaft centerline (i.e., *best-case* scenario). Under a *worst-case* scenario (possible), the axial restraining force would be centered at the outer radius of the cylinder ($R \cong 3.5$ in., 89 mm). For a representative bending moment calculation, the intermediate value of $R/2$ was used. Shaft compressive force was accordingly calculated to yield a shaft bending moment of 40,250 in. lb ($= 1.75 \times 23,000$), 4554 N m.

As illustrated in Figure 10.16, the bending moment was calculated to cause a 3.8 mil (0.097 mm) transient thermal bow of the shaft. This result was initially not believed by the client. So as a prudent next step, the client shop tested the plant's spare feed pump rotor on a rotor balancing machine with specially installed locally placed heaters on the shaft sleeves. This test confirmed the author's contention shown in Figure 10.16. That led to a shaft-sleeve retaining nut modification retrofit, by interposing a compressible gasket under each shaft-sleeve retaining nut at both ends of the pump shaft. This low-cost retrofit more evenly distributes the compressive force circumferentially, and freely allows the inherent cyclic differential thermal expansion while maintaining the nominal sleeve assembly compressive force. This retrofit is now installed on all four of the plant's 50% feed water pumps, with total success in eliminating the vibration problem's root cause, as reported by Adams and Gates (2002).

10.6 Power Plant Boiler Circulating Pumps

Some fossil-fired boilers for steam power plants incorporate boiler circulation pumps. This measure significantly reduces the size of the boiler, compared to free-convection boiler designs for the same capacity and thus significantly reduces the boiler first cost. Such boiler circ pumps come in a range of sizes matched to the output capacity of the boiler. The boiler circ pump discussed here is for quite older generating units rated at 130 MW. Each boiler is supplied with three such pumps, but can operate at full capacity with only two pumps operating, so pump maintenance and rebuilds can take place without backing off on unit generating power output.

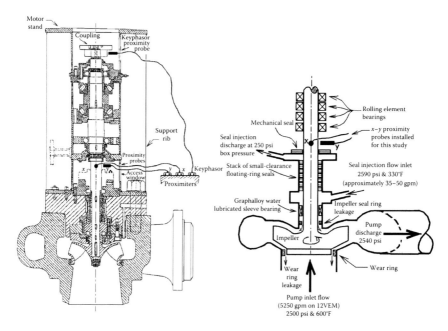

FIGURE 10.17 Boiler circulation pump: (a) cross section and (b) seal injection.

Figure 10.17a shows a cross section of the boiler circ pump in this case study. These pumps have a vertical centerline of rotation, as illustrated.

The author was retained to do a thorough top-to-bottom investigation of these pumps because they each had, for many years, required rebuilds approximately once per year, based on the floating ring seal clearance wearing open enough to exceed the capacity of the feed water system to supply seal injection water from a point upstream of the feed water manifold, that is, at a pressure sufficiently higher than boiler feed water pressure (see Figure 10.17b).

To gain potentially valuable diagnostic information, the author devised special fixtures so that shaft vibration measuring probes could be installed on one of the operating boiler circ pumps, at the only readily accessible location, the coupling. Figure 10.17a schematically illustrates that installation of x and y proximity probes. A spectrum of one of the shaft vibration displacement signals is plotted in Figure 10.18. It was fully anticipated that the vibration levels at the coupling would be relatively small. Also, since the rotational centerline is vertical, the appearance of a modest level of $N/2$ subsynchronous component was not surprising.

But what was surprising was the high level of the $6N$ impeller vane-passing shaft vibration level so far away from the impeller. This finding led to the identification of the root cause of the consistently high wear rate of the floating rings seal clearance, but not before some additional detective work.

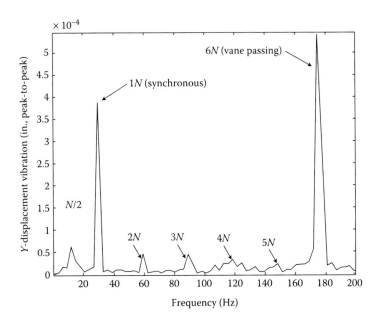

FIGURE 10.18 Shaft coupling vibration displacement spectrum.

It was discovered by the author that when one of these boiler circ pumps is removed to the rebuild shop, the boiler remains in operation. Furthermore, during the time the pump is away for rebuild, the only thing that separates the inside 600°F 2500 psi (316°C 170 bars) boiler water from the surrounding area is a single isolation valve, hopefully with a good seat! So naturally, no one in their right mind is going to stick his (or her) head down into the impeller vacated area to take an inside micrometer reading of the impeller-eye wear ring ID. When the simultaneous opportunity of unit shut down and boiler circ pump removal coincided, the impeller-eye wear ring ID was measured. The measurement showed that the impeller wear ring radial clearance was several times what it should be, worn open from years of unattended inspection and use.

The author made calculations which showed that the backflow through the impeller wear ring clearance to the impeller inlet (eye) was easily as high as 30% of the pump's rated capacity. This surely explained why these pumps were never quite delivering rated capacity, even just after a rebuild. Upon consultation with the boiler circ pump OEM's lead hydraulics engineer, it was revealed that such a high rate of wear ring backflow would significantly disrupt the impeller inlet flow velocity distribution, with the likely outcome of a quite high dynamic hydraulic impeller force at the vane passing frequency. The root cause *smoking gun* was finally found. The impeller wear rings need to be replaced when their clearance wears open to twice the as-new value. The pump OEM also has available a zero leakage mechanical seal retrofit that eliminates the floating ring seals and thus eliminates the need for seal injection flow. However, even with the superior OEM seal retrofit mechanical seal, the excessive 6N vane passing vibration must still be kept in check by proper wear ring replacement when needed.

10.7 Nuclear Plant Cooling Tower Circulating Pump Resonance

This case deals with a nuclear power plant boiling water reactor (BWR) generating unit rated at 1300 MW. It has three cooling tower circulation pumps, requiring at full load only two of these pumps operating during the winter season, but all three operating during the summer season. These are quite large vertical rotational centerline pumps with a rotor speed of 325 rpm (5.5 Hz). Over a period of several years, at least one of these pumps moved sufficiently out of plumb to need re-plumbing. That was accomplished during the winter season when this pump was taken out of service. Following this effort the unit was test run and found to have excessive levels of vibration at rotor speed frequency (5.5 Hz). The plumb shimming used at the base reduced base floor contact, called "soft foot." A structural resonance frequency moved right into the spin frequency. The result was a high level of structural resonance vibration, a rocking mode with maximum motion at the top of the unit (motor).

The ultimate long-term fix is to eliminate the *soft foot*. A cost-effective intermediate fix is to design and install a tuned vibration absorber attached to the top of the motor. The author and his staff designed a vibration absorber for this application. A tuned vibration absorber replaces the preexisting resonance with two side-band resonances, one below and one above the preexisting resonance frequency. The absorber is simply a spring–mass (Figure 1.1) tuned to the preexisting natural frequency, $\omega_n = \sqrt{k/m}$. The greater the absorber mass, the greater the frequency separation between the two resulting resonance peak replacements. This is a common cost-effective fix in the field. Figure 10.19 illustrates the design for this case.

10.8 Generator Exciter Collector Shaft Critical Speeds

The collector stub shaft of a 250 MW 3600 rpm steam turbine generator had a long history of high brush wear rate. It also needed frequent grind and polish operations on the collector rings to avoid "flash-over" arcing,

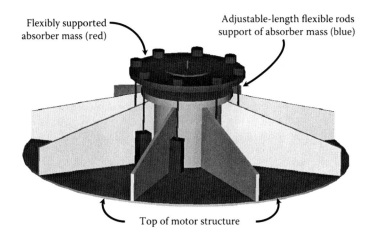

FIGURE 10.19 (See color insert following page 262.) Vibration absorber (blue and red) atop pump motor.

which occurs when the collector brushes do not follow (i.e., remain in contact with) the collector ring radial excursions. Such excursions come from rotor vibration and any collector-ring run-out from circumferentially nonuniform wear imposed (e.g., via the rotor vibration-modulated brush rub-contact forces; see Section 12.3, Chapter 12). The collector shaft is shown in Figure 10.20. It did not have any shaft targeting noncontacting displacement proximity vibration probes installed on it when the author

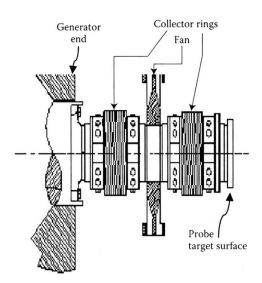

FIGURE 10.20 Collector shaft.

Forced Vibration and Critical Speed Case Studies

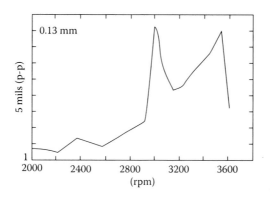

FIGURE 10.21 Measured 60 Hz vibration.

was first retained to troubleshoot this problem. So the first step by the author was to retrofit x and y (90° apart) proximity probes targeting the outboard end of the collector shaft. In addition, x and y velocity pickups were installed on the proximity probe holders so that absolute as well as relative collector shaft orbital vibration signals could be continuously measured. The recorded collector shaft vibration measurements are plotted in Figure 10.21. Utilizing the RDA rotor vibration analysis software provided with this book, a rotor unbalance vibration computer model simulation confirmed the vibration peak at 3000 rpm to be the *generator second critical speed* and the peak at 3550 rpm to be a *collector shaft critical speed*.

A heavier outboard disk on the collector shaft was substituted in the computer model and predicted that the addition of a heavier disk at the outboard end of the collector shaft would push the collector shaft's critical speed down by 300 rpm below 3550 rpm. This analysis result provided the basis for a practical cost-effective fix.

10.9 Summary

Critical speed troubleshooting case studies were selected for this chapter to highlight a number of important considerations. The large steam turbine cases demonstrate how rigidly coupled drive lines with more than two radial bearings are susceptible to nonstationary vibration characteristics resulting from statically indeterminate bearing load shifting. These turbine cases, along with those in Chapter 12, support the author's belief that in cases of excessive vibration problems in large steam turbine generators, journal bearing static load changes, such as from bearing alignment shifting, are most often one of the contributing factors to the problem.

Four feed pump cases were selected for this chapter to stress the considerable challenges in developing good predictive rotor vibration models and making correct problem diagnoses for high-energy density pumps. These challenges arise for two generic reasons. The first reason is the multiplicity of liquid-filled annular rotor–stator small-clearance radial gaps that dominate the vibration characteristics of such machines, combined with the dimensional variability of these small radial gaps from ID and OD manufacturing tolerances and in-service wear. Second, the potentially large and uncertain hydraulic radial static impeller forces, which vary with a pump's operating point over its head-capacity curve (see Section 6.1), introduce considerable uncertainty in radial bearing static loads. Since a bearing's rotor dynamic characteristics are a strong function of its static load, the inherent uncertainty of impeller static radial forces adds to the uncertainty of rotor vibration modeling and problem diagnoses. These pump cases demonstrate the diligent persistence required to isolate the root cause(s) in cases where simply rebalancing the rotor does not solve the problem.

The boiler circulating pump case study unfortunately typifies many long-standing problems that can exist for decades at a plant, incurring high upkeep costs and being a major nuisance for the operators until some new person arrives at the plant one day and simply asks why. The service and refurbishment companies are not necessarily going to ask the "why" because the plant's problem is their "honey basket." The cooling tower pump case study surely demonstrates what can easily happen when the in-house individuals in charge of a major task, like replumbing a massive piece of equipment, do not know what they are doing.

Bibliography

Adams, M. L. and Gates, W., *Successful Troubleshooting a Nuclear Feed Water Pump Vibration Problem*, Proceedings of ASME-NRC Annual Conference, Washington, DC, June 2002.

Adams, M. and Adams, M., *On the Use of Rotor Vibration Analysis and Measurement Tools to Cure Power Plant Machinery Vibration*, IFToMM Seventh International Conference on Rotor Dynamics, Vienna, Austria, September 25–28, 2006.

Giberson, M. F., *Evolution of the TRI Align-A-Pad® Tilt-Pad Bearing Through 20 Years of Solving Power Plant Machinery Vibration Problems*, Electric Power Research Institute (EPRI) Symposium on Trouble Shooting Power Plant Rotating Machinery Vibrations, San Diego, CA, May 19–21, 1993.

McCloskey, T. H. and Adams, M. L., *Troubleshooting Power Plant Rotating Machinery Vibration Problems Using Computational Techniques*, Fifth IMechE International Conference on Vibration in Rotating Machinery, Bath, England, September 1992.

11
Self-Excited Rotor Vibration Case Studies

11.1 Introduction

As discussed in Section 9.8 of Chapter 9, dynamic instability leading to self-excited rotor vibration can originate from several different sources. Modern turbomachinery is probably where self-excited rotor vibration is most often encountered, because of the high-power transfers and the attendant fluid dynamical interaction phenomena that abound inside turbomachinery. In most cases, self-excited rotor vibration can lead to excessively high vibration levels, and therefore when it is encountered the mandatory objective is its elimination from the operating zones of the machine. Paraphrasing Professor Stephen Crandall (1983), the available rotational kinetic energy in a machine is typically several orders of magnitude greater than the energy storage capacity of a destabilized rotor-whirling mode, and thus only a miniscule portion of the rotor kinetic energy channeled into an unstable mode can readily cause a failure. Even with the best of design practices and most effective methods of avoidance, many rotor causes of dynamic instability are so subtle and pervasive that incidents of self-excited rotor vibration in need of solutions continue to occur. Three interesting case studies from the author's troubleshooting experiences are presented in this chapter, all involving large steam turbo-generators. Each of these three cases is unique and thus individually informative.

11.2 Swirl Brakes Cure Steam Whirl in a 1300 MW Unit

As described in Section 6.3 of Chapter 6 and Section 9.8 of Chapter 9, steam whirl is a subsynchronous self-excited vibration, typically of the lowest natural frequency forward-whirling rotor mode. As previously explained, *steam whirl* differs from its close relative *oil whip* in that steam whirl has an instability threshold dictated by increasing power output, not increasing rotor speed. While steam whirl forces are present in all stages of steam turbines, when steam whirl occurs it is always in the high-pressure turbine, as explained in Section 6.3 of Chapter 6.

The case study presented in this section pertains to a 1300 MW cross-compound steam turbo-generator. As explained in Section 10.2 of Chapter 10, cross-compound units have two drive lines, each providing approximately 50% of the unit's rated power output. In the unit of this case, each drive line is rated at 650 MW. One drive line has a double-flow HP turbine, two double-flow LP turbines, a generator, and a drive-line-mounted exciter. The other drive line has a double-flow IP turbine, two double-flow LP turbines, a generator, and a drive-line-mounted exciter. The power plant in this case houses two of these 1300 MW generating units, each having it own coal-fired steam boiler. After being in service for about 15 years, one of the two units developed a subsynchronous vibration of excessive magnitude in its HP turbine section. At the point where the author became involved in this problem, steam whirl was already deemed the likely root cause responsible for the excessive subsynchronous (28 Hz) vibration on this 60 Hz machine. It had been established that the vibration kicked in at about 900 MW as the load was increased on the machine. The unit was temporarily derated to 900 MW pending a solution to the problem. The organization owning the power plant therefore incurred a 300 MW loss in generating capacity with the attendant lowered fuel efficiency of the unit at the reduced power output. At that point in time, the owning organization also had some of its nuclear-powered generating units under a temporary U.S. Nuclear Regulatory Commission (NRC) mandated shutdown pending resolution of regulatory concerns. The search for a solution to the problem was intense. The author was directed to this troubleshooting case by the EPRI, who sponsored the author's work on it.

As in the case presented in Section 10.2 of Chapter 10, the author developed a calibrated model for the entire HP drive line. That is, not to guess whether, or how much, the two LPs, the generator, and the exciter affect the vibration problem, but instead to include the complete rigid-coupled rotor drive line in the model.

The complete HP drive line is supported in six journal bearings. Figure 11.1 shows only the HP turbine portion of the HP drive line, which is where "swirl brakes" were retrofitted. As described in Section 6.3 of Chapter 6, the destabilizing effect known as steam whirl is actually the sum of two effects: (1) the Thomas-Alford forces due to variation of circumferential torque distribution (see Figure 6.2b) and (2) the leakage steam pressure distribution effect within the annuluses between the labyrinth strips of the blade tip seals (see Figure 6.6d), which is strongly abetted by the corotational preswirl of steam entering the seals. Swirl brakes work to negate the second of these two contributions, which is approximately twice as strong as the first contribution. Swirl brakes are axially oriented flow straightening stationary vanes installed just upstream of the annular tip seals. In this case, the rotor vibration model was used to determine instability thresholds (see Section 4.3 of Chapter 4) and thereby evaluate how many of the

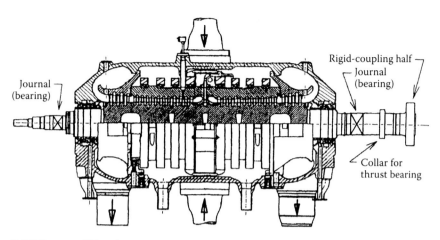

FIGURE 11.1 HP turbine section of the HP drive line of a 1300 MW 3600 rpm cross-compound steam turbo-generator.

first HP stages should be retrofitted with swirl brakes, which drastically reduce seal inlet corotational preswirl. The destabilizing effect of steam prerotation ahead of a seal varies approximately with the gas density and thus the largest stabilizing influence yielded by swirl brakes is in tip seals of the highest pressure stages.

As is typical, labyrinth tip seals in this HP turbine have multiple annular sealing strips at each stage. To make space for swirl brake axial strips, the first annular strip was removed at each stage and retrofitted with swirl brakes. This reduces the efficiency of the turbine; hence so an evaluation was made on the annual incremental fuel cost increase for each stage retrofitted with swirl brakes. Stage-by-stage cross-coupled (skew symmetric) bearing-like stiffness coefficients were incorporated into the RDA model at each HP turbine stage, with and without swirl brakes. The resulting analyses with the model indicated that swirl brakes installed in the rotor mid-plane seals and in the first three stages (both flow legs) would produce most of the stabilization influence that could be accomplished. Not only is the steam pressure (density) highest in this axial central region of the HP turbine, but the mode shape of the destabilized mode has its largest receptiveness (i.e., magnitude) in the axial central region. The model-computed HP unstable mode shape (Figure 11.2) clearly shows this, and it also shows that the steam whirl self-excited vibration in this case is primarily in the HP turbine. That is, the rest of the rotor sections rigidly coupled to the HP turbine do not participate vigorously in the unstable mode's self-excited vibration. This is consistent with the monitored vibration measurements from this machine, which showed the significant subsynchronous vibration component concentrated in the HP rotor.

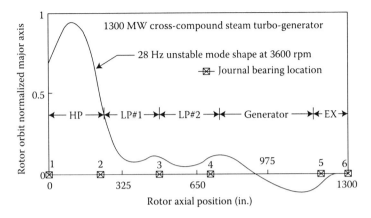

FIGURE 11.2 Steam whirl excited unstable mode of HP drive line.

The fix implemented at the power plant on this unit was therefore to install swirl brakes in the rotor mid-plane seals and in the first three stages of both flow legs. The unit was then able to operate free of steam whirl up to 1250 MW. Some subsequent adjustments to journal bearing vertical alignments provided the additional stabilizing influences that allowed the unit to be operated at its rated 1300 MW capacity, free of steam whirl.

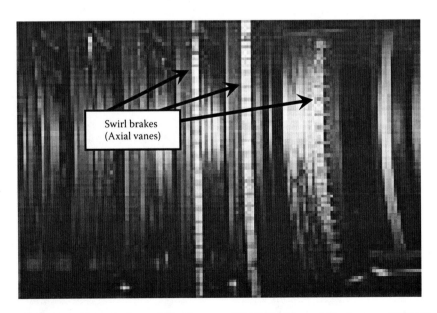

FIGURE 11.3 **(See color insert following page 262.)** HP turbine swirl brakes on a 1300 MW steam turbine.

A photo of the retrofitted swirl brakes is shown in Figure 11.3. Further details of the analyses for this case are given in the EPRI (1993) Symposium Short Course Proceedings and by McCloskey and Adams (1992).

11.3 Bearing Unloaded by Nozzle Forces Allows Steam Whirl

The steam turbo-generator unit in this case study is a 650 MW 3600 rpm tandem compound configuration (i.e., one drive line). It is one of five such units housed at the same plant. At about nine months prior to the unit's next scheduled outage for inspection and overhaul, the unit started to exhibit a large amplitude subsynchronous 27 Hz vibration concentrated in its HP–IP turbine section, 25 mils (0.64 mm) p.p. at journals. The initially diagnosed cause of the vibration source was *steam whirl*, because the associated threshold of instability was machine power output dependent. At the point in time when the author became involved in troubleshooting this problem, the subsynchronous vibration kicked in at about 500 MW as the load was increased on the machine.

Upon the strongest of recommendations by the author, the machine was temporarily derated to 500 MW. The power output where the subsynchronous vibration kicked in progressively lowered as the machine's scheduled outage approached, indicating a progressive worsening of the root cause. Accordingly, the unit was progressively derated in increments to about 300 MW as its scheduled outage was reached. When the unit was operable at its full 650 MW rated capacity, the power company owner of this plant had about 1000 MW of excess power from its most efficient generating units to sell to other power companies, such as the neighboring organization owner of the generating unit covered in Section 11.2, which was significantly short of capacity at the time because of its nuclear units in regulatory shutdown. Understandably, the quite significant income reduction from derating the machine to operate free of the high-amplitude subsynchronous vibration precipitated a tug-of-war between production financial management and plant engineering. Nonetheless, with the author's help, the engineers prevailed over the "bean counters" (not an everyday event!) and the unit was derated as necessary to operate free of the subsynchronous vibration. Naturally, the search for a solution to this vibration problem was intense.

A review of operation and maintenance records for this machine disclosed a history of excessive HP-turbine impulse (control stage) nozzle erosion for the unit. This discovery strongly suggested to the author that uneven static radial steam forces on the HP–IP rotor (at its impulse stage, see Figure 6.7) produced a net static radial rotor force that partially unload journal bearing static loads enough so that the bearings' normal squeeze-film capacity to suppress steam whirl was significantly diminished. The HP–IP journal bearings for this unit are of a 6-pad tilting-pad configuration with the rotor weight vector directed into the bottom pad's pivot location. In this scenario, the impulse stage's static radial load on the HP–IP rotor slowly increases over time because of the progressive closing of an impulse-stage control nozzle as some other nozzles' steam-flow areas enlarge due to erosion. This control nozzle closing must of course occur in order to maintain the steam power input to the machine within its rated capacity.

To test this hypothesis, the author developed a finite-element based total drive-line lateral rotor vibration RDA model calibrated for this unit's vibration symptoms, as had previously been done in successfully troubleshooting the unit in the Section 10.2 case study. Net static radial loading conditions on the HP–IP rotor were calculated as a function of slowly progressing time-dependent nozzle wear. Several analysis cases were undertaken by incorporating these progressive HP–IP rotor load changes into the journal bearing static loads for the drive line. The resulting stiffness and damping coefficients for all the journal bearings plus the skew-symmetric bearing-like stiffness coefficients for the HP section steam whirl forces were incorporated into the rotor vibration model of the unit. As reported by McCloskey and Adams (1992), the computed model results correlated well with the actual time line progression of the continuously worsening subsynchronous vibration problem. The conclusion was therefore drawn, with a high degree of confidence, that the vibration was in fact steam whirl and that progressive nozzle wear in the HP turbine impulse stage was the root cause. Therefore, a new nozzle plate replacement, with a change in material to provide improved resistance to erosion wear, was immediately ordered and ready for installation by the time of the scheduled outage. During the scheduled outage, the old nozzle plate was inspected and as expected was found to have considerable nozzle erosion wear. The unit was put back into service with the new nozzle plate and operated free of steam whirl up to its full rated 650 MW capacity. The unstable mode's shape is shown in Figure 11.4 and it is quite similar to that shown in Figure 11.2 for the steam whirl case in Section 11.2. The model-computed mode shape in Figure 11.4 shows that the rest of the rotor sections rigidly coupled to the HP–IP rotor do not participate vigorously in the steam whirl vibration. This is consistent with the monitored vibration measurements from this machine at the time of the problem. The experience gained from this problem by the owning company of this unit was especially valuable given the fact that the plant houses five such 650 MW units. In the same time frame of this problem, the author was consulted by another power company having a similar problem on a 620 MW version of the same design machine.

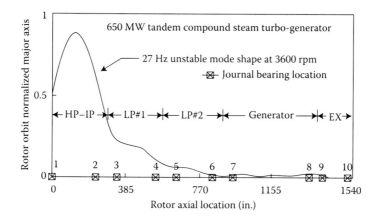

FIGURE 11.4 Steam whirl unstable mode of complete drive line.

11.4 Misalignment Causes Oil Whip/Steam Whirl "Duet"

The steam turbine generator unit in this case study is a 430 MW 3600 rpm tandem compound configuration (i.e., one drive line). It is the largest of four generating units housed at its plant. The author was consulted because the unit was in a derated mode of operation due to a strong subsynchronous 28.5 Hz vibration in the HP–IP rotor at loads above 390 MW. Based upon prior experiences, such as with the cases presented in Sections 11.2 and 11.3, it appeared to be a clear-cut case of steam whirl. However, the manufacturer of this machine claimed that this design did not have any history of steam whirl at other plants where the same design had been installed and no information was found at these plants to refute OEM's claim.

In a close collaborative effort with the OEM, the author developed a finite-element-based total drive-line rotor-bearing RDA model of the machine from detailed drawings and other information supplied by the OEM and the power company owner of the machine. Based on then recent bearing elevation measurements, for both "cold" and "hot" machine conditions, the rotor static sag-line was computed and the journal bearing static loads were thereby determined for both cold and hot conditions. The difference in journal bearing static loads between cold and hot elevation readings was large enough to critically affect the loads on the HP–IP bearings no. 1 and no. 2. Based on the so determined bearing static loads, bearing stiffness and damping coefficients were computed (see Chapter 5).

Under the hot operating condition, bearing alignments at the HP–IP end of the machine were significantly lowered relative to the rest of the machine as compared to the cold condition. Consequently, HP bearing

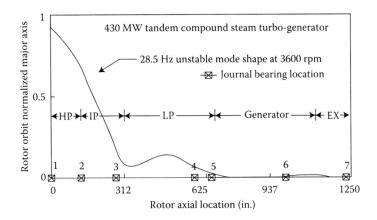

FIGURE 11.5 Oil whip/steam whirl unstable mode of complete drive line.

no. 1 (outboard) was about 90% unloaded. In fact, the HP–IP rotor was operating nearly in a condition of being cantilevered off the rest of the machine. It was amazing to the author that the rugged construction of the rotor had allowed this operating mode without structural fatigue damage to the shaft. However, as the model analyses showed, this allowed bearing no. 1 to contribute significant oil whip forces to help the steam whirl occur. All of the machine's journal bearings were of the two-lobe configuration, sometimes referred to as the "lemon" bore design, a metaphor particularly appropriate for this plant's installation. The unstable mode shape for this case is shown in Figure 11.5 and it is significantly different than the two unstable mode shapes shown in Figures 11.2 and 11.4 for the two other steam whirl cases presented in this chapter. Figure 11.5 reflects that bearing no. 1 was almost completely unloaded.

Further computer analysis studies singled out an optimum solution to stabilize the 28.5 Hz mode, which resulted in stable operation up to the machine's rated output. Bearing no. 1 was replaced in the model with a four-pad tilting-pad journal bearing with a range of preload factors, of which 0.5 provided the best compromise. As a result of the author's analyses, the plant replaced bearing no. 1 with a four-pad tilting-pad journal bearing having a preload factor of 0.5. This retrofit enabled the full load operation of the machine.

11.5 Summary

These cases and those of Chapter 10 demonstrate convincingly that adroit use of computer modeling drastically increases the probability of correctly diagnosing and curing difficult rotating machinery vibration problems that are not solved by routine maintenance actions or trial-and-error approaches.

Bibliography

Crandall, S. H., *The Physical Nature of Rotor Instability Mechanisms* (M. L. Adams, Ed.), Proceedings of the ASME Applied Mechanics Division Symposium on Rotor Dynamical Instability, AMD Vol. 55, pp. 1–18, 1983.

Electric Power Research Institute (EPRI) *Short Course,* Proceedings, EPRI Symposium on Trouble Shooting Power Plant Rotating Machinery Vibrations, San Diego, CA, May 19–21, 1993.

McCloskey, T. H. and Adams, M. L., *Troubleshooting Power Plant Rotating Machinery Vibration Problems Using Computational Techniques,* Fifth IMechE International Conference on Vibration in Rotating Machinery, Bath, England, September 1992.

12
Additional Rotor Vibration Cases and Topics

12.1 Introduction

The commonly identified rotor vibration root causes and symptom descriptions specific to each cause are summarized in Section 9.8. Chapters 10 and 11 present specific troubleshooting case studies that fall into two of the most frequently identified problem categories and that are drawn from the author's own troubleshooting experiences. This chapter is more a potpourri of rotating machinery vibration problem topics not readily grouped into a broad generic category, some taken from the author's own experience.

12.2 Vertical Rotor Machines

The topic of *vertical machines* warrants special treatment. The author gained valuable experience and insights on vertical machines (while employed at the Westinghouse R&D Center) from bearing and rotor vibration research on PWR primary coolant pumps, both of the shaft-sealed type for commercial nuclear plants and the canned-motor type for naval propulsion systems. That these types of pumps are vertical is dictated by the piping layout constraints of a typical PWR primary flow loop. Concerning rotor-bearing mechanics, vertical machines are fundamentally more difficult to analyze and understand than horizontal machines primarily because the radial bearing loads are not dead-weight influenced, the rotor weight being carried by the axial thrust bearing. Radial bearing static loads in vertical machines are therefore significantly less well defined and more nonstationary than bearing static loads in horizontal machines. Given the strong dependence of journal-bearing rotor dynamic characteristics on bearing static loads, the rotor vibration characteristics of vertical rotor machines are typically quite uncertain and randomly variable, far more than horizontal machines. While the author's Westinghouse experience was still fresh, he disclosed his insights on vertical machines into the EPRI publication by Makay and Adams (1979), which delineates important design and operational differences between vertical machines and horizontal machines.

A shaft-sealed reactor coolant pump (RCP) for a PWR nuclear power plant is illustrated in Figure 12.1. This pump is approximately 25 ft (7.6 m) high. The motor and pump shafts are rigidly coupled, which enables the entire coupled-rotor weight plus axial pump hydraulic thrust to be supported by one double-acting tilting-pad thrust bearing. This is the standard arrangement supplied by the U.S. RCP manufacturers. A major European pump manufacturer employs a flexible coupling, necessitating two thrust bearings, one for the pump and one for the motor. In the

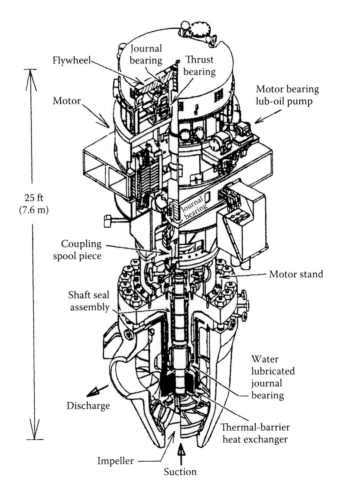

FIGURE 12.1 RCP of 100,000 gpm capacity and speed of 1200 rpm; typical PWR primary loop conditions are 2250 psi (153 bars), 550°F.

Figure 12.1 configuration, a large flywheel mounted at the top of the rotor is approximately 6 ft (1.8 m) in diameter and 15 in. (0.38 m) thick. It provides a relatively long coast-down time to insure uninterrupted reactor coolant water flow during the transition to emergency backup power in a pump power interruption. The use of a spool piece in the rigid coupling is to allow inspection and repair of the pump shaft seals without having to lift the motor. The pump impeller OD is approximately 38 in. (0.97 m). The rigidly coupled rotor shown in Figure 12.1 is held by three journal bearings, two quite narrow oil-lubricated tilting-pad journal bearings in the motor and one water-lubricated graphite-composition sleeve bearing located just above the thermal barrier. The water-lubricated bearing operates at primary loop pressure (~2250 psi, 153 bars) and thus this hydrodynamic bearing runs free of film rupture (cavitation). The attitude angle between the static load and the journal-to-bearing radial line-of-centers is therefore 90° over the full range of operation. This case is illustrated in Figure 6.6b. As a consequence, such RCPs usually exhibit a *half-frequency whirl* (i.e., half rotational speed) component in the rotor vibration signals. More detailed information on this pump and similar designs of different manufacturers is given in the Oak Ridge National Laboratory report of Makay et al. (1972).

RCP configurations in most U.S. and several foreign nuclear power plants have the rigid-coupled three-journal-bearing arrangement typified by the pump shown in Figure 12.1. From a rotor vibrations perspective, this presents possibly the most challenging type of system on which analysis-based predictions are made. That is, the journal-bearing static forces are not only devoid of dead-weight biasing, but they are also statically indeterminate. All this combines to make journal-bearing static loads, and thus rotor vibration characteristics, highly variable as related to manufacturing tolerances, assembly variations, pump operating flow point as well as normal wear at close clearance radial gaps. Given the absence of dead-weight journal-bearing loads, the primary source of radial static load is the static radial impeller force that changes with pump operating flow (see Section 6.1). However, given the three-bearing rigid-coupled configuration of RCPs, unless the three journal bearings are perfectly aligned on a straight line, there will be additional journal-bearing static loads from the bearings inadvertently preloading each other.

Jenkins (1993) attests to the considerable challenge in assessing the significance of monitored vibration signals from RCPs, and focuses on possible correlation of vibration signal content and equipment malfunction as related to machine age. He presents the "Westinghouse approach" in identifying vibration problem root causes and corrective changes for these aging Westinghouse RPC machines. In one of Jenkins' case studies, what appeared to be a sudden unfavorable change in monitored rotor vibration orbits was in fact eventually traced to a combined malfunction and faulty installation of the eddy current proximity probe system. This led to the conclusion that eddy current probe displacement systems are vulnerable to deterioration over time in the hot and radioactive environment around RCPs, and thus need to be replaced or at least checked at regular intervals. This "false alarm" case also emphasized the importance of closely following the proximity probe vibration instrumentation manufacturer's instructions regarding permissible part number with allowable target material combinations to avoid errors in probe-to-target scale factors, upon which all monitored vibration signals are based. For rotor vibration monitoring of RCPs, an x–y pair of proximity probes are installed 90° apart just below the coupling on a short straight low run-out section of the shaft.

A second case study presented by Jenkins (1993) pertains to an RCP of the model illustrated in Figure 12.1. It is one of the three identical pumps for a specific reactor. It developed a large half-frequency ($N/2$) rotor vibration whirl. With a cavitation-free water-lubricated sleeve bearing on a vertical centerline, there is nearly always some $N/2$ vibration content observed in the monitored rotor vibration signals of these pumps, but at tolerable levels when the pumps are operating "normally." In this case, the drastically increased level of $N/2$ vibration led to an investigation to determine the likely root cause(s) and the proper corrective action(s). Based on both the drastic increase in monitored $N/2$ vibration component (changed from 2 to 6 mils p.p. at coupling) and on a shift in static centerline position as indicated by the proximity probe DC voltages, it was diagnosed that the pump (water lubricated) journal-bearing clearance had significantly worn open. Some motor-bearing alignment adjustments allowed the $N/2$ vibration component to be held within levels deemed operable, pending a replacement of the pump bearing at the next refueling outage, or sooner if the monitored vibration developed a subsequent upward trend.

RCPs are not the only vertical pump applications. Some fossil-fired steam boilers in electric power-generating plants are designed with *boiler circulating pumps* (Section 10.6 of Chapter 10), which are incorporated into the design to make the boiler physical size much smaller than it would otherwise have to be if relying on free convection alone. Boiler circulating pumps have vertical centerlines as dictated by suction and discharge piping constraints. Steam turbine power plant *condensate pumps* are another example of vertical centerline machines. Marscher (1986) presents a comprehensive experience-based treatment to vibration problems in these vertical pumps. Most hydroelectric turbines and pump turbines are vertical.

12.3 Vector Turning from Synchronously Modulated Rubs

The propensity for rotor rubs to cause thermal rotor bows or local distortions that significantly increase synchronous vibration levels is greatest when the operating speed is close to a rotor critical speed with significant modal participation at the rub site. This tendency was recognized early in the era of modern rotating machinery as evidenced by the published works of Taylor (1924) and Newkirk (1926). More recently, Muszynska (1993) has provided an approximate computational model for this problem. The author became familiar with this problem firsthand when consulted by an electric power company in 1991 to help diagnose, explain, and cure the root cause of a serious exciter vibration problem that the company was experiencing on one of its fossil-fired 760 MW 3600 rpm steam turbine generator units. The unit had just been retrofitted with a brush-type exciter to replace its OEM-supplied brushless exciter with which the power company had had a long history of unsatisfactory experience. Both the original brushless exciter and the replacement brush-type exciter were configured to be direct connected to the outboard end of the generator shaft, see Section 10.8 of Chapter 10. The retrofitted brush-type exciter was custom designed and built by an OEM, which was not the turbine generator unit's OEM. Because of the serious vibration problem initially experienced with the newly retrofitted brush-type exciter, it was temporarily removed from generator outboard end and the turbine generator unit was then operated with an off-mounted exciter. To "add insult to injury" the power company had to lease (at a quite high daily rental rate) the off-mounted exciter from the turbine generator unit's OEM.

To a high degree of certainty, the exciter vibration problem was initiated by rub-induced friction heating at the sliding contact between the exciter brushes and collector rings. The power company's engineers working on this rub-induced vibration problem were surprised by a fundamental difference between a major symptom on this problem and the corresponding symptom of the previous rub-induced vibration problems that they had seen, that is, with rubs at packings, oil deflectors, interstage seals, and end seals. Specifically, in their prior experiences with rub-induced vibration, the vibration signal polar plot of amplitude versus phase angle (see Figure 9.8) exhibited a *counterrotational* slowly precessing "vibration vector." In the exciter vibration problem at hand, the vibration vector slowly precessed in the *corotational* direction, taking approximately 3 h per 360° vector turn, as illustrated in Figure 12.2. To understand and thereby properly diagnose the exciter vibration problem, the author developed a simplified model that explained the *corotational* direction precession of the exciter vibration vector. The presentation that follows on the author's simplified-model explanation is extracted from its first presentation, by Adams and Pollard (1993).

The simplified linear model has only 2-DOF (x and y) and treats the rotor as a single lumped mass, as shown in Figure 12.3. Furthermore, the radial stiffness and damping characteristic is assumed to be isotropic (i.e., same in all radial directions). Residual rotor unbalance is represented by the standard synchronous rotating force. The two equations of motion for

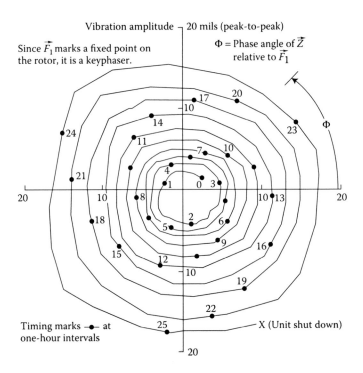

FIGURE 12.2 Polar plot of exciter journal vibration (Z) over 1 day.

this system are therefore given as follows (see Equation 2.1):

$$m\ddot{x} + c\dot{x} + kx = F_1 \cos \omega t$$
$$m\ddot{y} + c\dot{y} + ky = F_1 \sin \omega t \tag{12.1}$$

The steady-state vibration obtained from the particular solution to Equation 12.1 is as follows:

$$x_1 = Z_1 \cos(\omega t - \phi)$$
$$y_1 = Z_1 \sin(\omega t - \phi) \tag{12.2}$$

Assume that a rotor rub is initiated or at least modulated by this vibration. Consequently, a localized or cyclic heating of the shaft (hot spot) produces a thermal bowing or local distortion on the shaft such that an additional synchronous run-out and therefore additional unbalance-like force are added colinear with the vibration "vector" Z_1. The incremental change to the total vibration vector will lag the incremental unbalance force by the same characteristic phase angle, ϕ. Thus, the total unbalance

Additional Rotor Vibration Cases and Topics

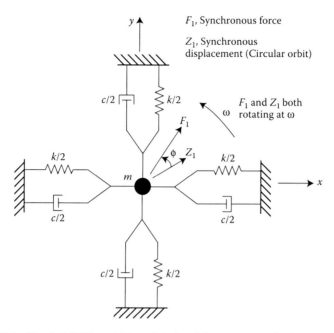

FIGURE 12.3 Simple 2-DOF model for rub-induced "vector turning" vibration.

force vector and the total synchronous vibration will be composed of the appropriate vector additions given in the following equations:

$$x = x_1 + x_2 = Z_1 \cos(\omega t - \phi) + Z_2 \cos(\omega t - 2\phi)$$
$$y = y_1 + y_2 = Z_1 \sin(\omega t - \phi) + Z_2 \sin(\omega t - 2\phi)$$
$$Z \equiv \sqrt{x^2 + y^2} = \sqrt{Z_1^2 + Z_2^2 + 2Z_1 Z_2 \left[1 - 2\sin^2\left(\frac{\phi}{2}\right)\right]}$$
(12.3)

Equation 12.3 is used to explore four cases that explain why the aforementioned exciter vibration problem was characterized by a *corotationally precessing vibration vector*, in contrast to plant engineers' prior experience on rub-induced vibrations at packings, oil deflectors, and seals.

Case 1: Stiffness-modulated rub with $\omega < \omega_{cr}$

If the rub is a "single-point" localized "hard" rub, or more generally stiffness modulated all around the shaft, then the incremental unbalance force (F_2) will be in phase with Z_1. Further, if the rotor speed (ω) is somewhat less than the critical speed (ω_{cr}), then the characteristic phase angle will be less than 90° (see Figure 1.5b). This case is illustrated in Figure 12.4a.

That F_2 is proportional and in phase with Z_1 is based on the notion that the rub contact pushes back on Z_1 approximately proportional to Z_1

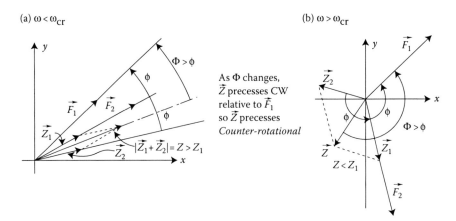

FIGURE 12.4 Stiffness modulated simple 2-DOF isotropic model.

(i.e., stiffness modulated such as rubs against packings, oil deflectors, interstage seals, and end seals, etc.). Two important observations can be made for this case from Figure 12.4a. First, the incremental effect of the rub-induced rotor bow is to increase the total synchronous vibration (i.e., $Z > Z_1$). Second, the phase lag (Φ) between F_1 (a point fixed on the rotor, i.e., keyphaser) and the total vibration Z is increased, which means the rotor high spot slowly precesses opposite the rotor spin direction. In other words, as time proceeds, the phase lag and the vibration amplitude will both slowly increase, because Z will produce a new incremental synchronous unbalance force colinear with Z and thus produce an additional incremental synchronous component lagging Z by ϕ (not added to Figure 12.4a).

Case 2: Stiffness-modulated rub with $\omega > \omega_{cr}$

In this case, using the same step-by-step approach, Figure 12.4b clearly shows that a stiffness-modulated rub at a speed somewhat above the critical speed also produces a total vibration vector that precesses opposite the rotor spin direction. However, in contrast to Case 1, the rub-induced rotor bow does not automatically tend to continuously increase the total vibration magnitude, an obvious consequence of ϕ being between 90° and 180°.

Case 3: Inertia-modulated rub with $\omega < \omega_{cr}$

Figure 12.5 is a visualization aid to explain the difference between *stiffness-modulated* and *inertia-modulated* rotor rubs. The stiffness-modulated rub (Figure 12.5a) pertains to a rotor-to-stator contact in which the normal contact force increases the more the rotor displaces into the rub site. In contrast, the inertia-modulated rub model presumes that the nonrotating contact rub surface is comprised of masses (e.g., exciter brushes) that are soft-spring preloaded against the shaft to prevent loss of contact, as

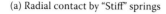
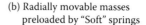
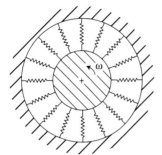
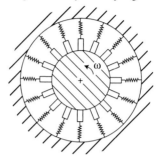

FIGURE 12.5 Rub force modulation models: (a) stiffness modulated and (b) inertia modulated. (Adams, M. L. and Pollard, M. A., *Rotor Vibration Vector Turning Due to Rotor Rubs*, Proceedings of the EPRI Symposium on Trouble Shooting Power Plant Rotating Machinery Vibrations, LaJolla, CA, May 19–21, 1993.)

illustrated in Figure 12.5b. Thus, the dynamics of inertia-modulated rotor rubs produce a normal dynamic contact force, which is proportional to the radial acceleration of the moveable stator masses. Therefore, for a synchronous circular shaft vibration orbit, the dynamic component of the normal contact force is 180° out of phase with the vibration displacement signal, as is obvious from the following equations: Given

$$x = X\cos(\omega t - \theta), \quad \text{then } \ddot{x} = -\omega^2 X\cos(\omega t - \theta) \qquad (12.4)$$

An inertia-modulated rotor rub thus tends to produce a maximum contact force (and thus "hot" spot) which is 180° out of phase with the displacement vibration vector ("high" spot). Therefore, in this case ($\omega < \omega_{cr}$, $\therefore \phi < 90°$) the result can be illustrated as shown in Figure 12.6a, which clearly indicates that the high spot (Z) will slowly precess in the *corotational* direction of shaft spin. Figure 12.6a also indicates that the total vibration amplitude is less than that from the initial mass unbalance alone, so in this case the vibration vector is not as likely to spiral out of control as in the next case.

Case 4: *Inertia-modulated rub with* $\omega > \omega_{cr}$

This case differs from Case 3 in that the characteristic phase angle (ϕ) is between 90° and 180°. Using the same type of vector diagram illustration as for the previous three cases, Figure 12.6b is constructed for this case. It shows that in this case the total vibration vector will also slowly precess in the *corotational* direction of shaft spin as with Case 3. However, in contrast to Case 3, the total vibration vector (Z) is shown to be more likely to spiral outward, and assuming insufficient heat removal capacity, can readily spiral out of control.

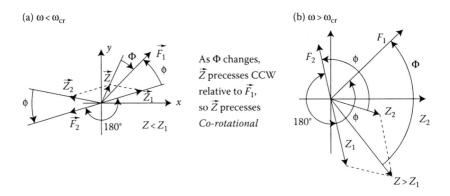

FIGURE 12.6 Inertia-modulated simple 2-DOF isotropic model.

Table 12.1 summarizes the four cases that are delineated with the model in Figure 12.3. The specific rub-induced vibration problem on the 760 MW unit referenced at the beginning of this section falls into Case 4. The successful fix implemented by the supplier of the custom-designed brush-type exciter retrofit entailed design changes to raise the exciter critical speed to a safe speed margin above 3600 rpm, thus moving it into Case 3 category.

Taylor (1924), Newkirk (1926), and Muszynska (1993) all sort of treat Case 1. Taylor and Newkirk also sort of treat Case 2. However, none of these publications mentions anything like the Adams and Pollard (1993) Cases 3 and 4 presented here.

TABLE 12.1
Four Cases of Rub-Induced Vibration Delineated by Simple Model

Rotor Speed	Stiffness Modulated Rub	Inertia Modulated Rub
	Case 1	Case 3
$\omega < \omega_{cr}$	• Slow precession of vibration vector is counterrotational	• Slow precession of vibration vector is corotational
	• Spiraling to high vibration levels is *more likely*	• Spiraling to high vibration levels is *less likely*
	Case 2	Case 4
$\omega > \omega_{cr}$	• Slow precession of vibration vector is counterrotational	• Slow precession of vibration vector is corotational
	• Spiraling to high vibration levels is *less likely*	• Spiraling to high vibration levels is *more likely*

Additional Rotor Vibration Cases and Topics

In all four cases, the thermal distortion attempts to make the hot spot become the high spot, and thus the position of the high spot will slowly change circumferentially and thereby continue to hunt for an equilibrium state, but never finding it. So the vibration phase angle continuously changes at a slow rate. The leverage that the thermal bow has on the vibration will be more amplified, the closer the running speed is to a critical speed with high modal participation at the rub site. The simple model indicates that the likelihood of severe vibration is much greater with situations that essentially fall into Cases 1 and 4. The plausible explanation of why such a phenomenon can reach a limit cycle is probably due to non-linear components in the heat removal mechanisms at work. That is, the incremental increase in heat removal near the hot spot region becomes progressively larger than the incremental decrease in heat removal near the cold spot. However, the existence of a limit cycle is not a guaranteed line of defense against a major failure since the limit cycle vibration may be larger than the vibration level sufficient to initiate failure.

For brush-type exciters, brush wear does not alleviate the rotor rub intensity since the brushes (rods of impregnated carbon) are kept in contact with the rotating collector rings by soft preload springs. Conversely, it is reasonable to hope that initial rub-induced thermal distortion rotor vibrations at packings, oil deflectors, and seals will eventually attenuate by rub alleviation through wear at the rub site. This may not be realized in specific configurations, especially when the wearing open of a radial clearance at the rub-site component appreciably reduces the component's otherwise significant contribution to total damping, for example, centrifugal pump wear rings.

12.4 Air Preheater Drive Structural Resonances

The boilers of fossil-fired steam turbine generator units are accompanied by several machines such as primary air fans, induced draft fans, forced draft fans, and air preheaters. The last one is a quite large carrousel that rotates about 3 rpm and supports hundreds of perforated metallic heat trays. The large carrousel slowly rotates through two wide sealed slots, one in the outgoing exhaust air duct and one in the incoming primary combustion air duct, thereby transferring heat from the outgoing hot exhaust air to the incoming primary combustion air. This is basically a heat recovery step that adds efficiency to the overall power generation process.

To drive the large air preheater carrousel at 3 rpm requires a three-stage speed reduction gear box, the last stage employing a worm-gear drive. The plant of this case study houses three tandem compound 3600 rpm steam

FIGURE 12.7 (See color insert following page 262.) Air preheater drive/platform with overhang support bars.

turbine generating units of the same configuration, each rated at 850 MW. The boiler for each generating unit has two air preheaters, for a total of six air preheaters for the three-unit plant. In an overall plant improvement audit, the drive and speed reduction gear box were identified as high maintenance systems due to a need for frequent rebuilds. This equipment deficiency apparently existed since the plant was commissioned several years previously. The plant's longstanding assessment (tribal knowledge) of the root cause was *high vibration*. The author was retained to determine how this vibration could be greatly attenuated and the air preheater drives thereby not needing frequent rebuilds. Figure 12.7 shows one of the plant's six air preheater drives in place and driving its air preheater carrousel at 3 rpm.

The author and his staff attached 11 channels of laboratory-grade piezoelectric accelerometers to a selected grid covering the drive support platform. The signals were simultaneously recorded with a multichannel digital tape recorder for subsequent analysis. This test setup was repeated on all six drive units in one 14-h work day visit. The recorded vibration measurements were analyzed for frequency content and found to have three main frequency components dominating the overall platform vibration,

Additional Rotor Vibration Cases and Topics

TABLE 12.2

Summary of Preheater Drive Vibration Reduction Improvements

	Original Structure		Modified Structure		Improvements	
Frequency No.	Frequency (Hz)	Vibration (mils p.p.)	Frequency (Hz)	Vibration (mils p.p.)	Vibration Twist and Flex Reduction	
1	6.3	2	8.9	0.5	4:1	Strong reduction
2	12.4	2	12.4	1.5	25%	Strong reduction
3	29.9	1.1	29.8	0.1	10:1	Strong reduction

6.3, 12.4, and 29.9 Hz. For each of these three harmonics, the signal amplitude and relative phase angles were extracted for the 11 simultaneously recorded locations. That information was then used to construct motion animations for each of the three dominant components of the platform vibration, which was almost entirely in the vertical direction.

The vibration motion animations for the three dominant vibration frequency components made it abundantly clear that the drive support platform simply needed more structural support of its cantilevered configuration. An inexpensive addition to the drive support structure is shown in Figure 12.7. After the new support bars were retrofitted to the drive platform, the author's staff returned to the plant and took vibration measurements on two of the units, to quantify the vast attenuation of the platform vibration. Even before these postfix measurements were taken, it was obvious even to one's finger tips that the platform vibration levels had been reduced significantly. Table 12.2 summarizes the vibration reduction improvements. Following installation of the added support bars to each of the six drive platforms, the need for frequent drive system rebuilds stopped. Success is fun, failure is not.

12.5 Aircraft Auxiliary Power Unit Commutator Vibration-Caused Uneven Wear

Figure 12.8 illustrates the auxiliary power unit (APU) for a commercial regional jet aircraft. The DC starter generator functions to start the small gas drive turbine, and then switches over to generate 10 kW of DC power. This is a standard piece of equipment on modern jet power aircraft.

Significantly uneven radial wear patterns on several commutator copper surfaces of 10 kW DC starter-generator units of APUs of a specific commuter jet led to an intensive engineering investigation to solve this

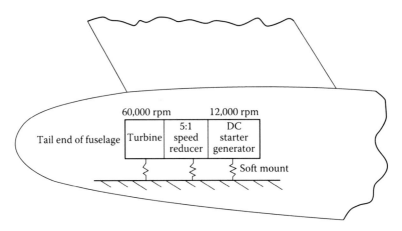

FIGURE 12.8 Commuter jet APU.

problem. In service, the circumferentially uneven radial wear typically progressed to a degree where the brushes could no longer track the commutator radial profile to maintain continuous rubbing electrical contact with the commutator, rendering the APU DC starter-generator inoperable. The author developed a commutator-brush dynamics-wear model and a corresponding new computer code to simulate commutator wear as a function of operating time. The code was formulated to simulate several design and operating parameters for this APU. This new analysis code was then used to simulate commutator time-dependent radial wear patterns around the commutator 360° arc of contact, specifically to study the influences of various parameters, especially (a) initial as-manufactured commutator radial run-out and (b) imposed unit vibration. The commuator rotor is illustrated in Figure 12.9 and carbon brush pair in their holder is shown in Figure 12.10.

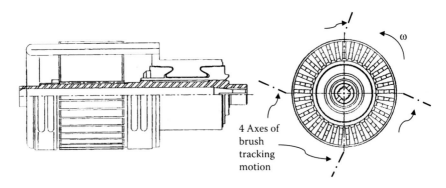

FIGURE 12.9 DC starter-generator rotor.

Additional Rotor Vibration Cases and Topics

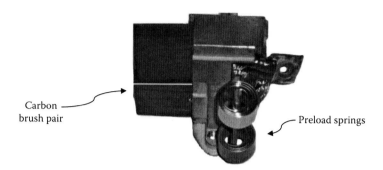

FIGURE 12.10 (See color insert following page 262.) Carbon brush pair in holder with preload springs.

The analysis development for simulated commutator wear is detailed by Adams (2007). The dynamics model of the spring-loaded brush is illustrated in Figure 12.11. Applying Newton's Second Law to the model illustrated in Figure 12.11 yields Equation 12.1 for the time-dependent contact force on the brush as the result of $x(t)$ tracking of the commutator profile.

$$F_c = \frac{m\ddot{x}\cos\alpha - (\dot{x}/|\dot{x}|)\mu_2 F_{sy} + F_{sx}}{[(\cos\alpha - \mu_1 \sin\alpha) + (\dot{x}/|\dot{x}|)\mu_2(\sin\alpha + \mu_1 \cos\alpha)]} \qquad (12.5)$$

It is well established that the wear rate for the rubbing brushes in this type of application is opposite of what occurs for a classical Coulomb friction wear. That is, the larger the normal contact force between brush and commutator, the smaller the wear rate, and vice versa. This is because the primary wear mechanism is neither abrasive nor adhesive wear, but is dictated by the current conduction effects across the rubbing contact. That is, the higher the contact force, the less the spark erosion wear of

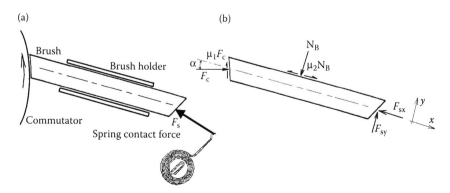

FIGURE 12.11 (a) Brush, holder, and spring. (b) Force free-body diagram of brush.

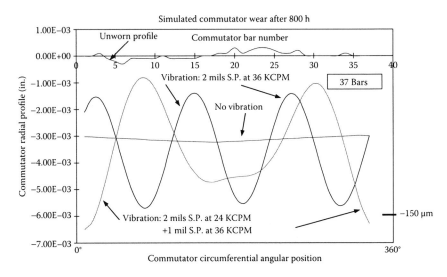

FIGURE 12.12 (See color insert following page 262.) Simulated commutator radial wear for APU DC starter-generator.

the brush. The wear model implemented in this analysis followed the well-established approach embodying the applicable wear phenomenon described by Adams (2007). Rubbing friction wear is also included, but its influence is secondary.

A sample of simulated wear results is shown in Figure 12.12. Note that the initial unworn profile (from factory measurements) is eliminated, as anticipated with the wear rate model. Similarly, for the case of no imposed rotor vibration, the wear model does not perpetuate but in fact works to eliminate initial manufacturing run-out. On the other hand, imposed vibration upon the rotor is shown by the simulation to produce the type of circumferentially uneven commutator wear patterns that were occurring in the fleet. Therefore, these dynamic wear simulation results strongly suggested that the root cause of the uneven commutator wear problem was excessive unit vibration imposed by the small 60,000 rpm gas turbine that connects to the APU's 12,000 rpm DC starter-generator through a planetary-gear 5:1 speed reducer (Figure 12.8). Subsequent vibration information from the turbine manufacturer confirmed this conclusion and thereby eliminated the commutator wear problem by curing the root cause of the turbine rotor vibration. The turbine manufacturer's fix was a change in material for the turbine rotor axial retaining bolt in order to eliminate a severe bolt material high-temperature creep phenomenon that allowed the turbine rotor to become loose in operation. Implementation of this turbine fix in the fleet resulted in the elimination of the commutator uneven wear problem.

12.6 Impact Tests for Vibration Problem Diagnoses

A laboratory impact test method for determining radial seal and bearing rotor vibration characteristics was discussed in Section 5.3 of Chapter 5 (see Figure 5.5). The major limitation of that experimental approach is the difficulty in getting sufficient energy into such highly damped dynamical components to enable retrieval of an adequately strong signal-to-noise response to the applied impact force.

To execute a comparable impact-based modal test on operating centrifugal pumps is an even bigger challenge than the laboratory experiment illustrated in Figure 5.11. Centrifugal pumps are highly damped dynamical systems. Although impact-based modal testing is a quite useful diagnostic tool for many low-damped structures, it was long considered not a practical or feasible diagnostic test method for centrifugal pumps in operation. In addition to having fluid-film bearings and radial seals, centrifugal pumps also internally generate a broadband set of dynamic forces emanating from the various internal unsteady flow phenomena (see Section 6.1 of Chapter 6). Furthermore, these pump unsteady-flow dynamic forces change during operation with changes in flow, head and speed, and change over time with internal component wear. In addition, the rotor dynamic properties of the bearings and radial seals of the typical centrifugal pump change significantly as a function of operating conditions and wear over time, making the rotor vibration natural frequencies nonstationary as well. Thus, the prospect of employing the quite powerful diagnostic test technique called *modal analysis* was basically not an option for centrifugal pumps until recent years. Marscher (1986) pioneered an impact method for centrifugal pumps in which multiple impacts are applied to the rotor (e.g., at the coupling) with impact magnitudes within ranges that are not injurious to a pump or its driver.

The key to the success of Marscher's method is the use of time averaging over several hundred impact strikes. By time averaging over several hundred impacts, only those vibration components that are the response to the impacts will be magnified in the time-averaging process. The time-averaged pump internally generated vibration and signal noise that do not correlate with the controlled impact strikes are progressively diminished as the number of time-averaged signal samples is increased. Marscher (1986) shows test results that provide convincing proof of the significant change in pump natural frequencies that can occur over the parameter changes within a pump's normal operating range.

Since Marscher first developed his test procedures, the field of data acquisition and signal analysis has advanced considerably (see Chapter 9). With the use of current generation multichannel high-sampling-rate digital tape recorders and companion analysis software that runs on a lap-top

PC right at the test site, Marscher's method can now be applied much more expeditiously and cheaply than when it was first implemented. Correctly diagnosing particularly troublesome pump vibration problems can be greatly facilitated by employing this modal test approach. The impact-based modal test method can greatly facilitate the development of "calibrated" computer models for pump vibration problem analyses (see Chapter 10) as useful in doing "what if" studies in search of the best corrective actions.

12.7 Bearing Looseness Effects

Vibration symptoms for mechanically loose connections are covered in Section 9.8. Probably every plant maintenance engineer and mechanic have their own long list of past cases where the root cause of a vibration problem was discovered to be looseness at the bearings. Virtually no rotating machinery type is immune to vibration problems when bearing or bearing support looseness is present. A short but informative paper by Bennett and Piatt (1993) documents three case studies that focused on looseness at journal bearings in power plant rotating machinery. Their three case studies are summarized in this section.

12.7.1 350 MW Steam Turbine Generator

This case study is in fact a continuation of the case study presented in Section 10.2 of Chapter 10, where the author's rotor unbalance computer model analyses on this 350 MW cross-compound steam turbo-generator indicated that bearing no. 2 (in the HP turbine) was not providing proper load support for the rotor. The excessive synchronous vibration peak (20 mils p.p.) of the HP turbine through its critical speed was not good for the turbine internal clearances and efficiency of the unit. The author's analyses further indicated that employing a modest amount of preload on bearing no. 2 (four-pad tilting-pad bearing) would reduce critical-speed peak vibration levels of the HP rotor to approximately half the experienced levels on run-ups and coast-downs. Based on the author's analyses, the indicated bearing preload was employed at the plant, and the result was as predicted: the HP critical-speed vibration peak level was more than halved. However, after only 6 months of operation, the problem reoccurred, indicating that the OEM bearing was not maintaining the setup.

Upon further investigation by the electric power company's engineers, it was uncovered that a number of deficiencies of the OEM HP turbine journal bearings contributed to the problem. Both HP journal bearings (no. 1 and no. 2) had developed looseness, which was the primary cause of the excessive critical-speed vibration peak level in the HP turbine. The looseness was found to be primarily an inherent characteristic of the OEM bearing design. Bearing no. 1 did not have bolts to hold together the top and bottom halves of the inside bearing support ring. Also, the bearings were designed to rely on differential thermal expansions to create the necessary "pinch" on the inner bearing ring by the bearing outer housing. Several attempts to create adequate pinch to properly secure the inner bearing rings of both HP bearings failed.

Because of the inherent design deficiencies of the OEM HP turbine bearing design, the electric utility company's corrective course of action was to find a retrofit replacement for the HP journal bearings that would not have the inherent deficiencies. Accordingly, a superior non-OEM six-pad tilting-pad bearing was retrofitted into the existing cylindrical bore bearing fits. Figure 12.13 shows both the original OEM bearing configuration and the non-OEM retrofit. After the outage to install the non-OEM bearings, excessive critical speed vibration in the HP turbine did not occur and that success of the fix has continued for several years. The electric utility company lists the following items as crucial to this success story: (1) use of horizontal joint bolts to insure adequate "pinch" on the inner bearing ring, (2) bearing pad preload, (3) and high-quality control over materials and construction details. They also recommend controlling the steam-valve sequencing so that the HP turbine bearings are always loaded (see Section 6.3.2 of Chapter 6).

12.7.2 BFP 4000 hp Electric Motor

During a routine maintenance vibration survey of plant machinery not instrumented with continuous vibration monitoring sensors, quite high vibration levels were detected on this 4000 hp 3600 rpm feed pump drive motor. Electrical problems were eliminated as the root cause because the vibration was dominated by the synchronous frequency component ($1N$) as well as the $2N$ and $N/2$ frequency components. The pump–motor set was removed from service and found to be out of alignment by about 9 mils. Furthermore, the drive-end bearing housing-to-endbell fit had 8 mils clearance instead of the zero to 1 mil pinch specified. After properly aligning the set and providing the proper bearing pinch, the unit was returned to service and exhibited an overall vibration level on the motor bearings of less than 1 mil p.p. One conclusion drawn from this case by the plant engineers is that bearing pinch is vitally important. The first of two feed pump case studies in Section 10.3 also demonstrates the importance of bearing inner shell pinch.

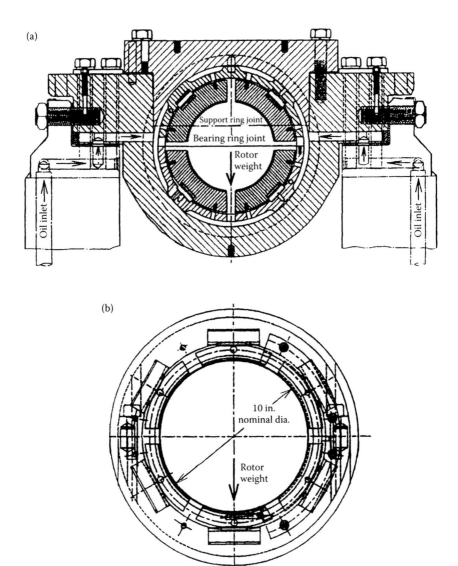

FIGURE 12.13 HP turbine tilting-pad journal bearings for a 350 MW unit.

12.7.3 LP Turbine Bearing Looseness on a 750 MW Steam Turbine Generator

This 3600 rpm unit has a similar rotor and bearing rigid-coupled configuration to that indicated in Figure 11.4 for a unit of the same manufacturer. Just following a major overhaul, high subsynchronous vibration (19.8 Hz) was experienced. Bearing or support-structure looseness was considered

Additional Rotor Vibration Cases and Topics 401

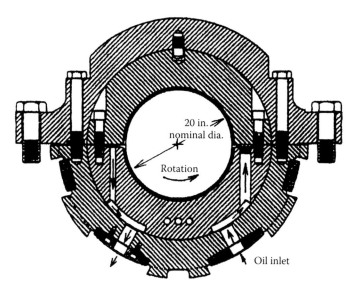

FIGURE 12.14 LP turbine bearing of a 750 MW 3600 rpm unit.

the most likely root cause or also a seal rub. This subsynchronous vibration was highest at bearing no. 6 (generator side of LP no. 2) and somewhat less on bearings no. 5 and no. 7 (see same drive line arrangement in Figure 11.4).

The LP turbine bearing (no. 6) was inspected during a short outage. An illustration of the LP turbine bearing configuration for this unit is shown in Figure 12.14. The spherical-seat pinch between the bearing housing and inner bearing halves was found to be zero, and the side alignment pad was loose. After adjusting the side adjustment pad and restoring pinch to the spherical-seat fit, the unit was returned to service with drastically reduced overall vibration levels at bearings no. 5 through no. 7. The lesson learned was that close attention must be paid to properly set the bearings during overhauls. Virtually any power plant engineer involved with major overhaul outages is unfortunately accustomed to debugging "new" problems caused by inadvertent mistakes and oversights when large machines are reassembled.

12.8 Tilting-Pad versus Fixed-Surface Journal Bearings

The tilting-pad journal bearing (also called pivoted-pad journal bearing, PPJB) has a proven history of avoiding the self-excited rotor vibration "oil whip," often encountered with fixed-surface cylindrical-bore journal bearings (CBJB). As explained in Section 2.4 of Chapter 2 and Section 9.8

of Chapter 9, the nonsymmetric portion of the bearing displacement-reaction dynamic force component perpendicular to radial displacement is a nonconservative destabilizing force. As explained by Adams and Makay (1981), PPJBs are basically immune to oil whip.

PPJBs were first introduced for vertical-rotor machines (see Section 12.2) because fix-surface CBJBs are most likely to cause self-excited subsynchronous rotor vibration when unloaded or lightly loaded. The success of PPJBs on vertical machines prompted designers to employ them on many horizontal-rotor machines where combinations of light bearing static loads and high rotational speeds made CBJBs highly susceptible to oil whip. Wide use of PPJBs has clearly shown that PPJBs are not a cure-all for fundamentally poor rotor dynamic design. Also, there are several ways in which PPJBs can be inadvertently designed, constructed, or applied to cause their own problems (see Sections 5.2.3 and 12.7.2), as Adams and Makay (1981) describe. Furthermore, PPJBs are somewhat more complicated and first-cost more expensive than CBJBs. Fixed-surface non-CBJBs (e.g., multilobe bearings) are a suitable improvement over CBJBs, at significantly less cost and complication than PPJBs. Each new rotating machinery design employing fluid-film journal bearings should be carefully analyzed before jumping to the conclusion that PPJBs are required for good rotor dynamic performance.

On large steam turbo-generators, Adams (1980) and Adams and McCloskey (1984) show that PPJBs are far superior to fixed-surface CBJBs under conditions of very large rotor unbalance such as those from loss of large turbine blades (see Section 2.5.2 of Chapter 2). Specifically, PPJBs more readily suppress subharmonic resonance from developing into catastrophically large amplitudes. On the other hand, Adams and Payandeh (1983) show that statically unloaded PPJB pads can incur a subsynchronous self-excited "pad-flutter" vibration that can lead to pad surface material fatigue damage (see Section 2.5.3 of Chapter 2).

12.8.1 A Return to the Machine of Section 11.4 of Chapter 11 Case Study

Discussion of the troubleshooting case study in Section 11.4 of Chapter 11 ended with the confirmation that replacing bearing no. 1 with a four-pad PPJB allowed that 430 MW unit to operate rotor dynamically stable up to its full rated capacity without excessive vibration. That retrofit of a PPJB to replace the original fixed-surface journal bearing was determined to be the least expensive and most readily implemented option available to the electric power company. The root cause of the problem was the severe shifting of journal-bearing support structures all along the machine's drive line, and correcting that root cause was deemed cost prohibitive. About 4 years after this retrofit was successfully implemented, this machine began to again exhibit some of the same subsynchronous self-excited vibration it had previously experienced, necessitating the unit to again be derated pending solution of the recurring problem. Apparently, shifting of journal-bearing support structures (the problem root cause) had continued to slowly worsen.

Again, the electric power company contracted the unit's OEM to retrofit a four-pad PPJB, (this time at bearing no. 2) to replace the original fixed-surface journal bearing, as the OEM had already done at bearing no. 1. This time the author's involvement was to independently check the OEM's rotor vibration computer model analysis results, because the OEM's predicted results surprised the power company's engineers. The OEM's computer analyses predicted that retrofitting a PPJB at bearing no. 2, to augment the PPJB they had previously retrofitted at bearing no. 1 4 years earlier, did more than just "push" the unstable mode above the rated power range of the unit. In fact, the OEM's analyses predicted that the unstable mode would totally "disappear." That is, the OEM predicted that the additional retrofit of a four-pad PPJB at bearing no. 2 effectively removed the offending mode from the rotor dynamical system of the unit.

The OEM's prediction suggested that the proposed retrofit would do more than provide the needed additional stabilizing damping capacity, that is, it suggested to the author that the proposed retrofit would radically alter the modal content of the system, eliminating the offending mode in the process. Augmenting the author's prior model of this unit to include the proposed bearing no. 2 PPJB retrofit, the author confirmed this "surprise" result predicted by the OEM. Bearing no. 2 retrofit was then implemented, which yielded the desired result. The unit was returned to service with restored operability to full rated capacity. However (the terrible "however"), this unit has more recently been having excessive synchronous vibration problems, also in the HP turbine. In this the author has analyzed the unit to assess potential benefits of employing additional balancing planes to those already used in plants balancing the unit's drive line. This is further discussed in Section 12.10.

12.9 Base-Motion Excitations from Earthquake and Shock

An important topic for which the author first gained appreciation during his Westinghouse experience is *earthquake and shock* inputs to rotating machinery. To study this topic, a good place to start is the keynote address paper of Professor Hori (1988) (University of Tokyo), in which he reviews published literature on the analysis of large power plant rotating machinery to withstand major earthquake events. The topic of base-motion excitations from earthquake and shock inputs to rotating machinery is one of the applications formulated by the author in his 1980 *Journal of Sound and Vibration* paper. More recently, the author has analyzed the stable nonlinear limit cycle of oil-whip rotor vibration, and confirmed interesting computational findings with laboratory tests, as provided in Adams and Guo (1996). Specifically, this research shows that a machine operating stable-in-the-small (i.e., below the linear threshold-of-instability speed) but above its "saddle node" speed can be "kicked" by a large dynamic disturbance into a high amplitude stable nonlinear limit cycle vibration with potentially catastrophic consequences.

12.10 Parametric Excitation: Nonaxisymmetric Shaft Stiffness

The rotor vibration consequences of anisotropic bending stiffness in a rotating shaft has been analyzed over the course of several decades by many investigators, with the earliest English citation attributed to Smith (1933). However, the German fluid mechanics specialist Prandtl (1918) appears to be the first investigator to publish a treatment of the problem. The fundamental problem has shown a considerable appeal to the more mathematically inclined mechanics theoreticians, providing a rich variety of possible vibration outcomes even for relatively simple configurations such as a Jeffcott rotor (see Figure 2.3) with axially uniform anisotropic shaft bending stiffness.

The fundamental problem did not attract the attention of rotating machinery designers until the post World War II period with the dramatic increases in maximum size of two-pole AC generators driven by large compound steam turbines. Figure 12.15, fashioned after that of Bishop and Parkinson (1965), illustrates the relative progressive change in rotor maximum physical size and power rating of two-pole steam turbine-powered AC generators from the 1940s to the 1960s. As the length-to-diameter proportions shown in Figure 12.15 indicate, this 20-year change from 120 MW to 750 MW generators has been accomplished by making the rotors longer but not appreciably larger in diameter, because the diameter is limited by allowable stress considerations. The progressive increase in slenderness led to lower generator critical speeds with the attendant increased propensity for oil-whip vibration. Also, the static deflection of generator center line under its own weight became a primary problem because of the inherent anisotropic shaft stiffness of two-pole generator rotors. Basically the rotor of a two-pole generator is a large rotating electromagnet with north

Additional Rotor Vibration Cases and Topics

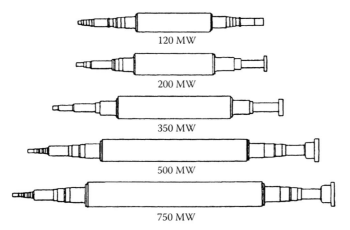

FIGURE 12.15 Maximum two-pole generator rotor sizes from the 1940s to the late 1960s.

and south poles on opposite circumference sides of the rotor. As illustrates in Figure 12.16, there are axial slots in the rotor into which copper conductors are embedded to provide a rotating magnetic field from the DC exciter current fed to the rotor windings, usually through brushes rubbing on exciter collector rings (see Section 12.3). This construction makes the rotor's two principal bending-area moments-of-inertia different, thus the rotor has a maximum and a minimum static deflection line. Without proper bending stiffness equalization measures (see Figure 12.16), a large two-pole generator rotor slowly rotating about its centerline would cycle between maximum and minimum static beam deflections, twice each revolution. Operation at full speed (3600 rpm on 60 Hz systems, 3000 rpm on 50 Hz systems), without adequate bending stiffness equalization measures, would

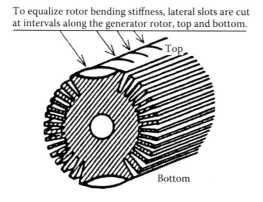

FIGURE 12.16 Rotor construction for two-pole generators.

produce quite high rotor vibration levels at a frequency of twice the running speed (2N). Lateral slots cut at intervals along the generator rotor is now the standard design approach to sufficiently reduce the rotor bending stiffness anisotropy so that the residual 2N generator vibration is much smaller than the residual synchronous vibration.

The 2N rotor vibration exhibited by two-pole generators falls into a vibration generic category called *parametric excitation*. A comprehensive theoretical treatment of this type of rotor vibration as well as the design ramifications of it for generators are given by Kellenberger and Rihak (1982).

12.11 Rotor Balancing

As succinctly stated by NASA's Dr. David Fleming (1989), "A rotor is said to be unbalanced if its *mass axis* does not coincide with its axis of rotation." A *mass axis* is the locus of the distributed *mass-center* along the rotor length. Rotor balancing is the most important and frequently addressed day-to-day operation in achieving smooth running rotating machinery. It starts with the basic machine design process coupled with the construction details of the rotor, including shop rotor balancing of new and repaired rotors using a *balancing machine*. For some machinery types and applications, this is all that may be required. However, in-service rebalancing of some machinery types is periodically needed to reduce their residual unbalance-driven vibration to within acceptable levels (e.g., see Figures 7.1 and 8.1). There is a sharp distinction between *shop balancing* a rotor in a balancing machine and *in-service balancing* a rotor in an assembled machine. Since this book is aimed more toward the troubleshooter than toward the machine designer, the emphasis here is more on in-service balancing. To that end, a general purpose computer code is furnished here for determining balance correction weights. A recent advancement for balancing in operation is a new type of rotor-mounted automatic real-time balancing system, which is described later in this section.

That the subject of rotor balancing warrants its own book was rectified by Rieger (1986), whose book is the most complete and comprehensive treatise on the subject to date. In addition to fundamental theory and application details for different balancing methods and balancing machines, Rieger also provides a historical perspective on rotor balancing and a summary of balancing specifications for the different classes of machines. Thus, Rieger's book provides coverage of the field both as needed by the designer/builder of rotating machinery as well as the in-service user/maintainer of rotating machinery. The emphasis here is focused on the needs of the user/maintainer of rotating machinery. If one is just beginning to study rotor balancing, it is helpful to delineate between so-called *static unbalance* and *dynamic unbalance* as well as distinguish between so-called *rigid rotors* and *flexible rotors*.

12.11.1 Static Unbalance, Dynamic Unbalance, and Rigid Rotors

The simplest rotor unbalance condition is characterized by the rotor mass center being eccentric to the rotor's geometric spin axis. This is called *static unbalance*. A static unbalance can be likened to an unbalance mass (m_s) at some nonzero radius (r_s) superimposed (in the radial plane of the rotor's mass center) on an otherwise perfectly balanced rotor, as illustrated in Figure 2.7. Such a concentrated static unbalance clearly acts on the rotor like an equivalent synchronous corotational force ($F_u = m_s r_s \omega^2$) that is fixed in the rotor. Thus, a purely static unbalance on a simple rotor configuration like in Figure 2.7 can theoretically be corrected by a single balance correction with the same magnitude ($m_s r_s$) and in the same radial plane as the initial static unbalance, but positioned 180° from the initial unbalance. That is, static unbalance is theoretically correctable by adding a balance correction mass in a *single plane*, that is, in the plane of the unbalance. There are many examples where single-plane balancing produces an adequate state of rotor balance quality.

Dynamic unbalance refers to rotor unbalance that acts like an equivalent radial corotational moment fixed in the rotor. Referring to Figure 2.7 and using its nomenclature, the equivalent corotational moment of a concentrated dynamic unbalance has magnitude $M_d = m_d r_d l \omega^2$. If rotor flexibility is not a significant factor to unbalance vibration response, then the "rigid rotor" assumption can be invoked. Then the total dynamic unbalance of a rotor is theoretically correctable by adding two equal-magnitude ($m_c r_c$) corrections (separated by 180°), one at each of two planes axially separated by l_c (where $m_c r_c l_c = m_d r_d l$). The two $m_c r_c$ corrections are positioned in the plane of the initial dynamic unbalance, but 180° out of phase with the initial dynamic unbalance. The initial dynamic unbalance is thereby theoretically negated since the corotational moment produced by the two correction masses has the magnitude $m_d r_d l \omega^2$ of the initial dynamic unbalance, but 180° out of angular position with the corotational moment produced by the initial dynamic unbalance. Since a static unbalance can be negated by two in-phase correction weights appropriately placed in the same two planes as the dynamic unbalance correction masses, it is clear that a complete rotor balance (static + dynamic) of a "rigid" rotor can be accomplished by adding correction masses in only two planes. Since a general state of rotor unbalance is a combination of both static and dynamic unbalance, the correction weights at different axial locations will generally be neither at the same angular position nor separated exactly by a 180° in their relative angular positions.

The defining property for so-called rigid rotors is that rotor flexibility is not a significant factor to unbalance vibration response. Therefore, the *two-plane* balance procedure for a rigid rotor can be performed at a speed lower than the operating speed of the rotor. In practical terms, this means the rotor may be balanced using vibration or dynamic force measurements at balancing spin speeds substantially lower than the rotor's in-service operating speed.

12.11.2 Flexible Rotors

As all inclusively stated by Dr. Neville Rieger (1986), "A flexible rotor is defined as being any rotor that cannot be effectively balanced throughout its speed range by placing suitable correction weights in two separate planes along its length." Synonymous with this definition is that a so-called flexible rotor has an operating speed range that closely approaches or encompasses one or more bending critical speeds whose rotor flexural bending contributes significantly to the corresponding critical speed mode shape(s) and unbalance vibration responses. Table 2.1 provides an introductory composite description of the increased rotor dynamic complexity produced when rotor flexibility is significant to unbalance vibration characteristics.

In contrast to a *rigid rotor*, adequate balancing of a dynamically *flexible rotor* often requires placement of correction weights in more than two separate planes along the rotor length. What is an adequate number of balancing planes and what are their optimum locations along the rotor are factors dictated by the mode shape(s) of the critical speed(s) that significantly affect the rotor's unbalance vibration response. The first three flexure mode shapes of a uniform simply supported beam are illustrated in Figure 4.13, and provide some insight into proper axial locations for balance correction weights in balancing flexible rotors. That is, for a rotor with critical speed mode shapes similar to those in Figure 4.13, a midspan balance plane clearly has maximum effectiveness on the first mode. Similarly, the 1/4 and 3/4 span locations have maximum effectiveness on the second mode, and the 1/6, 1/2, and 5/6 span locations have maximum effectiveness on the third mode.

A flexible rotor is also definable as one whose dynamic bending shape changes with rotational speed, and this speed-dependent dynamic bending may alter the state of balance. Ideally, a flexible rotor should therefore be balanced at full in-service rotational speed and at speeds near critical speeds within or near the operating speed range. This point is clearly demonstrated by an example from Fleming (1989) illustrated in Figure 12.17, which shows a uniform shaft in stiff bearings unbalanced by a single mass attached at the axial center of the shaft. Figure 12.17 also illustrates that the shaft has been rebalanced using the low-speed rigid-rotor

Additional Rotor Vibration Cases and Topics

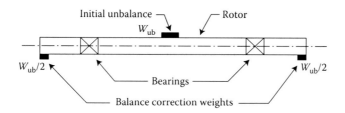

FIGURE 12.17 Simple uniform diameter flexible rotor.

approach by adding a correction weight at each end of the shaft. As long as the shaft speed is significantly below its first bending critical speed, it will remain essentially straight and thus will remain in balance. But as its rotational speed approaches its first bending critical speed, it deforms as illustrated in Figure 12.18 (illustrated deflection grossly exaggerated). As is clear from Figure 12.18, at speeds near its first bending critical speed the shaft illustrated in Figure 12.17 has its initial unbalance and both unbalance correction weights acting together to excite the first bending critical speed. If this experiment were performed, one would find that the vibration near the first bending critical speed is worse (higher) with the two low-speed-balancing correction weights attached than without. In this simple example, the initial unbalance is known to be concentrated at the midspan location and thus it is a trivial case. In a general case with manufacturing and assembly tolerances, the unbalance is of an unknown distribution along the rotor, such as that illustrated in Figure 12.19.

Balancing a dynamically flexible rotor is a considerably more involved process than low-speed two-plane balancing of rigid rotors. Each type of flexible rotor has its own preferred number and location of balance planes. Many multistage machines require a component-balance of each impeller or blade disk assembly before mounting and balancing the fully assembled rotor.

There are some dynamically flexible rotors that can be adequately well balanced like a rigid rotor, that is, on a low-speed balancing machine with correction masses placed in only two planes along the rotor length.

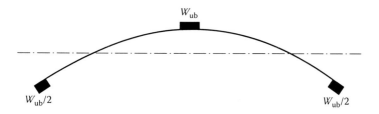

FIGURE 12.18 Unbalance vibration mode shape of a first bending critical speed.

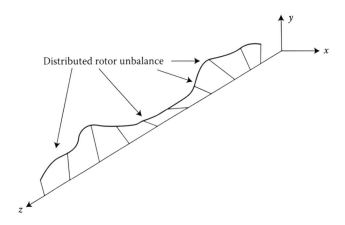

FIGURE 12.19 Isometric view of a general rotor unbalance axial distribution.

Such a rotor is characterized by having most of its unbalance concentrated at a known axial region of the rotor. A prominent example represents large double-suction power plant centrifugal fans, where the single impeller (midspan) essentially dominates the assembled rotor's unbalance distribution.

As described by Rieger (1986), there are historically a number of competing methods for balancing flexible rotors, the two most recognized being the *modal method* and the *influence coefficient method* (ICM). Both of these methods assume that the rotor dynamic system is linear. Significant nonlinearity can be tolerated, but it will likely increase the number of balancing iterations needed to achieve the required quality of rotor balance. The modal method requires detailed modal information (mode shapes) for all the critical speeds that significantly influence the rotor's unbalance vibration characteristics over its entire speed range. To the extent that critical-speed modal characteristics are a function of radial bearing dynamic characteristics, the bearings in a modal balancing machine need to match the dynamic characteristics of the actual machine's bearings, and this is often not practical. Although the modal method is considered in some circles to be theoretically a more effective approach than the ICM for balancing flexible rotors, in practice the ICM is used in most applications, being less complicated and more practical than the modal method. The strong preference for the ICM is particularly true for in-service rebalancing of rotors in assembled machines, where a correction is often limited to one plane.

12.11.3 Influence Coefficient Method

The ICM does not require critical-speed mode shapes, but approximate mode shapes can be helpful in the design process to prescribe where

Additional Rotor Vibration Cases and Topics

balance planes and unbalance vibration measurements are best located. However, after a rotor is installed in its machine, accessible planes for rebalancing are often limited to locations near the axial ends of the rotor, for example, at a coupling. Thus, all the potential benefits of multiplane rotor balancing are only of academic interest to the person in the power plant who must implement a "balance shot" during a short outage of a machine. A general summary of the ICM fashioned after Fleming (1989) is presented here, followed by some examples using the ICM balancing code **Flxbal.exe** contained in the directory **Balance** on the diskette supplied with this book.

The ICM is based only on the assumption of linear dynamic characteristics, so nonplanar modes are automatically accommodated. Utilizing the linearity assumption, the rotor vibration response can be given as the superposition or sum of individual vibration responses from an unbalance at each of the selected balance planes, as expressed in the following equation:

$$V_j = \sum_{k=1}^{N_p} A_{jk} U_k, \quad j = 1, 2, \ldots, N_m \quad k = 1, 2, \ldots, N_p \qquad (12.6)$$

where N_m is the number of independent vibration observations = no. of locations × no. of speeds, N_p is the number of balance planes, V_j is the vibration response from jth measurement $\equiv \mathbf{V_j} e^{i(\omega t - \theta_j)}$ (complex), U_k is the unbalance at kth balance correction plane $\equiv \mathbf{U_k} e^{i(\omega t + \phi_k)}$ (complex), and A_{jk} is the influence coefficient $N_m \times N_p$ Array $\equiv \mathbf{A_{jk}} e^{i\alpha_{jk}}$ (complex).

Vibration measurements need not be taken at the same locations as the balance correction planes. Also, any and all of the three basic vibration sensor types may be used, that is, accelerometer, velocity pick-up, and displacement proximity probe. Furthermore, vibration measurements may be made on adequately responsive nonrotating parts of the machine (e.g., bearing caps). For in-service rebalancing of machines with displacement proximity probes installed (typically mounted on the bearings targeting the shaft), the rotor vibration relative to the bearing(s) may be used and is recommended.

The influence coefficients are experimentally obtained by measuring the incremental change in each of the measured vibration responses to a *trial mass* individually placed at each balance correction plane. With the influence coefficients known, balance corrections for each correction plane can be computed. After the correction masses ($W_j, j = 1, 2, \ldots, N_p$) are installed, the residual rotor vibration for all the specified observations (locations and speeds) are expressible as follows:

$$V_j = V_j^{(0)} + A_{jk} W_k \qquad (12.7)$$

where $V_j^{(0)}$ are the measured vibration responses before adding the balance correction masses.

If the number of observations ($N_m = N_p$) is equal to the number of balance correction planes ($N_m = N_p$), then the influence coefficient array A_{jk} yields a square matrix which is presumably nonsingular by virtue of making N_m linearly independent vibration measurement observations. Using Equation 12.7, unbalance vibration at the observation locations and speeds can then theoretically be made zero by using balance corrections given by the following equation:

$$\{W\} = -[A]^{-1}\{V^{(0)}\} \tag{12.8}$$

It is widely suggested that better balancing often results if the number of observations exceeds the number of correction planes, that is, $N_m > N_p$. Since it is then mathematically impossible to make all the observed vibrations go to zero, the approach generally taken is to base the balance correction masses on minimizing the sum of the squares of the residual observed vibrations.

What the ICM can theoretically achieve is best understood by considering the following. If the system were perfectly linear and the vibration observation measurements were made with zero error, then for the case of $N_m = N_p$, the observed vibrations (at location–speed combinations) are all made zero by the correction masses. Similarly, for the case of $N_m > N_p$, the sum of the squares of the residual observed vibrations can be minimized. However, there is no mathematical statement for unbalance vibration amplitudes at any other location–speed combinations. By choosing balance speeds near all important critical speeds and at maximum operating speed, and balance planes where actual unbalance is greatest, as well as choosing vibration measurement points that are not close to critical speed mode-shape nodal points, smooth running over the full speed range is routinely achievable.

12.11.4 Balancing Computer Code Examples and the Importance of Modeling

Rotor balancing examples are presented here to demonstrate the use of the PC code **Flxbal.exe** contained in the directory **Balance** that accompanies this book. **Flxbal.exe** is based on the *ICM*. It is written by NASA's Dr. David Fleming and provided as part of his invited two-lecture presentation on "Balancing of Flexible Rotors," given regularly to the author's graduate class on Rotating Machinery Dynamics at Case Western Reserve University. **Flxbal.exe** is demonstrated here using some of the **RDA99.exe** unbalance sample cases in Section 4.2 of Chapter 4. That is, the **RDA99.exe** computational models are treated here as "virtual machines" for balancing

Additional Rotor Vibration Cases and Topics

data "input" and "output." This way, the balance corrections computed by **Flxbal.exe** can be investigated at midspan rotor locations where rotor vibration measurements are typically not available with in-service machinery.

The "virtual machine" provided by an **RDA99.exe** model is equivalent to a hypothetical machine that is perfectly linear and on which there is zero error in the balancing vibration measurements. **Flxbal.exe** runs in the DOS environment just like **RDA99.exe** and thus all input and output files in use during a run must reside in the same directory as **Flxbal.exe**. To initiate it, simply enter "Flexbal" and the code then prompts the user, line by line, for input options and data, as will be demonstrated by the following examples. There are a few key factors in using this balancing code in conjunction with RDA99 models for virtual machines, itemized as follows:

- RDA99 vibration phase angles are *leading* since $x_{RDA} \equiv X \cos(\omega t + \theta)$.
- Flxbal vibration phase angles are *lagging* since $x_{Flxbal} \equiv X \cos(\omega t - \theta)$.
- Both RDA99 and Flxbal use *leading* angles for unbalance placement.

Therefore, when transferring data between RDA99 and Flxbal, the sign on the indicated vibration phase angle(s) must be reversed, but the indicated angles for placement of unbalance trial weights and correction weights are the same. The nomenclature for Equation 12.6 is defined consistent with Flxbal.

- Flxbal correction weights are based on trial weight(s) being first removed.

Case-1: *3 Mass Rotor, 2 Bearings, 1 Disk, Unbalance and Correction Same Plane*
This numerical balancing experiment example uses the first model in Section 4.2 of Chapter 4 and is a trivial case since the balance correction can be automatically seen. As detailed in Section 4.2 of Chapter 4 for this model, a single unbalance of 0.005 in. lb, located at 0° phase angle, is attached to the disk at the axial center of the rotor. With a 0° phase angle, it becomes the angular reference point fixed on the shaft. The rotor will be balanced at 1700 rpm (near its first critical speed) using the RDA99 x-displacement of 16.388 mils s.p. at −108.1° (from first sample case tabulation in Section 4.2) at the axial center of the rotor where the disk is located. Thus, this example is really a case of static unbalance correction since the correction is to be placed in the same plane as the initial unbalance. Furthermore, a trial

"weight" of unbalance magnitude 0.0025 in. lb will be used also at the midspan disk and at 0° phase angle. Thereby, the Flxbal inputs and the sought answer for the "correction weight" are already obvious.

- Begin by entering Flxbal. The prompt reads ENTER DESCRIPTIVE LINE TO IDENTIFY RUN:

  "Balancing example Case-1" is entered here.

- The next prompt is ENTER NAME OF FILE FOR OUTPUT DATA:

  "Case-1" is entered here.

- The next prompt is ENTER NUMBER OF PROBES:

  (NUMBER OF PROBES) TIMES (NUMBER OF SPEEDS)<=50:

  The number "1" is entered here.

- The next prompt is ENTER NUMBER OF SPEEDS:

  The number "1" is entered here.

- The next prompt is ENTER NUMBER OF BALANCING PLANES:

  The number "1" is entered here.

- The next prompt is ENTER CALIBRATION FACTOR FOR PROBE 1:

  The number "1" is entered here.

- The next prompt is DO YOU WANT TO ENTER LOW SPEED RUNNOUT? (Y/N):

  The letter "N" is entered here for "no."

- The next prompt is DO YOU WANT TO ENTER NEW INFLUENCE COEFFICIENTS?

  The letter "Y" is entered here for "yes."

- The next prompt is ENTER BALANCING SPEED 1 IN RPM:

  "1700." is entered here.

- The next prompt is ENTER AMPLITUDE AND PHASE ANGLE FOR PROBE 1:

  "16.388 108.1" is entered here.

- The next prompt is ENTER SIZE AND ANGULAR LOCATION OF TRIAL WEIGHT:

  "0.0025 0.0" is entered here.

Additional Rotor Vibration Cases and Topics

In this last entry, one may either enter a "weight" or an "unbalance" magnitude (in any system of units), provided the usage is consistent throughout the exercise.

- The next prompt is ROTOR SPEED 1700 RPM
  ENTER AMPLITUDE AND PHASE FOR PROBE 1:

  "24.582 108.1" is entered here.

  By inspection for this last entry, the addition of 0.0025 (in-lb) to the initial unbalance of 0.005 (in-lb), both at 0° phase angle, simply increases the total vibration by 0.5 times the initial unbalance magnitude ($1.5 \times 16.388 = 24.582$) while leaving the phase angle unchanged at 108.1°.

- The next prompt is DO YOU WANT TO SAVE THESE INFLUENCE COEFFICIENTS?

  The letter "N" is entered here for "no."

The output file **Case-1** is automatically written to the same directory (folder) in which **Flxbal.exe** has been executed. The following is an abbreviated list from the output file **Case-1**.

CORRECTION WEIGHTS

PLANE	WEIGHT	ANGLE, DEG.
1	0.5000E-02	180.0

RESIDUAL VIBRATION AFTER BALANCING

PROBE	SPEED	AMPLITUDE	PHASE
1	1700.0	0.2719E-06	108.1

The complete **Flxbal.exe** generated output file for this case is contained in the subdirectory **BalExpls**, along with the output files for all the other subsequent rotor balancing examples presented here. The correction "weight" shown in the abbreviated output provides the obvious correct answer of 0.005 (in. lb) at 180°, which directly cancels the initial unbalance. Thus, the residual vibration amplitude (mils s.p.) after balancing is essentially zero.

Case-2: Same as Case-1 Except Trial Weight Angle is More Arbitrary
In Case-1, the trial weight is placed at 0° (the same angle as the initial unbalance), and so the total resulting vibration with the trial weight is deducible from the original **sample01.out** results simply by multiplying the vibration amplitude by 1.5. Here in Case-2, the same trial weight unbalance is used but is placed at a different angle than the original unbalance. This is to demonstrate the correct interpretation of weight placement angles.

A trial weight unbalance of 0.0025 (in. lb) is placed at 30° on the disk located at the axial center of the rotor (station no. 2 in RDA99 model). The total unbalance at station no. 2 of the RDA99 model is therefore the vector sum of 0.005 (in. lb) at 0° plus 0.0025 (in. lb) at 30°. This vector sum gives 0.00727382 (in. lb) at 9.896091°, which is implemented in the RDA99 input file **ubal02tw.inp**. The RDA99 computed response with this input file is contained in the output file **ubal02tw.out**, and shows that the x-vibration computed for station no. 2 at 1700 rpm is 23.838 mils s.p. at a phase angle of $-98.2°$. Thus, for the Flxbal input the trial weight is 0.0025 in. lb placed at 30° and the resulting total vibration is 23.838 at $+98.2°$.

The Flxbal output file for this case, named **Case-2**, shows the same unbalance correction as determined in the Case-1 example, that is, same magnitude as initial unbalance, but 180° from the initial unbalance. It is not necessary to confirm this result with an RDA99 run with the trial weight removed and the correction weight added because the net unbalance is obviously zero.

Case-3: Case-1 Model with Measurement & Correction at Rotor End
This next example also uses the first model in Section 4.2 of Chapter 4 with the same single midspan initial unbalance as in the first two examples. It also uses the same 1700 rpm balancing speed near the critical speed. But this case is less trivial than the first two examples. In this case, the correction weight is placed at one end of the rotor (station no. 1). Also, the y-displacement signal at station no. 1 is used as the single vibration "measurement." Using the **Flxbal.exe** results, the specified correction is added to the RDA99 model and the RDA99 code is used to compute the unbalance vibration amplitudes at all the stations. Thus, this example will demonstrate the overall results that a single rotor-end "shot balance" in the field might produce. In an actual machine, the change produced by a "balance shot" cannot generally be measured at the important midspan axial locations. However, a "calibrated" RDA99 model for an actual machine can provide a reliable estimate of midspan vibration after the balance correction is implemented. This is demonstrated in the next subsection from a case study on a 430-MW steam turbo-generator.

The RDA99 output file named **sample01.out**, summarized in Section 4.2 of Chapter 4, shows the initial station no. 1 unbalance y-vibration at 1700 rpm as 1.897 mils s.p. at a phase angle of 140.4°. The first step in this exercise is to use RDA99 to compute the unbalance response with a trial weight added at station no. 1. Accordingly, the RDA99 input file named **ubal03tw.inp** reflects the addition of a trial weight of 0.01 (in. lb) at 0° placed at station no. 1. The corresponding RDA99 output file, **ubal03tw.out**, shows that the y-response at station no. 1 with this trial weight added is 2.210 mils s.p. at 145.1° phase angle. Therefore, the Flxbal inputs are 1.897 at $-140.4°$ for the initial station no. 1 y-vibration at 1700 rpm. The Flexbal inputs after the trial weight is added are 2.210 at −145.1°. An abbreviated Flexbal output is listed as follows:

CORRECTION WEIGHTS

PLANE	WEIGHT	ANGLE, DEG.
1	0.5341E-03	149.3

The next step is to remove the trial weight, add the Flxbal indicated correction weight, and then compute the unbalance response of the rotor with the balance correction in place, using RDA99. The RDA99 input file named **ubal03cw.inp** reflects the Flxbal computed balance correction. The corresponding RDA99 output file, **ubal03cw.out**, shows that the 1700 rpm y-response at station no. 1 is essentially zero with the correction added. However, a quite important observation is made by observing the unbalance response at all other rotor locations and at other speeds. Clearly, the "balance shot" did exactly what it was mathematically programmed to do, that is, make the 1700 rpm vibration at station no. 1 become zero through the addition of a correction weight at station no. 1. The general unbalance response was not overall improved, but in fact became worse after addition of the correction.

Case-4: Rotor-End Measurement, But Mid-Plane Correction
Case-3 shows the potential pitfall of adding a "balance shot" correction weight at the end of a rotor. The next case is a variation of Case-3. The vibration measurement is still taken at the rotor end (station no. 1) where proximity probes, velocity pickups or accelerometers can generally be placed on actual machinery. But the trial weight and subsequent correction weight are placed at the rotor midspan location (station no. 2) where the initial unbalance is concentrated. This is to demonstrate a typical situation where the vibration measurement cannot be made at the rotor midspan location, but the balance weights can be added at the midspan location when a midspan access plate has been designed into the casing so that the rotor midspan plane is easily accessible.

The trial weight RDA99 unbalance vibration here can be taken from the Case-2 output file **ubal02tw.out** for vibration at station no. 1 with the trial weight 0.0025 (in. lb) at 30° placed at station no. 2. This output is 2.759 (in. lb) at 150.3°. From the first case tabulated in Section 4.2, the initial midspan (station no. 2) 1700 rpm y-direction vibration is 1.897 mils s.p. at 140.4°. Thus, the Flxbal inputs are as follows (refer to Case-2):

- Balance speed, 1700 (rpm)
- Initial y-vibration data, 1.897 –140.4°
- Trial weight, 0.0025 30°
- y-vibration data with trial weighted added, 2.759 –150.3°

The Flxbal file (**Case-4**) correction weight data are 0.005 at 180°, the exact cancellation of the initial unbalance and thus zero vibration everywhere.

A comparison of Cases 3 and 4 demonstrates the critical importance of proper balance plane(s) selection. It also demonstrates that the vibration measurement point is not as critical provided that measurement point is adequately responsive to the initial unbalance distribution and the added unbalance trial weight(s). That is, the measurement point should not be near a nodal point of any of the important critical speed modes nor near a nodal point of the rotor response shape at operating speed. These two cases clearly demonstrate the value of employing a calibrated rotor unbalance response computer model in concert with standard balancing procedures to predetermine whether a quick "balance shot" during a short outage will actually reduce the rotor vibration at the important midspan rotor locations. An unbalance response computer model used in this manner basically "measures" the midspan vibration reduction from a "balance shot."

With the example Cases 1–4 provided here, interested readers can create additional interesting examples using any of the other RDA99 unbalance response sample cases presented in Chapter 4. Of course, the primary reason for this is to prepare interested readers to generate new RDA99 models of rotor–bearing systems for machinery, important to their respective organizations.

12.11.5 Case Study of 430 MW Turbine Generator

The machine in this case study is the same machine described in the self-excited subsynchronous vibration case study of Section 11.4. After this machine was reassembled at the end of a recent major planned outage for the unit, it exhibited excessive synchronous rotor vibration concentrated in the HP and IP rotors (see Figure 11.4 for the schematic layout). Successive attempts by a quite competent industry recognized specialist in balancing such machines were unsuccessful in reducing the journal-to-bearing vibration at bearing no. 1 to less than 10 mils p.p. Since the author already had modeled this machine to solve the problems described in Section 11.4 of Chapter 11 and Section 12.8, the author was retained to employ the model to determine whether using additional balancing planes (not typically used for in-service rebalancing) could potentially reduce the synchronous vibration from 10 mils p.p. to possibly 5 or 6 mils p.p.

The influence of employing various multiplane balancing combinations was computationally researched using an RDA99 unbalance response model. The model included the entire rigidly coupled drive line, including the HP, IP, LP, generator, and exciter rotors, all supported on seven journal bearings, as sketched in Figure 11.5. In the process of conducting this work, the author's computed unbalance responses were compared

Additional Rotor Vibration Cases and Topics   419

to the incremental responses produced by the trial weights used on the actual turbo-generator during the most recent attempt to balance the machine. These comparisons are summarized as follows, to demonstrate the expected accuracy of a properly devised RDA99 model for such a machine.

Plane-1: HP rotor between bearing no. 1 and HP end seals; 488 g at 180° at 7.58 in. radius, for 8.135 in. lb; vibration (mils p.p.), phase angle° (lagging)

Incremental vibration from measurement at bearing no. 1: 3.1 at 290°

Incremental vibration from RDA99 model at bearing no. 1: 3.3 at 272°

Plane-10: IP end of LP, just outside of LP last-stage blades; 950 g at 255° at 19.68 in. radius, for 41.13 in lb

Incremental vibration from measurement at bearing no. 3: 1.7 at 52°

Incremental vibration from RDA99 model at bearing no. 3: 2.8 at 27°

Given the complexity of the machine and the uncertainty of actual journal-bearing static loads, these comparisons are remarkably close. This excellent comparison added to the author's own confidence in applying computer model simulations to aid in troubleshooting vibrations problems, even on such large complicated machines with the inherent uncertainties.

Unusually large balance correction weights indicated by the analyses in this case led the author to conclude that the root cause of the excessive vibration was not unbalance but more likely an improper setting of a rigid coupling between two of the rotors at the scheduled outage reassemble, probably at the coupling between the HP and IP rotors. Power generation revenue considerations dictated this unit to be deemed "operable" until the next opportunity to remove the turbine covers for general inspection of internals and to check for rotor-to-rotor run-out at the turbine couplings.

12.11.6 Continuous Automatic In-Service Rotor Balancing

No up-to-date discussion on rotor balancing would be complete without mentioning the latest and most advanced product in automated real-time continuous rotor balancing. Figure 12.20 shows this in a cutaway view of the newest automatic balancing product from the Lord Corporation. The rotor-mounted portion houses two equally unbalanced counterweight/stepping-motor rotors, separately indexed in 5° increments relative to the rotor. Power and control is through magnetic couplers.

Conventional rotor-mounted automatic balancing devices are designed to minimize residual rotor mass unbalance so that the rotor vibration level is maintained within a given application's requirements. Precision machine tool spindles, especially for grinding, are a major application for such devices since successful high-volume high-precision grinding requires continual automatic adjustment of balance correction weights on the rotating assembly as grinding wheel material is removed. The conventional devices available for such automated balancing are configured to change the correction weight magnitude and angular location based on many successive incremental moves that reduce the monitored vibration (usually measured with an accelerometer attached to the spindle housing). However, such conventional systems do not "know" the magnitude or angular location of the continuously changed correction weight nor are they able to execute a "command" to perform a specified incremental change to the correction weight.

The Lord product shown in Figure 12.20 has significantly advanced the field of automatic rotor balancing by tracking the magnitude and angular location of the instantaneous balance correction. The author designed and constructed a flexible-rotor test rig in his laboratory at Case Western

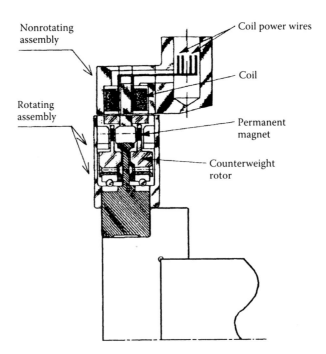

FIGURE 12.20 Lord Corporation automatic rotor mass balancer.

Reserve University, which is configured with two of these Baladyne balancing devices. The software supplied with these two matched devices executes real-time two-plane automatic rotor balancing, with the controlling algorithm based on the ICM. The software also permits manual control of counterweight placement and magnitude (through a host PC controller). That feature plays prominently in current ongoing model-based monitoring and diagnostics research in the author's laboratory, see Adams and Loparo (2000). That is, by being able to impose a known incremental change to the state of unbalance (i.e., active probing of the dynamical system), a continuous real-time comparison can be made between how the actual machine incrementally responds and how an observer model tracking the machine's vibration responds.

The author believes this new type of real-time automatic balancing system can be a quite cost-effective method for minimizing rotor vibration levels on flexible-rotor machines that currently necessitate considerable compromises between various important critical speeds and operating speeds that each individually have somewhat unique optimum balance correct weight placements. The author is familiar with some large steam turbo-generator configurations now in service that would benefit considerably from such a system.

12.11.7 In-Service Single-Plane Balance Shot

The most frequent rotor balancing job is the in-service quick balance correction. Rotating machinery in power plants, process plants, and in machine tool spindles are typical examples. The machine is in service and vibration has increased above allowable maximum amplitude levels (see Chapter 8). A balance correction weight is placed on the rotor at a readily accessible location on the machine. The objective of such single-plane in-service balance shots is to reduce the maximum vibration levels. It is not intended nor is it feasible that such a single-plane balance shot provide the high degree of rotor balance quality that is achievable when the removed bare rotor is factory balanced in a precision balancing machine.

Refer to Section 12.11.4 (Case-4) of this chapter. When the vibration measurement and added correction weight locations are limited to access points of an assembled machine, the vibration levels at critical internal inaccessible rotor locations may actually increase, even while the vibration levels at accessible vibration measurement points has been significantly reduced. Thus, the small internal radial clearances between rotating and nonrotating components at midspan locations (e.g., at turbine and compressor blade tips and pump sealing wear rings) may not benefit from such a balance shot. For a detailed treatment of single-plane balance shots, see Adams (2007).

12.12 Summary

This chapter is a potpourri of rotating machinery vibration problems. The emphasis is on solving real problems in real machines, with actual case histories being the primary basis of the material presented. This chapter and Chapters 10 and 11 form Part 4 of this book on Case Studies. Since the author's own troubleshooting experience has been, and continues to be, heavily focused on power plant rotating machinery, the case studies here are primarily from power plants. However, the particulars of each case study have much broader value in guiding problem solution investigations in many other different industrial applications of rotating machinery.

Bibliography

Adams, M. L., "Non-linear dynamics of flexible multi-bearing rotors," *Journal of Sound & Vibration*, 71(1):129–144, 1980.

Adams, M. L. and McCloskey, T. H., *Large Unbalance Vibration in Steam Turbine-Generator Sets*, Proceedings of the Third IMechE International Conference on Vibrations in Rotating Machinery, York, England, 1984.

Adams, M. L. and Makay, E., *How to Apply Pivoted-Pad Bearings*, Power, McGraw-Hill, New York, October 1981.

Adams, M. L. and Payandeh, S., "Self-excited vibration of statically unloaded pads in tilting-pad journal bearings," *ASME Journal of Lubrication Technology*, 105: 377–385, 1983.

Adams, M. L. and Pollard, M. A., *Rotor Vibration Vector Turning Due to Rotor Rubs*, Proceedings of the EPRI Symposium on Trouble Shooting Power Plant Rotating Machinery Vibrations, LaJolla, CA, May 19–21, 1993.

Adams, M. L., Adams, M. L., and Guo, J. S., *Simulations and Experiments of the Non-Linear Hysteresis Loop for Rotor-Bearing Instability*, Proceedings of the Sixth IMechE International Conference on Vibration in Rotating Machinery, Oxford University, Oxford, England, September 1996.

Adams, M. L. and Loparo, K. A., *Model-Based Condition Monitoring From Rotating Machinery Vibration*, Final Report, EPRI Project WO3693-04, 2000.

Adams, M. L., "Rotor balancing and unbalance-caused vibration," in *Handbook of Noise and Vibration Control*, Chapter 62, M. J. Crocker, (Ed.), Wiley, New York, 2007.

Bennett, M. G. and Piatt, B. J., *Experience with the Effects of Bearing Looseness*, EPRI Symposium on Trouble Shooting Power Plant Rotating Machinery Vibrations, LaJolla, CA, May 19–21, 1993.

Bishop, R. E. D. and Parkinson, A. G., "Second order vibration of flexible shafts," *Philosophical Transactions, Royal Society*, 259(A. 1095):1965.

Fleming, D. P., *Balancing of Flexible Rotors*, Invited-lecture handout for graduate course in Rotating Machinery Vibration, Case Western Reserve University, 1989, pp. 11.

Hori, Y., *Anti-Earthquake Considerations in Rotordynamics*, Keynote address paper, Fourth IMechE International Conference on Vibration in Rotating Machinery, Edinburgh, Scotland, September 1988.

Jenkins, L. S., *Troubleshooting Westinghouse Reactor Coolant Pump Vibrations*, EPRI Symposium on Trouble Shooting Power Plant Rotating Machinery Vibrations, San Diego, CA, May 19–21, 1993.

Kellenberger, W. and Rihak, P., *Double-Frequency Vibration In Large Turbo-Generator Rotors—A Design Problem*, Proceedings of the IFToMM International Conference on Rotordynamics, Rome, Italy, 1982.

Makay, E., Adams, M. L., and Shapiro, W., "Water-Cooled Reactor Pumps Design Evaluation Guide," Oak Ridge National Laboratory, ORNL TM-3956, November 1972, pp. 412.

Makay, E. and Adams, M. L., *Operation/Design and Evaluation of Main Coolant Pumps*, Final Report, EPRI-NP-1194, September 1979.

Marscher, W. D., *The Effect of Fluid Forces at Various Operating Conditions on the Vibrations of Vertical Turbine Pumps*, Seminar by the Power Industries Division of IMechE, London, England, February 5, 1986.

Muszynska, A., "Thermal rub effect in rotating machines," *ORBIT, Bently Nevada Pulication*, 14(1): 8–13, 1993.

Newkirk, B. L., "Shaft rubbing, relative freedom of rotor shafts from sensitiveness to rubbing contact when running above their critical speeds," *Mechanical Engineering*, 48(8):830–832, 1926.

Prandtl, L., "Beiträge zur Frage der kritischen Drehzahlen," *Dinglers Polytechn. Journal*, 333:179–182, 1918.

Rieger, N. F., *Balancing of Rigid and Flexible Rotors*, The Shock and Vibration Information Center, Book No. SVM-12, U.S. Department of Defense, 1986, pp. 614.

Smith, D. M., "The Motion of a Rotor Carried by a Flexible Shaft in Flexible Bearings," Proceedings of the Royal Society, (London) A 142, 1933, pp. 92–118.

Taylor, H. D., *Rubbing Shafts Above and Below the Critical Speed*, General Electric Review, April 1924.

Index

A

Acceleration value, 298
Accelerometers, 281–283
   sensitivity, 283
   unidirectional, 282
Addition of nonstructural mass and inertia to rotor element, 62–63
Adiabatic formulation, 199
Aerodynamic drag forces, 335
AGMA. *See* American Gear Manufacturers Association (AGMA)
Air preheater drive structural resonances, 391–393
Aircraft auxiliary power unit vibration-caused wear, 393–396
Alford turbine forces, 265
Aliasing, 293
Altitude wind tunnel high-capacity two-stage fan, 130–132
American Gear Manufacturers Association (AGMA), 300
American National Standards Institute (ANSI), 300
American Petroleum Institute (API), 300
American Society of Mechanical Engineers (ASME), 300
Amplitude conventions. *See* Vibrations
Angular contact, 230
Angular momentum, time-rate-of-change of, 46–47
Angular momentum equation, 47
Anisotropic model, 205, 214
Anisotropic rotor stiffness, 316
Annular seals, 212–230
   brush, 227–229
   bulk flow model, 218, 219
   circumferentially grooved, 224–225
   gas seals
      seal LRV-coefficient uncertainties, 229–230
      steam whirl and oil whip, 226–227
      typical configurations, 227–229
   honeycomb, 227–229
   liquid, 215–223
   seal dynamic data and resources, 215
   tapered, 217
   ungrooved, 215–223
ANSI. *See* American National Standards Institute (ANSI)
API. *See* American Petroleum Institute (API)
Apparent viscosity, 222
ASME. *See* American Society of Mechanical Engineers (ASME)
Average amplitude, 295
Axial contact, 230
Axial flow compressors, 270–272
Axial momentum equation, 220–222

B

Backward whirl, 75, 76, 78, 79, 80, 81, 160, 161, 312, 339
Balancing, 406
   430 turbine generator case study, 418–419
   computer code and modeling, 412–418
   computer code examples and modeling importance, 412–418
   continuous automatic in-service rotor, 419–421
   drum, 194, 255
   dynamic balance, 407

425

Balancing (*continued*)
   flexible rotors, 36, 408–410
   influence coefficient method (ICM), 410–412
   in-service single-plane balance shot, 421
   machine, 406
   multi-plane, 411, 418
   rigid rotors (two plane), 407–408
   static unbalance, 407
Balancing drum (piston). *See* Boiler feed water pumps
Ball bearings, 230, 231
Bandwidth filter. *See* Filters
Band-pass filter, 293–294
Base-motion excitations, 403–404
Basic rotor finite element in RDA, 55–57
Beam deflection formulas, 53
BearCoef, 198
Bearing and seal rotor dynamics, 183
   annular seals, 212
     annular gas seals, 225–230
     circumferentially grooved, 224–225
     seal dynamic data and resources, 215
     ungrooved, 215–223
   compliance surface foil gas bearings, 243–246
   dynamic coefficients, experiments to measure, 201
     instability threshold-based approach, 210–212
     mechanical impedance method with harmonic excitation, 203–208
     mechanical impedance with impact excitation, 208–209
   liquid-liquid fluid-film journal bearings
     computer codes, 199
     fundamental caveat of LRV analyses, 199–200
     Reynolds lubrication equation (RLE), 184–187
     stiffness and damping data and resources, 196–198
     stiffness and damping formulations, 187–192
     tilting-pad mechanics, 192–196
   magnetic bearings, 239
     short comings, 241–243
     unique operating features, 240–241
   rolling contact bearings (RCB), 230–235
   squeeze-film dampers, 235
     with centering springs, 236–237
     without centering springs, 237–238
     Reynolds-equation-based solutions, 238–239
Bearing cap vibration guidelines, 298–299
Bearing coefficients
   connect rotor directly to ground, 67–68
   connect to an intermediate mass, 68–70
Bearing looseness effects. *See also* Troubleshooting
   BFP electric motor, 399
   LP turbine bearing looseness, 400–401
   steam turbine generator, 398–399
Bearing pinch, 399
Bearing stiffness and damping (coefficients), 187–192, 233, 234
   angular displacement, 17, 40
   concept development, 195
   data resources, 196–200
   dynamic models, 93, 191, 240
   insights, 68–83
   measurement, 202, 206
   misalignment, 82–83
   non-symmetries, 71–77
   numerical accuracy considerations, 191
   tensor transformation property, 190–192
Bearing support models, 17, 64–70

Bearings
  hydrodynamic (fluid film), 97, 173
  magnetic, 239–243
  rolling contact, 230–235
BEP. See Best efficiency point (BEP)
Bernoulli equation, 267–268
Best efficiency point (BEP), 175
BFM. See Bulk flow model (BFM)
BFP. See Boiler feed water pump (BFP)
Bifurcation diagram. See Chaos
Blade tip clearance effects. See Thomas-Alford forces
Blade tip-on-casing rubs, 339
Bode diagram, 317–318
  and plot, 318
Boiler circulating pumps, 384
Boiler feed water pump (BFP), 194, 255
  critical speeds, 354–361
    case study 1, 354–358
    case study 2, 358–360
    case study 3, 360–361
  electric motor, 399
  multistage, 213
Branched systems, 126–127
  complete equations of motion, 129–130
  flexible connections, 129
  rigid connections, 127–129
Brush seal, 227, 228–229.
  See also Annular seals
Bulk flow model (BFM), 218, 219.
  See also Annular seals
Bump-type foil bearings, 244

C

Cal Tech tests, 259
Calibrated model, 229
Calibration factor, 286, 287
Campbell diagrams, 163–165
Canned-motor pump model, 79
Cantilever beam, 53
Capacitance-type displacement transducer, 285
Cascade plot, 318–321, 319, 320
  forward and backward movements, 320–321
Catastrophic failures, 84, 86, 87

Catcher bearings, 241. See also Ball bearings
Caveat of rotor vibration analyses, 199–200
CBJB. See Cylindrical-bore journal bearings (CBJB)
Centering springs. See Squeeze-film dampers (SFD)
Centrifugal compressors, 260
Centrifugal pump impellers, 251, 397
  dynamic radial force, 255–259
  static radial force, 251–254
Chaos, 99–100, 325–330
  in rotor dynamical systems, 99–100
  signal analysis tools, 325–330
Characteristic determinant, 22
Circumferential momentum equation, 219
Circumferential pressure distribution, 266, 267
Circumferentially grooved annular seals for liquids, 224–225
Clearance, 89, 95, 183, 195, 200, 212, 213–214, 216, 223, 227, 232, 245–246, 264–265, 301, 355–356, 421
Coaxial same-speed coupled rotors, 119–120
COJOUR journal bearing code, 199, 351, 359
Collector shaft, 368
Collocation error and active magnetic bearings, 242
Combined radial and misalignment motions, 82–83
Combustion gas turbines, 270
Complete equations of motion
  for branched systems, 129–130
  unbranched systems, 124–126
Complete free–free rotor matrices, 63–64
Completed RDA model equations of motion, 70
Complex eigenvalue problem, 28–30
Complex plane representation, 145, 150–151
Complexity categories for lateral rotor vibration, 37

Compliance surface foil gas bearings, 243–246
    types, 243, 244
Compressed Air and Gas Association, 300
Compressor bearing failure, 101–104
Computer codes, journal bearing, 199
Concentrated disk mass properties, 56–57
Conservative force field, 73, 192
Consistent mass matrix, 17, 59–61
Contact forces, 231, 232
Continuity equation, 184, 218, 219, 220
Continuous automatic in-service rotor balancing, 419–421
Control room display, 291
Coriolis acceleration, 272
Coulomb damping, 9
Coupled rotors, 119–120
    branched systems, 126–127
        complete equations of motion, 129–130
        flexible connections, 129
        rigid connections, 127–129
    coaxial same-speed coupled rotors, 120
    unbranched systems, 121–122
        complete equations of motion, 124–126
        flexible connections, 124
        rigid connections, 122–124
Cracked shafts, 342. *See also* Troubleshooting
Critical speed, 36
    and forced vibration, 349–370
        boiler feed pumps critical speeds, 354–361
        boiler circulating pumps, 364–367
        circulating pump resonance, 367
        generator exciter collector shaft critical speeds, 367–369
        HP steam turbine passage through first critical speed, 350–352
        HP–IP turbine second critical speed, 352–354
        nuclear feed water pump cyclic thermal rotor bow, 361–364
Critically damped. *See* Damping
Crowned-cylindrical roller elements, 230
Curve-fit coefficients, bearing and seal rotor dynamic coefficients, 202
Cylindrical and tilting-pad journal bearings, comparison between, 192, 193
Cylindrical-bore journal bearings (CBJB), 402
Cylindrical-bore seals, 217

D

Damped natural frequency. *See* Natural frequency vibration
Damping, 8–10
    critical, 4, 5, 10–11
    linear, 8, 24
    matrix, 14, 173
    modal, 22–23, 26, 27
    over damped, 4, 5
    proportional, 26
    ratio, 25
    rotor internal, 334–336
    structural, 9
    under damped, 4, 5
Damping to critical damping ratio, 10
Data acquisition. *See* Vibration
Deflection shape functions, 60
Denoising filters, nonlinear, 294
Destabilizing forces. *See* Dynamic instability
Diagnostics. *See* Troubleshooting
Differential equations, 7
Digital tape recorders, 241
Direct $F = ma$ approach, 52–55
Direction cosines, 191
Displacement
    measurement
        proximity probes, 285–289
        shaft riders, 285
    transducers, 301
        background, 284–285

Index   429

inductance-type displacement, 285–289
value, 298
Displacement measurement proximity probes, 285–289
   calibration, 286
   carrier frequency, 287
   electrical runout, 287
   gap, 285–286
   inductance, 285
   modulator-demodulator, 285
   operating principle, 285
   vibration severity guidelines, 298–299
Distributed mass matrix, 17, 57, 116–117
Distribution of rotor unbalance. *See* Unbalance
Double-spool-shaft spindle, 206
Double-volute pump configurations, 254
Dynamic balance, 407
Dynamic coefficients, 201
   instability threshold-based approach, 210–212
   mechanical impedance method with harmonic excitation, 203–208
   with impact excitation, 208–209
Dynamic instability, 6
   complex eigenvalue problem, 28–30
Dynamic nonlinearity, 316
Dynamic radial force on pump impellers. *See* Centrifugal pump impellers
Dynamic radial hydraulic impeller forces, 255
   centrifugal compressors, 260
   interaction impeller forces, 257–259
   interactive force modeling similarities, 262–263
   stability criteria, 260–262
   unsteady flow dynamic impeller forces, 255–257
Dynamic rotor unbalance. *See* Unbalance
Dynamically stable, 7

E

Earthquake excitation. *See* Base-motion excitation
Eccentricity, 190, 199, 206, 217, 221, 264–265, 267
Eigenvalues
   and eigenvectors, 19
      categories table, 29
      complex, 28–30
      formulations, 30
      real, 22, 30
   and extraction, 54
   types and mode motion properties, 29
Electric motor, 399
Electric Power Research Institute (EPRI), 199, 255
Electrical runout. *See* Displacement measurement proximity probes
Elliptical orbits, 158–163
Elrod–Ng approach, 222
Energy conservation, 72
Energy nonconservation, 72
Energy per cycle, 24, 75, 76, 169, 171, 172–173, 210
Entrance loss coefficient. *See* Bulk flow model (BFM)
EPRI. *See* Electric Power Research Institute (EPRI)
Euler angles, 43
Exciters, 337, 391
Expert system, 279
   flow chart, 279

F

Fast Fourier Transform (FFT) algorithm, 278
   and filtered orbit, 314, 315
   and spectrum analysis, 308–310
Feedback speed control, to TRV, 114–115
FFT algorithm. *See* Fast Fourier Transform (FFT) algorithm
Film thickness, 92, 103, 185, 220, 221, 222, 237, 238, 288

Filters, 293–294
Finite element models, 25, 26, 41, 86
First critical speed, 272
Flexible connections
  branched systems with, 129
  unbranched systems with, 124
Flexible couplings, 111, 120
Flexible rotors, 408–410. *See* Balancing
  applications, 230
Floating ring seals, 214
Fluid-film bearings, 65, 85, 241. *See also*
  Liquid-liquid fluid-film
  journal bearings
Foil bearings. *See* Compliance surface
  foil gas bearings
Footprint and RCB, 230, 231
Forced resonance, 10
Forced systems decoupled in modal
  coordinates, 27
Forced vibration and critical speed,
  349–370
  boiler feed pumps critical speeds,
    354–361
    case study 1, 354–358
    case study 2, 358–360
    case study 3, 360–361
  circulating pumps, 364–367
  circulating pump resonance, 367
  generator exciter collector shaft
    critical speeds, 367–369
  HP steam turbine passage through
    first critical speed, 350–352
  HP–IP turbine second critical speed,
    352–354
  nuclear feed water pump cyclic
    thermal rotor bow, 361–364
Forward whirl, 76
Fourier transform, 278, 308–310
Four-square gear tester, 132–133
Free body diagrams, 14
Free–free rotor matrices, 63–64, 118–119
Free-free rotor model. *See* Rotor
  Dynamic Analysis (RDA)

G

Gas turbines, 35, 163, 229, 235, 245, 270,
  271, 339

Geared connections, TRV coupling, 121
General solution, 18
Generalized coordinates. *See* Lagrange
  equations
Generalized forces. *See* Lagrange
  equations
Generator exciter collector shaft critical
  speeds, 367–369
Generators, 35, 111, 134, 316, 337, 351,
  402, 404
Gyroscopic
  effect, 40, 43
  explanation of, 77–79
  matrix. *See* RDA
  moment, 44, 47, 49, 52–53, 73, 86

H

Half power bandwidth test method, 9
Half-frequency whirl, 383
Hamiltonian motion, 99
Hammer kits, 204
Harmonic. *See* Vibration
Harmonic excitation. *See also* Vibration
  of linear models, 27–28
High rotational Reynolds number fluid
  annulus, 267
High-capacity fan for large altitude
  wind tunnel, 130–132
High-pass filter, 293
  See also Filters
High-pressure steam turbines and gas
  turbines
  combustion gas turbines, 270
  partial admission in steam turbine,
    269–270
  steam whirl, 263–269
High-speed reclosure (HSR), 119
Homogeneous solution, 7
Honeycomb seal, 227–228, 261. *See also*
  Annular seals
HP steam turbine passage through first
  critical speed, 350–352
HP turbine, 373, 374
HP–IP turbine second critical speed
  through power cycling,
  352–354
Hydraulic Institute, 300

Hydrodynamic bearings, 97, 173
Hydrodynamic pressure
　　distribution, 186
Hydrostatic gas bearings, 243
Hysteresis loop, 96–97

I

ICM. *See* Influence coefficient method
　　(ICM)
IEC. *See* International Electrical
　　Commission (IEC)
Imbalance. *See* Unbalance
Impact approach for mechanical
　　impedance, 204
Impact excitation of radial seals
　　experimental setup, 209
Impact tests for vibration problem
　　diagnoses, 397–398
Impulse turbines, 269, 352, 353
Inductance-type displacement
　　transducer, 285–289
　　position sensing system, 286
　　for rotor orbital displacement
　　　trajectories measurement, 311
Inertia-modulated rub, 388–390
Influence coefficient method (ICM),
　　410–412. *See* Balancing
Initial conditions, 4, 7, 20
In-service single-plane balance
　　shot, 421
In-servicing balancing, 406
Insights into rotor vibration, 70–83
Instability, 6, 28–30, 76, 165–173,
　　331–336
Instability growth orbits, 167
Instability self-excited-vibration
　　threshold computations,
　　165–173
　　mass rotor and disk
　　　different, 172–173
　　　same, 166–172
Instability thresholds, 160, 165–166,
　　198, 332
Instability threshold speed, 178–179
Instability threshold-based approach,
　　210–212

vertical spindle rig for controlled
　　speed tests, 210
Integration and signal conditioning,
　　293
Interaction forces, dynamic radial
　　hydraulic, 255
Interaction impeller forces, 257–259
Interactive radial force vector, 201
Internal rotor damping, 335–336
International Electrical Commission
　　(IEC), 300
International Standards Organization
　　(ISO), 300
ISO. *See* International Standards
　　Organization (ISO)
Isotropic model, 38, 79–81, 205, 208,
　　209, 214, 217, 218, 221, 257
Isotropic tensor, 75, 80, 191

J

Japanese Society of Mechanical
　　Engineers (JSME), 198
Jeffcott rotor model, 39–41
Jet engine. *See* Combustion gas
　　turbines
Journal bearings. *See* Hydrodynamic
　　bearings
Journal of Sound and Vibration, 404
Journal vibration orbit, 315
Journal-bearing hysteresis loop, 96–97
Journal-bearing nonlinearity with large
　　rotor unbalance, 85–93
JSME. *See* Japanese Society of
　　Mechanical Engineers (JSME)

K

Keyphaser, 289–290, 312
Kinetic energy. *See* Lagrange equations
Kronecker's delta, 23, 25

L

Labyrinth seals, 213, 227, 373. *See also*
　　Annular seals
　　destabilizing forces, 266
Lagrange approaches, 44–52
Lagrange equations, 15

Large amplitude vibration sources, 84–85
Large steam turbo-generator sets, example, 134–135
Lateral rotor vibration (LRV) analyses, 35–37
  categories, 18
  complexity categories, 37
  linear LRVs, 70–71
    gyroscopic effect, 77–79
    isotropic model, 79–81
    physically consistent models, 82
    radial and misalignment motions, 82–83
    nonsymmetric matrices, 71–77
  nonlinear effects in rotor dynamics, 83–84
    chaos in rotor dynamical systems, 99–100
    damping masks oil whip and steam whirl, 100–101
    journal-bearing hysteresis loop, 96–97
    journal-bearing nonlinearity, 85–93
    large amplitude vibration sources, 84–85
    shaft-on-bearing impact, 97–99
    unloaded tilting-pad self-excited vibration, 94–95
  RDA code. *See* RDA code for LRV analyses
  RDA software formulations, 55
    addition of nonstructural mass and inertia, 62–63
    basic rotor finite element, 55–57
    complete free–free rotor matrices, 63–64
    completed RDA model equations of motion, 70
    radial-bearing and bearing-support models, 64–70
    shaft element consistent mass matrix, 59–61
    shaft element distributed mass matrix, 58–59
    shaft element gyroscopic matrix, 62
    shaft element lumped mass matrix, 57–58
    shaft element stiffness matrix, 61–62
  simple linear models
    Jeffcott rotor model, 39–41
    point-mass 2-DOF model, 37–39
    simple nontrivial 8-DOF model, 41–43
  versus TRV, 135–137
Least-squares linear regression, 205
Limit cycle, 84
Linear LRVs, 70–71
  gyroscopic effect, 77–79
  isotropic model, 79–81
  physically consistent models, 82
  radial and misalignment motions, 82–83
  systems with nonsymmetric matrices, 71–77
Linearity assumption, 4
Liquid-liquid fluid-film journal bearings
  caveat of LRV analyses, 199–200
  computer codes, 199
  Reynolds lubrication equation (RLE), 184–187
  stiffness and damping data and resources, 196–198
  stiffness and damping formulations, 187–192
  tilting-pad mechanics, 192–196
Log-decrement test method, 9
Lomakin effect, 216–218
Loose connection. *See* Troubleshooting
Lord Corporation automatic rotor mass balancer, 420
Low-pass filter, 293, 312. *See also* Filters
LP turbine bearing looseness, 400–401
LRV. *See* Lateral rotor vibration (LRV) analyses
Lubricant pressure distribution, 85
Lumped mass matrices, 16–17, 57, 115–116. *See also* Rotor Dynamic Analysis (RDA)

Index

M

Machinery management systems, 279
Magnetic bearings, 239, 244
  collocation error, 242
  feedback control, 240, 242
  load capacity saturation effects, 240, 242
  microprocessor controller, 241
  short comings, 241–243
  spillover 242
  unique operating features, 240–241
Makay charts, 343
Marscher's method, 397
Mass axis, 406
Mass matrix, 14, 15, 16, 57
Mass moment of inertia, 12, 47, 190, 191
Mass unbalance. *See* Unbalance
Mass–spring–damper model, 3
Material damping. *See* Damping
Matrix bandwidth, 16–18
Matrix zeros, 15, 16–18
Mechanical impedance hypothesis, 221–222
Mechanical Technology Incorporated (MTI), 199
Mechanically loose connections. *See* Troubleshooting
Misalignment, 377–378. *See also* Bearing stiffness and damping; Troubleshooting
  vibration symptoms, 339–340
Modal
  analysis, 397
  coordinates, 20
  damping, 24–26
  decomposition, 19–24
  decoupling, 20, 26
  method for balancing flexible rotors, 410
  testing, 26, 397–398
  vectors, 22–23
Mode shapes, 18, 19, 36, 86, 136, 169, 176, 255, 373, 378, 408, 410
Model-based
  condition monitoring, 278, 279, 281, 289, 290, 321, 328
  diagnostic approach, 280–281, 302
  statistical-correlation filtering, 280
Monitoring. *See* Vibration
Mother wave. *See* Wavelet transforms (WTs)
Motion equations, 13, 14, 20, 26
MTI. *See* Mechanical Technology Incorporated (MTI)
Multistage axial compressors, 271
Multi-DOF models, 13–30. *See also* Vibration
  dynamic instability, 28–30
  forced systems decoupled in modal coordinates, 27
  harmonic excitation of linear models, 27–28
  matrix bandwidth and zeros, 16–18
  modal damping, 24–26
  modal decomposition, 19–24
  standard rotor vibration analyses, 18–19
  two-DOF models, 13–16
Multi-stage centrifugal pumps, 204, 225

N

Narrow bandwidth matrices, 16
National Electrical Manufacturers Association (NEMA), 300
Natural frequency vibration, 10–11
Navier–Stokes (N–S) equation, 184–185
Negative damping, 6
NEMA. *See* National Electrical Manufacturers Association (NEMA)
Net destabilizing force, 264, 265
Newton's Second Law, 3
Nine-stage centrifugal pump model, 175–180
  instability threshold speed, 178–179
  unbalance response, 175–178
Nodal points, 27, 35, 178
Nonaxisymmetric shaft stiffness, 404–406
Nonconservative force fields, 74
Nonlinear contact forces, 231

Nonlinear damping masks oil whip and steam whirl, 100
   compressor bearing failure, 101–104
   oil whip masked, 100–101
   steam whirl masked, 101
Nonlinear effects in rotor dynamic, 83–84
   chaos in rotor dynamic, 99–100
   journal-bearing hysteresis loop, 96–97
   journal-bearing nonlinearity, 85–93
   large amplitude vibration sources, 84–85
   nonlinear damping masks oil whip and steam whirl, 100
      compressor bearing failure, 101–104
      oil whip masked, 100–101
      steam whirl masked, 101
   shaft-on-bearing impacting, 97–99
   unloaded tilting-pad self-excited vibration, 94–95
Nonlinear jump phenomena, 90, 91, 93, 316
Nonlinear limit cycles. *See* Limit cycle
Nonlinear rotor vibration, 83–104
Nonstructural mass, 62–63
Nonsymmetric matrices, systems with, 71–77
Nonsynchronous orbit frequency, 312
Nonsynchronous vibration. *See* Vibration
Notch filter, 294. *See also* Filters
NRC. *See* Nuclear Regulatory Commission (NRC)
N–S equation. *See* Navier–Stokes (N–S) equation
Nuclear feed water pump cyclic thermal rotor bow, 361–364
Nuclear plant cooling tower circulating pump resonance, 367
Nuclear Regulatory Commission (NRC), 372

O

Oak Ridge National Laboratory report, 383
Observers. *See* Model-based condition monitoring
OEMs. *See* Original equipment manufacturers (OEMs)
Oil whip, 96, 179, 331–332, 333
   masked, 100–101
   and steam whirl, 226–227
Oil-free bearings, 239, 244
One-degree-of-freedom model, 3–13, 238
   as an approximation, 11–13
   damping, 8–10
   linearity assumptions, 3–4
   self-excited dynamic-instability vibrations, 6
   steady-state sinusoidally forced systems, 7–8
   undamped model, 10–11
   unforced system, 4–6
1-DOF impedance test, 204–205
Orbital displacement signals, 202
Orbital trajectories, 73, 74, 103, 151, 154, 161, 165, 238, 311, 316, 326
Original equipment manufacturers (OEMs), 279, 353, 354, 367, 385, 398, 403
Oscillatory signal's frequency spectrum, 309
Over damped. *See* Damping
Over-the-shoulder hand-held vibration analyzers, 291

P

Pad radial film stiffness, 193
Parametric excitation, 234
   nonaxisymmetric shaft stiffness, 404–406
Parametric excitation. *See* Non-axisymmetric shaft stiffness
Partial admission in steam turbine impulse stages, 269–270
Partial derivatives and RLE, 188
Partial differential equation (PDE), 42

Index

Participation factor, 27
Particular solution, 7
PDE. *See* Partial differential equation (PDE)
Peak-to-peak amplitude, 295
Pendulums, 12–13, 15–16
Period. *See* Vibrations
Phase angle, 6, 7–8, 144, 149–152, 161, 287, 307, 317, 340, 389, 391
Phase angle explanation and direction of rotation, 149–152
Physical model-based statistical approach, 280–281
Physically consistent models, 82
Piezoelectric crystal, 281, 282
Pivoted-pad bearing, chaotic rotor vibration in, 329
Pivoted-pad journal bearings (PPJB). *See* Tilting-pad journal bearings
Planar double-compound pendulum, 15–16
Poincaré maps, 326, 327. *See also* Chaos
Point-mass 2-DOF model, 37–39
Polar plot, 317, 318, 319, 386
Portable machinery vibration analyzers. *See* Vibration
Potential energy. *See* Lagrange equations
Power plant boiler circulating pumps, 364–367
Predictive maintenance, 277, 278
Preload spring, 281, 282
Preventive maintenance, 277, 278
Principal stress, 190
Proportional damping method, 26
Proximeter, 285
Proximity probe, 285, 287–288, 311, 312, 313. *See also* Displacement measurement

Q

Quad volute pump configurations, 254
Quasiperiodic motion, 326–327

R

Radial centering force, 217
Radial contact, 230
Radial eccentricity, 193, 199, 206
Radial force, 237
Radial force signals, 202
Radial stiffness effect, 217
Radial-bearing, 121–and
  bearing-support models, 64–67
  bearing coefficients
    connect rotor directly to ground, 67–68
    connect to an intermediate mass, 68–70
Rankin's model, 272
RCB. *See* Rolling contact bearings (RCB)
RCP. *See* Reactor coolant pump (RCP)
RDA. *See* Rotor Dynamic Analysis (RDA)
RDA code for LRV analyses, 141–142
  additional sample problems, 173–180
    nine-stage centrifugal pump model, 175–180
    mass rotor and pedestals, 174
  instability self-excited-vibration threshold computations, 165–173
    mass rotor (different) and disk, 172–173
    mass rotor (same) and disk, 166–172
  unbalance steady-state response computations, 142–165
    Campbell diagrams, 163–165
    elliptical orbits, 158–163
    mass rotor and disk, 152–158
    phase angle explanation and direction of rotation, 149–152
RDA Sample problems, 173–180
  Campbell diagrams, 163–165
  elliptical orbits, 158–163
  mass rotor and disk, 145–149, 152–158

RDA Sample problems (*continued*)
    different, 172–173
    same, 166–172
    and 2-pedestals, 174
    nine-stage centrifugal pump model, 175–180
    phase angle explanation and direction of rotation, 149–152
RDA99, 141–144
Reactor coolant pump (RCP), 382–384
Real-time probabilities for defined faults and severity levels, 280
Resonance, 340. *See also* Vibration
Reynolds lubrication equation (RLE), 85, 173, 184–187
    assumptions, 184–185
    bearing static load, 186
    cavitation, 186
    journal bearings, 185–186
    laminar flow, 216
    limitations for squeeze-film dampers, 238–239
    long-bearing approximation, 186
    short-bearing approximation, 186
    single solution point, 187
Reynolds numbers. *See* Bulk flow model (BFM)
Rigid body's angular momentum, 46–47
Rigid connections
    branched systems with, 127–129
    unbranched systems with, 122–124
Rigid rotors, 407–408
Rigid-body mode, 114
RLE. *See* Reynolds lubrication equation (RLE)
Rolling contact bearings (RCB), 230–235, 241
    distribution of contact loads in, 231
Rolling-element bearings, 342–343
Rotary inertia, 57, 59, 112
Rotating coordinate systems, 191, 192
Rotational kinetic energy, 52
Rotor balancing. *See* Balancing
Rotor Dynamic Analysis (RDA)
    basic rotor finite element, 55–57
    completed RDA model equations of motion, 70

complete free–free rotor matrices, 63–64
nonstructural mass and inertia to rotor element, 62–63
radial-bearing and bearing-support models, 64–67
bearing coefficients, 67–70
shaft element
    consistent mass matrix, 59–61
    damping matrix, 26, 64, 77, 173
    distributed mass matrix, 58–59
    formulations, 55–70
    gyroscopic matrix, 62
    lumped mass matrix, 57–58
    stiffness matrix, 61–62
Rotor Dynamic Analysis (RDA) Finite Element PC software. *See* RDA
Rotor dynamic coefficients
    bearings. (*see* Bearing stiffness and damping)
    centrifugal pumps, 251–259
    seals. *See* Annular seals
    turbine steam-whirl forces. (*see* Thomas-Alford forces)
Rotor dynamic models, for chaos studies, 325, 326
Rotor Dynamics Laboratory, 205
Rotor mass unbalance vibration, 330–331
Rotor orbit
    and chaos-tool mappings for, 327
    growth, for unstable rotor mode, 332
    and inductance-type displacement transducer, 311
    trajectories, 310–317
Rotor slotting, 316, 342
Rotor unbalance, 7
Rotor vibration analyses (standard), 18–19
Rotor vibration case studies
    base-motion excitations, 403–404
    bearing looseness effects
        steam turbine generator, 398–399
        electric motor, 399
        LP turbine bearing looseness, 400–401

Index   437

impact tests for vibration diagnoses, 397–398
parametric excitation, 404–406
rotor balancing, 406
  balancing computer code, 412–418
  dynamic balance, 407
  flexible rotors, 408–410
  influence coefficient method (ICM), 410–412
  in-service rotor balancing, 419–421
  in-service single-plane balance shot, 421
  rigid rotors, 407–408
  static unbalance, 407
  turbine generator case study, 418–419
self-excited
  misalignment, 377–378
  steam whirl, 371–377
structural resonances, 391–393
tilting-pad versus fixed-surface journal bearings, 401–403
vector turning from modulated rubs, 384
  simplified linear model, 385–391
vertical rotor machines, 381–384
vibration-caused wear, 393–396
Rotor vibration measurement and acquisition
monitoring and diagnostics, 277–281
signal conditioning, 292
  amplitude conventions, 294–295
  filters, 293–294
vibration data acquisition
  applications, 292
  large multibearing machines, 289–291
  monitoring, 291–292
vibration signals and associated sensors
  accelerometers, 281–283
  displacement transducers, 284–289
  velocity transducers, 283–284
Rotor vibration orbit, 288, 316–317

Rotor-based spinning reference frames, 113
Rotor-relative-to-bearing (stator) vibration orbits, 154, 411
Rotor–stator rub-impacting, 316, 336–339. *See also* Troubleshooting
Rotor-to-stator position, 284
Rotor-to-stator vibration displacement, 302, 303
Rubbing. *See* Troubleshooting
Rub-impacting
  on flexible-rotor test rig, 338
  rotor–stator, 336–339

S

Seal flow analysis models, 218
Seal rotor dynamic coefficient data resources. *See* Annular seals
Seals. *See* Annular seals
Seismic sensor, 288
Self-excited dynamic-instability vibrations, 6
Self-excited instability vibrations, 331–336
  internal rotor damping, instability caused by, 334–336
  oil whip, 333
  steam whirl, 333–334
Self-excited rotor vibration case studies
  misalignment, 377–378
  steam whirl
    bearing unloaded by nozzle forces and, 375–377
    and swirl brakes, 371–375
Semidefinite systems, 130
Sensors. *See* Accelerometers; Velocity transducers; Displacement
Separation of variables method, 20–21
SFD. *See* Squeeze-film dampers (SFD)
Shaft element
  consistent mass matrix, 59–61
  distributed mass matrix, 58–59
  gyroscopic matrix, 62
  lumped mass matrix, 57–58
  properties, RDA, 55–56
  stiffness matrix, 61–62

Shaft riders, 284–285
  shortcomings of, 285
Shaft-on-bearing impacting, 97–99
Shaker test, mechanical impedance, 204
Shock. *See* Base-motion excitation
Shop balancing, 406
Signal amplitude conversion, 293
Signal conditioning, 292–295
  amplitude conventions, 294–295
  filters, 293–294
Simple linear models for LRVs
  Jeffcott rotor model, 39–41
  point-mass 2-DOF model, 37–39
  simple nontrivial 8-DOF model, 41–43
    direct approach, 52–55
    Lagrange approaches, 44–52
Simple nontrivial 8-DOF model, 41–43
  direct approach, 52–55
  Lagrange approaches, 44–52
Simply supported beam, 53
Single mass rotor model, 145–149, 152–158, 166, 172, 174, 342
Single uncoupled rotor, 113–115
  lumped and distributed mass matrices, 115–116
    distributed mass matrix, 116–117
    stiffness matrix, 117–119
Single-peak amplitude, 295
Skew-symmetric parts of coefficient matrices, 71–75, 82, 226
Sliding velocity term, 185
Smooth-bore seals, 213
Sommerfeld number, 197, 198, 211
Spectrum analyzer and FFT, 308
Spectrum cascade plots, 317–321
Squeeze-film dampers (SFD), 230
  with centering springs, 236–237
  without centering springs, 237–238
  Reynolds-equation-based solutions, 238–239
Squeeze-film term, 185
Stability of
  axial flow compressors, 270–272
  centrifugal compressors, 260–262
  centrifugal pumps. (*see* Troubleshooting)
  multi-degree-of-freedom models, 19
  rotor-bearing systems
    computations, 6, 36
  steam turbines. (*see* Troubleshooting)
Standard rotor vibration analyses, 18–19
Standards for machinery vibration severity. *See* Vibration
Static condensation method, 26
Static equilibrium, 199, 203
Static radial force, 314
Static radial force on pump impellers. *See* Centrifugal pump impellers
Static unbalance, 407
Static rotor unbalance. *See* Unbalance
Statically indeterminate contact forces, 231
Steady-state harmonic vibration. *See* Vibration
Steady-state responses, 89–92
Steady-state sinusoidally forced systems, 7–8
Steady-state unbalance response, 54
Steam turbines, 35, 65, 100, 113, 226, 263–270, 289, 316–317, 334, 350–352, 398–401
Steam turbine generator, 350, 398–399
Steam turbine power plant condensate pumps, 384
Steam whirl, 226–227, 263, 333–334
  bearing unloaded by nozzle forces and, 375–377
  blade shroud annular seal contribution, 265–269
  blade tip clearance contribution, 264–265
  and swirl brakes, 371–375
Steam whirl masked, 101
Stiffness and damping coefficients, 185, 186
  data and resources, 196–198
    tables of dimensionless coefficients, 198
  perturbation sizes, 189–190
    coordinate transformation properties, 190–192
    symmetry of damping array, 192

Index

Stiffness matrix, 62, 117–119
Stiffness-modulated rub, 387–388
Straight cylindrical roller elements, 230
Stresses, 20
Structural damping. *See* Damping
Sub-harmonic resonance. *See* Vibration
Subsynchronous rotor vibrations, 313
Sulzer tests, 259
Swirl breaks, 265, 269, 371–375
Symmetric mass rotor anisotropic bearings
  and disk
    different, 172–173
    same, 166–172
  and pedestals, 174
Symmetric parts of coefficient matrices, 71–75, 82, 226
Synchronous bandwidth filtering, 312
Synchronous electric motor, 112
Synchronous rotor vibration. *See* Vibration
System identification, 203
Systems with nonsymmetric matrices, 71–77

T

Tangential force, 237
Tapered roller elements, 230
Tapered-bore seals, 217
Tensor filtering, 192, 202
Tensor transformation, 190, 191
Thermal distortion, 391
Thomas–Alford forces
  coefficient, 264, 271–272
  in compressors, 271
  in steam turbines, 253, 264, 270
Thresholds of instability, 260, 264
Tilting-pad journal bearings, 192–196
  comparison with cylindrical journal bearings, 193
  and grinder spindle, 196
  HP turbine, 400
  load-direction vibration factors of, 194
  preloaded, 195
  three-pad inside-out, 197

Tilting-pad versus fixed-surface journal bearings, 401–403
Time marching computation, 85
Time–frequency localization, 322. *See also* Wavelet transforms (WTs)
Time-rate-of-change, 46–47
Time-varying bearing loads, 302
Time-varying hydraulic forces. *See* Dynamic radial hydraulic impeller forces
Top foil, 244
Torsional rotor vibration (TRV)
  analyses, 111–112
  coupled rotors, 119–120
    branched systems, 126–130
    coaxial same-speed coupled rotors, 120
    unbranched systems, 121–126
  finite element model, 114, 115
  flexibly coupled, 124, 129
  four-square gear tester, example, 132–133
  gear sets, 121, 124, 126, 127, 132, 133
  high-capacity fan, 130–132
  large steam turbo-generator, 134–135
  versus LRV, 135–137
  pulley-belt sets, 121, 127
  rigidly coupled, 127–129
  rotor-based spinning reference frames, 113
  self-excited instability vibration, 165–173
  semidefinite systems, 130
  single uncoupled rotor, 113–115
    mass matrices, 115–117
    stiffness matrix, 117–119
Total solution, 7
Tracking filter, 294, 312. *See also* Filters
Transverse rotor vibration analyses. *See* Lateral rotor vibration analyses
Trending, 277
TRI. *See* Turbo Research Inc. (TRI)
Tri-volute pump configurations, 254
Troubleshooting
  boiler feed pumps, 354–361
  cracked shafts, 342

Troubleshooting (*continued*)
    critical speed, 324, 333, 349–369
    exciters, 337
    loose connections, 341
    misalignment, 339–340
    motors, 355
    oil whip, 333
    reactor coolant pumps, 342
    resonance, 340
    rotor-stator rub-impacting, 316, 336–339
    self-excited instability vibrations, 331–333
    steam turbines, 263–270, 331
    steam whirl, 333–334
    vibrations, 330–331
TRV. *See* Torsional rotor vibration (TRV) analyses
Turbine generator
    case study, 418–419
    rotor vibration model, 353
Turbine pivoted-pad bearing configurations, 354
Turbo Research Inc. (TRI), 354
Turbo-machinery impeller and blade effects
    axial flow compressors, 270–272
    centrifugal pumps, 251
        static impeller force, 251–254
        dynamic impeller forces, 255
            centrifugal compressors, 260
            interaction impeller forces, 257–259
            interactive force modeling and pumps, 262–263
            stability criteria, 260–262
            unsteady flow dynamic impeller forces, 255–257
    high-pressure steam turbines and gas turbines
        combustion gas turbines, 270
        partial admission in steam turbine impulse stages, 269–270
        steam whirl, 263–269
Two-DOF bearing pedestal model, 67
Two-DOF impedance test, 205
Two-DOF models, 13–16
Two-DOF x–y model, 209
Two-plane rigid-rotor balancing. *See* Balancing

U

Unbalance
    excited rub-impact simulation model, 326–329
        Poincaré mapping of chaotic response, 328
        rotor orbits and chaos-tool mappings for, 327
    response of centrifugal pump, 175–178
    rotor mass, 7, 86, 92, 330–331
        distribution, 331, 406
        dynamic, 330, 407–408
        flexible rotors, 408–410
        force, 236, 256
        response computation, 142–165
        static, 407–408
    steady-state response computations, 142–165
        pedestals and disk, 155–158
        mass rotor model, 145–149, 152–155
        Campbell diagrams, 163–165
        elliptical orbits, 158–163
        phase angle and direction of rotation, 149–152
Unbranched systems, 121–126
    complete equations of motion, 124–126
    flexible connections, 124
    rigid connections, 122–124
Undamped model, 5, 10–11
Undamped natural frequency. *See* Natural frequency vibration
Underdamped system, 5, 8. *See also* Damping
Unforced, 1-DOF system, 4–6
Unforced underdamped system, 5–6
Ungrooved annular seals for liquids, 215–223, 227
    axial momentum equation, 220–222
    bulk flow model approach, 219

Index

circumferential momentum equation, 219
and journal bearings, comparison between, 222–223
Lomakin effect, 216–218
seal flow analysis models, 218
Uniform viscosity, 199
Unloaded tilting-pad self-excited vibration, 94–95
Unsteady flow dynamic impeller forces, 255–257
U.S. Navy and rotating machinery, 300

V

Vane-passing and blade-passing effects, 343
Variable viscosity, 199
Vector turning, 337
   synchronously modulated rubs, 384
   simplified linear model, 385–391
Velocity transducers, 283–284
   elementary scheme, 283
Vertical rotor machines, 381–384
Vertical shaker test, 204
Vibration, 3–31
   alarm levels, 278
   amplitude conventions, 294–295
   damped natural frequency, 11
   data acquisition, 289–292
   decay, 6
   frequency, 6–9
   harmonic excitation, 9, 27–28
   measurement, 277, 411
   monitoring, 99, 277, 287, 290, 291
   multi-degree-of-freedom models, 6, 10, 13–19
   natural frequency, 10–11
   one-degree-of-freedom model, 3–13
      as approximation, 11–13
      damping, 8–10
      linearity assumptions, 3–4
      self-excited dynamic-instability vibrations, 6
      steady-state sinusoidally forced systems, 7–8
      undamped model, 10–11
      unforced system, 4–6

portable analyzers, 291–292
resonance, 10, 19
severity acceptance codes, 300–301
subharmonic, 91, 316
subsynchronous, 100, 245, 264, 270, 372, 375, 376, 418
synchronous, 100, 256, 294, 330, 341, 342, 350, 388, 398, 418
trip levels, 278
Vibration absorber, 368
Vibration cues
   Bode diagram, 317–318
   cascade plot, 318–321, 319, 320
   chaos analysis tools, 325–330
   FFT spectrum, 308–310
   polar plot, 318, 319
   rotor orbit trajectories, 310–317
   symptoms and identification
      cracked shafts, 342
      mechanically loose connections, 341
      misalignment, 339–340
      resonance, 340
      rolling-element bearings, 342–343
      rotor mass unbalance vibration, 330–331
      rotor–stator rub-impacting, 336–339
      self-excited instability vibrations, 331–336
      vane-passing and blade-passing effects, 343
   vibration trending and baselines, 307–308
   wavelet transforms, 321–325
Vibration damping, 9
Vibration data acquisition
   applications, 292
   large multibearing machines, 289–291
   monitoring, 291–292
Vibration severity guidelines, 297
   acceptance criteria, 300–301
   bearing cap vibration displacement, 298–299
   shaft displacement criteria, 301–302

Virtual control rooms, 290
   vibration levels at bearings, 291
Virtual sensors, 302
Viscosity, 65, 186, 199, 200, 211, 221, 222
Viscous damping, 24

W

Water lubricated bearing, 383, 384
Wavelet transforms (WTs), 279, 321–325
Westinghouse approach, 383–384
WFT. *See* Windowed Fourier transform (WFT)
Whirl frequency ratio, 171
Windowed Fourier transform (WFT), 322–323
   time windowing, 322
Wind tunnel fan, 130–132
Wrist pin bearings, 103
WTs. *See* Wavelet transforms (WTs)

Z

Zeros, in matrices, 17–18